Lubricant Technology

In the rapidly evolving world of mechanical systems, lubricants play a critical role in minimizing friction, reducing wear, and improving energy efficiency. *Lubricant Technology: Fundamentals and Applications* delves into the multifaceted science of lubrication, bridging the gap between foundational principles and cutting-edge innovations.

Written by leading experts, this comprehensive volume explores:

- The fundamentals of friction, wear, and lubrication.
- The chemistry and properties of base oils and advanced lubricants.
- Revolutionary bio-lubricants, nanolubricants, and ionic liquid additives.
- Emerging trends such as superlubricity and smart lubricants.
- Sustainable practices in used-oil disposal and future industry directions.

With detailed studies, practical insights, and research-backed discussions, this book is an indispensable resource for senior undergraduate, graduate students, and academic researchers in fields including mechanical engineering, industrial and production engineering, manufacturing engineering, and materials science.

Whether you're exploring the basics of lubricant formulation or investigating advanced applications, *Lubricant Technology: Fundamentals and Applications* provides the knowledge you need to drive innovation and sustainability in lubrication science.

Application Tribology

Materials, Design, Manufacturing and Testing

Series Editors:
Jibin Thazhethil Philip, Indian Institute of Science, Bengaluru, Deepak Kumar, Lijesh, K.P.

This series comprehensively discusses the basics and fundamentals of friction, wear and lubrication to emerging and special topics, such as nano-tribology, bio-tribology, food tribology, tribology of fabrics, tribology of extreme environments, tribo-corrosion, wear modeling, monitoring, and testing of wear. It is aimed at senior undergraduate, graduate students, academic researchers and professionals and presents the latest research in the field of tribology. It also explores different aspects such as materials, design, manufacturing, and testing.

Lubricant Technology: Fundamentals and Applications
Edited by Upendra Maurya, Gourav Khajuria, and Velagapudi Vasu

Green Tribology and Industry 4.0: Fundamentals and Sustainable Advancements
Edited by Velagapudi Vasu, Upendra Maurya, and Gourav Khajuria

For more information about this series, please visit: https://www.routledge.com/Application-Tribology-Materials-Design-Manufacturing-and-Testing/book-series/CRCATMDMT

Lubricant Technology
Fundamentals and Applications

Edited by
Upendra Maurya, Gourav Khajuria,
and Velagapudi Vasu

CRC Press
Taylor & Francis Group
Boca Raton London New York

CRC Press is an imprint of the
Taylor & Francis Group, an **informa** business

Front cover image: © Upendra Maurya, Gourav Khajuria, & Velagapudi Vasu

First edition published 2025
by CRC Press
2385 NW Executive Center Drive, Suite 320, Boca Raton FL 33431

and by CRC Press
4 Park Square, Milton Park, Abingdon, Oxon, OX14 4RN

CRC Press is an imprint of Taylor & Francis Group, LLC

ISBN: 978-1-032-59333-3 (hbk)
ISBN: 978-1-041-07919-4 (pbk)
ISBN: 978-1-003-64284-8 (ebk)

DOI: 10.1201/9781003642848

Typeset in Sabon
by codeMantra

Contents

Preface

The demand for advanced lubrication technologies is surging as industries evolve to meet the challenges of energy efficiency, sustainability, and environmental goals. Lubricants are the lifeblood of mechanical systems, enabling efficient energy transfer, reducing friction and wear, and prolonging the service life of machinery. This book, *Lubricant Technology: Fundamentals and Applications*, aims to bridge the gap between traditional lubricant technologies and the cutting-edge advancements shaping the future of this vital field.

This book is divided into 13 comprehensive chapters, meticulously covering the foundational principles and the latest developments in lubrication science. Beginning with the fundamentals of friction, wear, and lubrication, the chapters delve into the diverse types of lubricants, their chemical compositions, and their performance characteristics. Special attention is given to the role of additives, surface protectants, and the significance of lubrication regimes in optimizing tribological performance.

As the world pivots toward sustainability, bio-lubricants, nanolubricants, and ionic liquids are highlighted as transformative solutions to reduce environmental impact without compromising efficiency. Cutting-edge topics such as superlubricity, next-generation nanolubricants, and advanced materials demonstrate the synergy between scientific innovation and industrial applications. Additionally, critical issues like used-oil disposal and the socio-economic impact of lubricant technologies are thoroughly addressed to provide a holistic perspective.

This book is a collaborative effort of experts in the field, offering insights and case studies that cater to researchers, industry professionals, and students alike. Whether you are seeking a deeper understanding of lubrication science or exploring the future trends in this domain, this book is designed to be a valuable resource.

I extend my heartfelt gratitude to the contributors, reviewers, and publishers who have made this endeavor possible. It is our hope that this book will inspire further research and innovation, driving the lubricant industry toward a sustainable and efficient future.

Upendra Maurya
Mechanical Engineering Department
Pimpri Chinchwad College of Engineering
Pune, India

Gourav Khajuria
Mechanical Engineering Department
Government College of Engineering
Jammu, India

Velagapudi Vasu
Mechanical Engineering Department
National Institute of Technology
Warangal, India

Acknowledgments

The successful completion of this book, *Lubricant Technology: Fundamentals and Applications*, has been made possible through the collective efforts, support, and dedication of numerous individuals and organizations. We extend our heartfelt gratitude to the contributing authors and researchers whose exceptional expertise and insights have shaped the foundation of this work. Their commitment to advancing the science and technology of lubrication has been invaluable.

We are deeply appreciative of the academic and industrial communities that supported this endeavor by providing access to critical resources, facilities, and knowledge-sharing platforms. The encouragement and collaboration we received from colleagues and peers have been instrumental in refining the scope and depth of this book. Special thanks go to the publishing team for their professionalism and guidance throughout the editorial process. Their meticulous attention to detail and commitment to excellence ensured that this book met the highest standards of quality.

Lastly, we extend our sincere gratitude to our families and friends for their unwavering encouragement, patience, and support. Their understanding and belief in our vision made this endeavor possible. This book is a testament to the collective passion and dedication of all involved, and we hope it serves as a valuable resource for researchers, students, and professionals working in the field of lubrication science.

Upendra Maurya
Mechanical Engineering Department
Pimpri Chinchwad College of Engineering
Pune, India

Gourav Khajuria
Mechanical Engineering Department
Government College of Engineering
Jammu, India

V. Vasu
Mechanical Engineering Department
National Institute of Technology
Warangal, India

About the editors

Upendra Maurya completed his Bachelor's degree in Mechanical Engineering from the University of Mumbai, India, in 2012 and M.Tech. in Industrial Tribology and Maintenance Management from the National Institute of Technology Srinagar, India, in 2016. During his Master's program, he worked as a visiting researcher at the CSIR-Indian Institute of Petroleum, Dehradun, India, studying the influence of molecular chain length on lubricant properties. He earned his Ph.D. from the National Institute of Technology Warangal (NITW), India, specializing in the development of hybrid nanolubricants using ionic liquids and nanoparticles. He is currently an Assistant Professor at the Mechanical Engineering Department, Pimpri Chinchwad College of Engineering, Pune, India. With over 12 publications in reputed international journals, his research focuses on the synthesis of ionic liquids, surface characterization techniques, and advanced lubricant formulations. His work has led to novel insights into the tribological performance of hybrid nano-oils and their industrial applications.

Gourav Khajuria is working in the Mechanical Engineering Department, Government College of Engineering and Technology, Jammu, Jammu and Kashmir, India. He completed his Bachelor of Technology (Mechanical Engineering) at Shri Mata Vaishno Devi University, Katra, Jammu, and his Master of Technology (Industrial Tribology and Maintenance Management) at the National Institute of Technology, Srinagar (Hazratbal, Kashmir), India. His research focuses on tribological and surface interactions of materials in the application of the automobile industry. He worked on the super alloy materials at high temperatures to investigate tribological behavior of worn surfaces and mechanism of *in situ* development of tribo-films under dry conditions. He has over five research work published in various international reputed journals.

Velagapudi Vasu is a Professor at the Department of Mechanical Engineering, National Institute of Technology, Warangal, India. He obtained his Bachelor of Mechanical Engineering from Nagarjuna University, India, in 1999, and a post-graduation in Manufacturing Engineering from Birla Institute of Technology & Science, Pilani, India in 2003. He obtained his Ph.D. from Jawaharlal Nehru Technological University, Hyderabad, India, in 2010. His doctoral dissertation explored eco-friendly nanolubricants as cutting fluid. He was awarded Heritage-Erasmus Mundus research fellowship in 2013 to visit Politecnico di Milano, Italy, as a Post-Doctoral Associate where he carried out wear and tribological studies of aluminum composites. He has published over 40 papers in refereed journals and refereed conference proceedings. He has research interests in the areas of tribology, nanomaterials, nanofluids, micro-nano machining mechatronics, and systems controls. He has completed several research projects funded by various agencies such as DST, SERB, NRCM, ISC Bangalore, UGC, IIT Bombay, and IE India and is currently handling the SERB project on "Synergistic Effect of Phosphonium, Ammonium Based Ionic Liquid and Nano Lubricants on the Tribological and Rheological Properties" (worth Rs. 17.64 Lakhs). He is the coordinator of "Smart factory: Industry 4.0 Lab, Seimens Center of Excellence" NIT Warangal Project (worth 200 Lakhs). He was conferred the "Outstanding Scientist Award 2015" in Nanotechnology by Venus International Foundation and the "Young Engineer Award 2010" from the Institute of Engineers, India.

Contributors

Kailash Arole
Department of Materials Science and Engineering
Texas A&M University
College Station, Texas

Amit Choudhari
Mechanical Engineering Department
Cleveland State University
Cleveland, Ohio

Sachin G. Ghalme
Department of Mechanical Engineering
Sandip Institute of Technology and Research Centre
Nashik, Maharashtra, India

Ashish Kumar Gupta
School of Mechanical and Aerospace Engineering
Oklahoma State University
Stillwater, Oklahoma

Manoj Kumar Gupta
School of Mechanical Engineering
SMEC, VIT-AP University
Amaravati, Andhra Pradesh, India

Mahesh Harne
Department of Mechanical Engineering
Amrutvahini College of Engineering
Sangamner, Maharashtra, India

Harsha Arakere Puttaswamy
Department of Mechanical Engineering
Indian Institute of Technology (Banaras Hindu University)
Varanasi, Uttar Pradesh, India

Om P. Khatri
CSIR-Indian Institute of Petroleum
Chemical and Material Sciences
Mohkampur, Dehradun, India

Abhishek Kumar
Department of Mechanical Engineering
Texas A&M University
College Station, Texas

Ajay Kumar
R&D Apar Industries Ltd
Navi Mumbai, India

Ashwani Kumar
Professor & Head Mechanical Engineering, Technical Education Department Uttar Pradesh
Kanpur, India

Department of Mechanical Engineering Graphic Era deemed to be University Dehradun, India

Avinash Kumar
Department of Mechanical Engineering
Indian Institute of Information Technology Design and Manufacturing (IIITDM)
Kancheepuram, Tamil Nadu, India

Ajay Pratap Singh Lodhi
Institute of Functional Surfaces
School of Mechanical Engineering, University of Leeds
Leeds, United Kingdom

Max Marian
Department of Mechanical and Metallurgical Engineering, School of Engineering, Pontificia Universidad Católica de Chile, Vicuña Mackenna 4860, Macul, Región Metropolitana 6904411, Chile

Institute of Machine Design and Tribology (IMKT), Leibnitz University Hannover, An der Universität 1, 30823 Garbsen, Germany

Harshal P. Mungse
Department of Chemistry, Adarsha Science
J. B. Arts & Birla Commerce College
Sant Gadge Baba Amravati University
Amravati, India

Biswadeep Pal
Assistant Professor in Chemical Engineering
Parul University
Vadodara, India

Sooraj Singh Rawat
Department of Mechanical Engineering
Indian Institute of Technology (Banaras Hindu University)
Varanasi, Uttar Pradesh, India

Andreas Rosenkranz
Department of Chemical Engineering
Biotechnology and Materials (FCFM), Universidad de Chile
Santiago, Chile

Krishnkant Sahu
Mechatronics Engineering Department
Sharad Institute of Technology College of Engineering
Ichalkaranji, Maharashtra, India

Vinay Saini
Department of Chemical and Petroleum Engineering
University of Calgary
Calgary, Canada

Neha Sharma
Academy of Scientific and Innovative Research (AcSIR)
Ghaziabad, Uttar Pradesh, India

Gananath D. Thakre
CSIR-Indian Institute of Petroleum
Advanced Tribology Research Centre
Dehradun, India

Sandeep Thorat
Department of Mechanical Engineering
MIT Arts, Design & Technology University's, MIT School of Engineering
Pune, Maharashtra, India

Adesh Kumar Tomar
Department of Mechanical and Metallurgical Engineering
School of Engineering, Pontificia Universidad Católica de Chile
Región Metropolitana, Chile

Chapter 1

Fundamentals of lubricants and lubrication

Harshal P. Mungse and Om P. Khatri

1.1 INTRODUCTION

Tribology is the field of science and technology that deals with the study of friction, wear, and lubrication when the interacting surfaces are in relative motion. The Greek word "tribos," which means rubbing, is the source of the word "tribology" and was coined by Professor Peter Jost [1]. Increasing transportation activities and use of machines for industrial and domestic purposes consume energy, and a substantial amount of energy is utilized to overcome the friction in moving components of mechanical systems [2]. The transportation sector produces ~18% of greenhouse gas emissions and consumes ~20% of the total energy worldwide [3]. The increase in greenhouse gas emissions is caused by burning of fuels, coal-fired power plants, and different industrial processes. This has resulted in air pollution and rise in global warming.

Increasing environmental concerns and energy demand have encouraged the use of clean and renewable energy sources along with the development of energy-saving technologies and equipment. Utilization of efficient lubricants is one of the potential ways to reduce energy consumption by lowering friction and material loss of interacting surfaces. It also increases the life of machinery parts and minimizes CO_2 emission, which are important concerns for environmental protection. Friction was first studied scientifically by Leonardo da Vinci in the 15th century [4]. Reynolds proposed the scientific theory of lubrication in the 19th century [5]. Since that time, a lot of research works have been carried out in order to grasp the fundamental aspects of wear, friction, and lubrication.

1.1.1 Friction

The friction, a force resisting the relative motion of two or more surfaces (Figure 1.1), results in the energy being released as heat. It is beneficial in some cases, such as walking, writing on a notebook, and use of break to stop the car. However, high friction during rolling, sliding, and rotating movements is not desirable in various parts of machineries. Friction is a

DOI: 10.1201/9781003642848-1

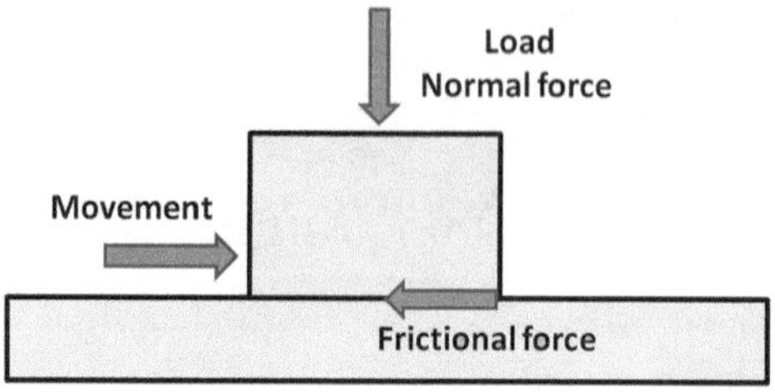

Figure 1.1 Block diagram representing the friction between two sliding bodies.

main cause for energy and material loss in several mechanical processes. To reduce friction in the various engine components, a passenger car uses about 33% of its total fuel energy [2]. Leonardo da Vinci made the discovery of the classic rules of sliding friction, although they were never published.

Guillaume Amontons rediscovered these rules in 1699, and they are as follows:

Amontons' First Law: The force of friction (F_f) is directly proportional to applied load (F_n), and it is expressed as follows:

$$F_f = \mu F_n \tag{1.1}$$

Herein, μ is the coefficient of friction

Amontons' Second Law: The frictional force does not depend on the apparent area of contact. These laws have limitations, particularly when dealing with polymers, but they are applicable to a variety of tribo systems, including metals, in both dry and boundary-lubricated conditions. In general, friction is categorized into two types: dry and fluid film friction. When two solid interacting surfaces are in relative motion and the energy dissipation is high, the friction is called dry friction. This is further subcategorized into static and kinetic friction. In fluid film friction, two or more tribo surfaces in relative motion are assisted by a thin liquid layer between the contact surfaces and energy dissipation is noted to be lower than in dry friction.

1.1.2 Wear

Wear results from material loss and deformation caused by mechanical actions (sliding, rolling, or rotating) between interacting surfaces (Figure 1.2). It typically happens at the points where surface asperities are in contact with each other and various factors such as load, speed, and lubricant effectiveness govern the wear of contact interfaces. The main reason for the failure of machinery parts is the loss of materials and surface damages during tribo events. Therefore, the use of effective lubricants is

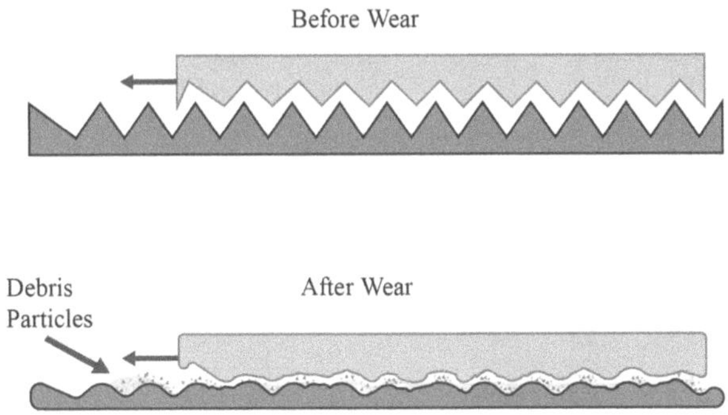

Figure 1.2 Block diagram expressing wear of tribo surfaces and the formation of debris particles.

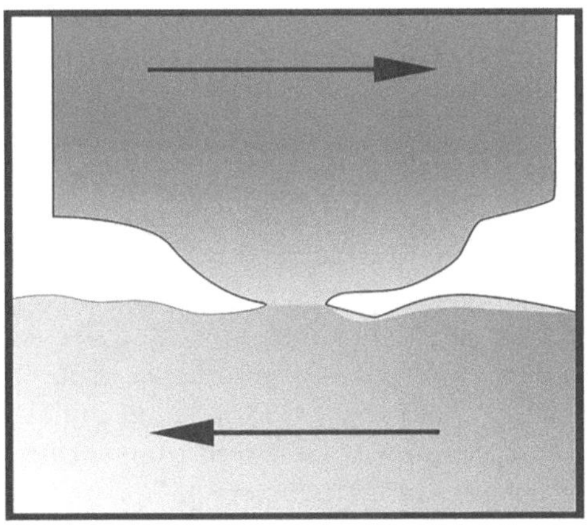

Figure 1.3 Block diagram representing adhesive wear between two contact bodies.

very important for minimizing friction and wear to conserve energy and prolong the lifespan of mechanical components. Various types of wear can arise depending on the interacting contact surfaces, environment, and operating conditions (Figure 1.3).

Adhesive wear is commonly known as scuffing, galling, or seizure. It occurs when two or more bodies/counterparts are in direct contact *via* asperities, which undergo plastic deformation and adhere to each other, causing severe wear. A large amount of adhesive forces may cause cold welding at the confined contact points. Adhesive wear occurs in bearing surfaces where lubricants are not present or not effective.

Abrasive wear occurs when hard counterparts remove materials from a softer counterpart. Scratches and grooves are the products of abrasive wear. Abrasive wear is facilitated by the hard particles that are present at

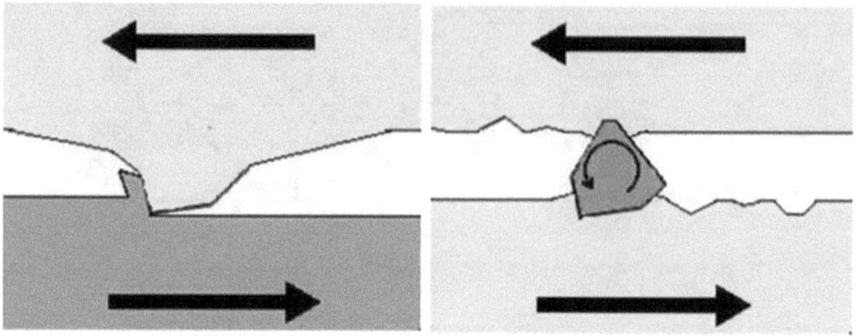

Figure 1.4 Schematic view representing abrasive wear (reprinted with permission from Ref. [6]).

the contact surface, either as contaminants or as a third body. The abrasive wear process involves both plastic flow and brittle fracture. As shown in Figure 1.4, abrasive wear happens in a two-body contact when a hard surface's asperities erode a softer surface and take materials away from its counterpart. However, in a three-body mode, the loose hard particles are unrestrained and free to roll and slide on a surface.

Erosive wear occurs by impacting particles, which are carried by a flowing liquid or gas. Fatigue wear can be subsurface- or surface-initiated. Subsurface-initiated fatigue occurs due to repeated stresses under sliding or rolling conditions. The materials weaken due to structural changes, leading to subsurface cracks that propagate and result in flaking, spalling, and peeling. The tribo surface fatigue is initiated from surface distress, causing micro-cracks, micro-spalls, and micro-pitting. Corrosive wear occurs when materials are exposed to the surrounding corrosive medium. Oxidative wear occurs when the atmospheric oxygen functions as a corroding agent [7,8]. Tribochemical wear occurs when solid surfaces react chemically with the lubricant or environment around them. The chemical reactions can weaken the surface layer and enhance the effect of other wear mechanisms (Figure 1.5).

1.1.3 Effect of friction and wear on energy and environment

Friction is the main cause of energy loss in the form of heat in various processes such as transportation, mining, power generation, and industrial machineries to domestic activities. In paper industries, 15%–25% energy is consumed to tackle the friction. Annually, 381,000 TJ of energy is utilized globally to overcome the friction. In a passenger car, energy loss due to friction is mostly found in engines (35%), breaks (15%), transmission (15%), and tires (35%) [2]. The gaseous emissions from transportation sectors add

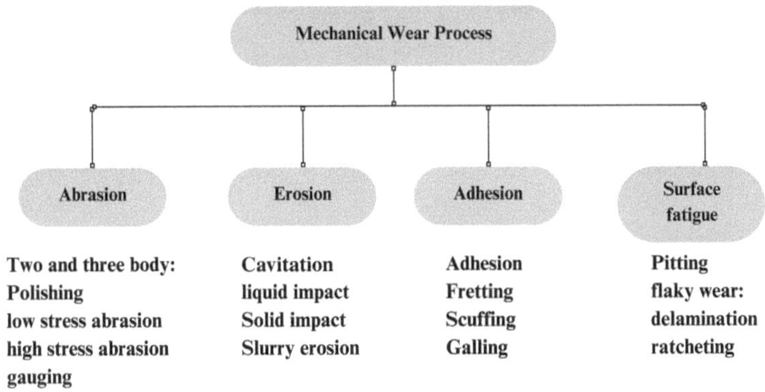

Figure 1.5 Classification of the wear process (reprinted with permission from Ref. [9]).

~18% to the greenhouse gas [3,10]. It is estimated that utilization of efficient lubricants in the engine of a four-seater car can result in a 1% reduction in friction. This reduction can lead to an annual saving of 1 billion liters of liquid fuel, thereby contributing to a remarkable reduction in CO_2 emission associated with fuel consumption [11]. The damage of engineering surfaces due to high wear leads to frequent failure of machinery parts. Therefore, for the long-term survival of machineries, friction and wear must be controlled judiciously. Lubrication serves as a successful method for managing wear and minimizing friction. An effective lubricant reduces the friction by creating a thin film of weak shear strength on the surfaces in contact, thus preventing direct interaction between them and reducing the wear of contact interfaces.

1.1.4 Lubrication

Lubricants are frequently used to furnish the lubrication between the contact surfaces to minimize energy dissipation and protect the tribo-interfaces. The surfaces of interacting bodies are often separated by the lubricant film, and hence, there is reduction of friction and wear.

The friction for lubricated surfaces is governed by the viscosity of lubricants, applied load, and speed. The Stribeck curve (Figure 1.6), representing correlations of friction with the viscosity of lubricants, applied load, and speed, was addressed to study frictional properties between lubricated surfaces [13]. The friction coefficient is plotted as a function of $\eta V/P$ in the Stribeck curve, and it reveals the dependency of the coefficient of friction on the dynamic viscosity (η), speed (V), and normal load (P). These parameters also govern the lubrication regimes, namely, boundary lubrication, mixed lubrication regime, and hydrodynamic lubrication. The concept of boundary lubrication was introduced by Hardy and Doubleday to explain friction and wear reduction in sliding of two solid surfaces [14]. In the

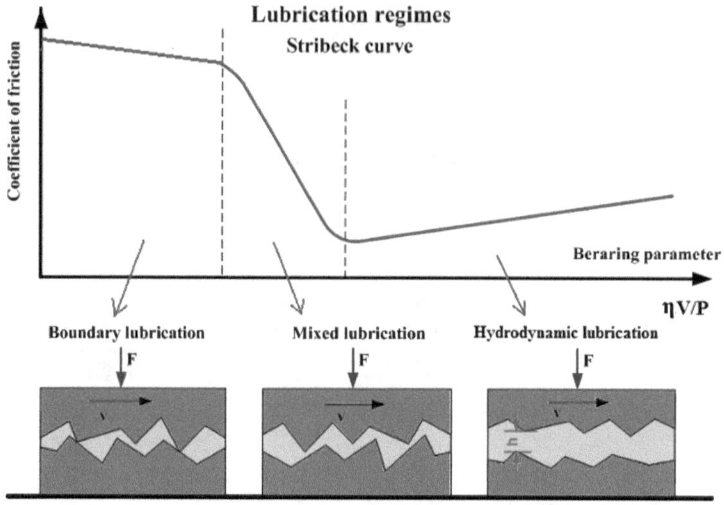

Figure 1.6 Stribeck curve demonstrating the dependence of the coefficient of friction on viscosity (η), speed (V), and normal load (P) for a lubricated tribo-system (reprinted with permission from Ref. [12]).

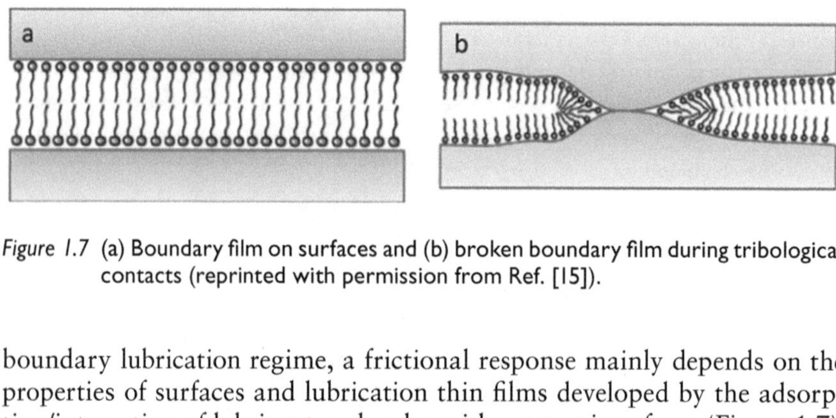

Figure 1.7 (a) Boundary film on surfaces and (b) broken boundary film during tribological contacts (reprinted with permission from Ref. [15]).

boundary lubrication regime, a frictional response mainly depends on the properties of surfaces and lubrication thin films developed by the adsorption/interaction of lubricant molecules with contact interfaces (Figure 1.7). The boundary lubrication is usually furnished by the use of a low viscosity lubricant during tribological contacts at low velocity and high load. This type of lubrication results in a high coefficient of friction due to an inadequate lubricant at the interface region, which causes a high degree of direct surface contact by breaking of the boundary thin film.

The mixed lubrication regime exhibits intermittent contact, in which a lubricant film exists between the moving surfaces with some asperity contact. As one moves from the boundary to elasto-hydrodynamic or hydrodynamic regimes, the mix or thin-film lubrication acts as a bridge. According to the hydrodynamic lubrication theory, as applied load increases, the lubricant film thickness decreases. As a result, direct asperity contacts are established at certain locations, leading to poor lubrication. In this regime, the load is shared by lubricant films and asperities [16]. A mix lubrication regime is an intermediate phase, where the principles of hydrodynamic lubrication are not entirely valid.

In a hydrodynamic lubrication, the lubricant film is sufficiently thick (1–1000 ηm) to avoid any direct interaction between the lubricated surfaces, and the load is entirely borne by the lubricant film. When the surfaces move relative to each other in the presence of fluid lubricants, a thick pressurized layer is formed. This thick layer minimizes the friction between the surfaces by creating a space between them. As a result, this regime has a low friction and negligible wear.

1.1.5 Lubricants

The most feasible and economical way to minimize friction and wear is through the use of lubricants. They reduce friction and wear by avoiding direct contact between rolling and sliding surfaces. Furthermore, supplementary components/additives present in lubricants play important roles in heat transfer and removal of contaminants and debris, expand the oxidation stability, control the viscosity and pour point, minimize the corrosion, etc., besides minimizing the friction and wear for their optimum performance. Lubricants are classified according to their physical state as solid lubricants, semi-solid lubricants, and liquid lubricants. The liquid lubricants are also classified based on type of base oil used for lubricant formulation such as synthetic/semi-synthetic, mineral, environmentally friendly, and vegetable oils. Automotive oils occupy the greatest portion of the market in terms of volume (57%), followed by industrial oils (26%), and the rest consist of process oils, metal working fluids, and greases. The worldwide lubricant consumption in 2016 was predicted to be around 40 million tones. The compound annual growth rate of lubricants is estimated to be 2.8% from 2019 to 2025.

1.1.5.1 Solid lubricants

The solid lubricants are materials which furnish lubrication under essentially dry conditions to the interacting surfaces under high loads. Solid lubricants are employed in two different forms: dispersed particles or surface thin film to reduce wear and friction. The use of self-lubricating surface coating has an advantage compared to traditional fluid lubricants. Solid lubricants are mostly applied in space technology and ultra-high vacuum, where the contamination by liquid lubricants must be avoided [17]. Liquid lubricants have some limitations, which restricts their use in specific conditions. They cannot be used in certain applications because of the decomposition problem at high temperatures and sealing issues. Liquid lubricants under corrosive

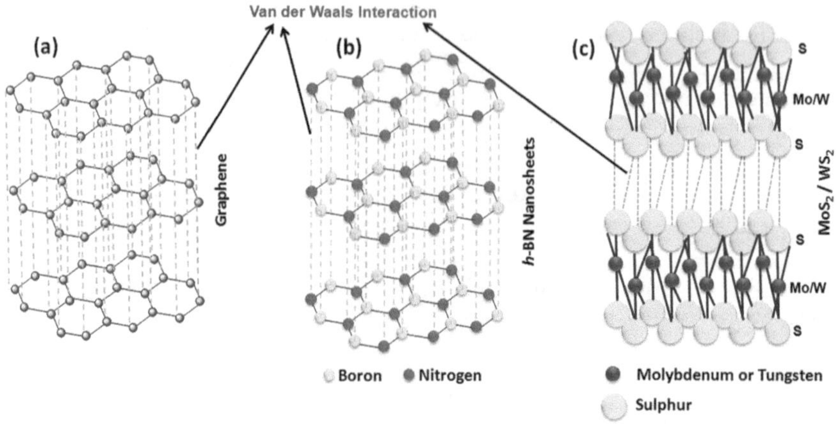

Figure 1.8 Schematic structure of (a) graphene, (b) *h*-BN, (c) MoS_2, and WS_2 nanosheets along with the illustration of the van der Waals interaction between their atomic/molecular lamellae (reprinted with permission from Ref. [18]).

environments are decomposed or may be contaminated. These limitations can be addressed by the use of solid lubricants. The most recognized solid lubricants are polymers, lamellar materials, soft metals, metal oxides, and their composites. Hexagonal boron nitride (h-BN), graphite, graphene, and transition-metal dichalcogenides (WS_2 and MoS_2) are the most prominent and lamellar solid lubricants. The low friction properties of these inorganic 2D materials are because of their layered structure. In these layered materials, the constituent atoms within layer/plane are bonded to each other by a strong covalent interaction, whereas a weak van der Waals interaction exists between adjacent atomic/molecular thick lamellae, as shown in Figure 1.8.

When the two surfaces slide over each other under tribological contact, these planes have an ability to align themselves in the direction of relative motion. This makes it possible to slide the parallel planes because of the low shear strength, which reduces friction. Molybdenum disulfide and graphite are frequently used as thermally stable extreme-pressure materials with excellent thermal stability.

The solid lubricant's shear strength at the sliding interfaces can be influenced by various factors such as the test environment, contact pressure, temperature, microstructure, and chemistry of the tribo-film. Additionally, friction can also be influenced by deposition or lubricant application methods as well as the presence of specific chemical adsorbates [19,20]. It has been noted that the friction and wear of MoS_2 rises as humidity in the surrounding air rises, but the opposite trend was observed in the case of graphite [21,22].

Soft metals like indium (In), gold (Au), lead (Pb), silver (Ag), etc., whether in their pure form or as coatings on relatively hard substrates, reduce the friction because of their weak shear strength. Soft metals exhibit a range of friction coefficients, typically ranging from 0.1 to 0.4. These variations in the friction coefficient depend on the testing conditions and the surfaces on which they interact. However, they generally exhibit a higher friction compared to lamellar materials [23]. The thickness of the lubricating film is a crucial factor in determining friction and wear in the contact area. A thicker film can lead to a larger actual contact area, resulting in increased friction. However, if the film thickness is too low, it may wear out rapidly, exposing the bare surfaces and causing a high level of friction and wear [23]. Hence, an ideal value of friction and wear can be achieved by optimizing the film/coating thickness, depending on nature of the film material and method used for their deposition such as self-assembly, electroplating, vapor deposition, and thermal spraying [24].

In recent years, polymeric materials have drawn significant interest as solid lubricants. Polytetrafluoroethylene (PTFE) is a commonly used polymeric material to minimize the friction for several applications [25]. The large molecular weight polyethylene (UHMWPE), nylon, polyimide, polychlorofluoroethylene, etc. have also been revealed as good lubricating materials for tribological applications. The efficient lubrication characteristics of these polymeric materials are linked to their structure. Most of the polymers contain long-chain molecules arranged parallel to each other with a poor bonding strength. Therefore, these chains can easily slide over each other, furnishing low resistance to shear under tribo-stress, resulting in less friction [26].

Certain metal oxides such as ZnO, MnO_2, TiO_2, and B_2O_3, when used as coatings, can improve anti-wear properties and reduced friction, even at higher temperatures. Because of the formation of soft and highly lubricious surfaces, these materials exhibit a decreased frictional response at a higher temperature [27]. Extensive research has been carried out on carbon-based lubricious materials such as diamond-like carbon compounds and graphene, which are commonly used as thin films or coatings for tribological components. The most ideal tribological condition, where the force of friction almost disappears between the solid interacting surfaces, is superlubricity, which is demonstrated by the fundamental studies of these materials and tribo surfaces. The presence of ultra-low friction can play a significant role in minimizing maintenance requirements, enhancing reliability, and decreasing energy consumption. This phenomenon is referred to as superlubricity. This is the most effective way to minimize the energy losses due to friction, and it has gained a lot of attention in the last few years.

Graphene, a carbon material with a honeycomb lattice structure, has gained a lot of attention due to its impressive mechanical, thermal, electrical, and optical properties [28,29]. Its high surface area ($2630\,m^2/g$), good mechanical strength (Young's modulus: 1 TPa; fracture strength: 130 GPa), and excellent thermal conductivity (5000 Wm/K) make it a classic material for tribological applications [30–32]. Graphene is the thinnest solid lubricant film, which can be coated on different surfaces and effectively reduces the adhesion and friction forces [33]. Because of its high mechanical strength, it can reduce wear significantly and protects the tribo surfaces. The superlubricity can be obtained by combining graphene with nanodiamond particles and diamond-like carbon in a synergistic manner. Under the sliding tribo-stress, the graphene sheets at a sliding interface wrapped the nanodiamonds, resulting in nanoscrolls of the reduced contact area. These nanoscrolls then slide over the diamond-like carbon surface, resulting in incommensurate contact, and significantly minimize the friction coefficient to ~0.004. This reduction signifies the achievement of macroscopic superlubricity [34]. Computational techniques have been employed to predict the superlubricity mechanisms of graphene at various scales. Ultra-thin graphene's capability to conformally coat tribological contacts at the nano to micro scale has established it as a very useful material for minimizing friction and wear of nanoelectromechanical/microelectromechanical (NEMS/MEMS) systems [35]. The extraordinary properties of graphene and its widespread applications have driven the emergence of a new group of materials known as 2D materials (Table 1.1).

1.1.5.2 Semi-solid lubricants

Semi-solid lubricants are gel-like materials such as greases, wax, and vaseline, which are used to minimize friction and wear between sliding or rolling surfaces when in contact with each other. They consist of a thickener agent

Table 1.1 2D materials family

Graphene family	Graphene	hBN "white graphene"	BCN	Fluorographene	Graphene oxide
2D Chalcogenides	MoS_2, WS_2, $MoSe_2$, and WSe_2	Semiconducting dichalcogenides: $MoTe_2$, WTe_2, ZrS_2, $ZrSe_2$, etc.		Metallic dichalcogenides: $NbSe_2$, NbS_2, TaS_2, TiS_2, and $NiSe_2$ Layered semiconductors: GaSe, GaTe, InSe, Bi_2Se_3, etc.	
2D Oxides	Micas, BSCCO	MoO_3 and WO_3	Perovskite-type: $LaNb_2O_7$, $(Ca, Sr)_2Nb_3O_{10}$, $Bi_4Ti_3O_{12}$, $Ca_2Ta_2TiO_{10}$, etc.	Hydroxides: $Ni(OH)_2$, $Eu(OH)_2$	
	Layered Cu oxides	TiO_2, MnO_2, V_2O_5, TaO_3, RuO_2, etc.		Others	

Source: Ref. [36].

in liquid lubricants. The greases are made by mixing oil with a thickener or soap. Grease is the most common semi-solid lubricant in which the base oil is the primary lubricant. A commercial grease typically consists of base oil (65–95 wt%), thickener (5–35 wt%), and additives (0%–10%) [37]. The base oil remains inside the thickener structure via van der Waals and capillary forces [38]. In the recent past, different types of nanomaterials have been considered as additives in grease to increase friction and anti-wear properties in boundary and mixed lubrication regimes. Fan et al. showed that adding 0.1 wt% graphene to the Bentone grease not only resulted in a decrease in friction and wear, but also led to a substantial enhancement in the load-bearing capacity [39]. Commercially available lithium grease of NLGI grade 2 containing 2% graphene demonstrated a significant reduction in both friction and wear of tribopair [40].

1.1.5.3 Liquid lubricants

Lubricants are formulated products comprising base oil and different specialty additives, which are designed for targeted performance. The base oil acts as a solvent for the additives, which control the bulk and surface properties. Additives in a lubricant range from 1 to 25 wt% depending on their applications [41]. The lubricants are generally used to minimize friction and wear, cool components, and remove wear debris. The base oil, originating from various sources such as minerals or synthetic, must possess a viscosity that allows it to sustain a lubricant film in the specific operating conditions. It should also have a fluidity to effectively dissipate heat and to prevent power loss due to viscous drag. Furthermore, it should also exhibit stability when subjected to thermal and oxidative stresses, have low volatility, and possess inherent capabilities to regulate friction and wear by itself. Also important is the ability of the base oil to dissolve/disperse the additives.

Mineral oils are the most commonly used base stocks in the lubrication. They are composed of either straight or branched carbon chains with 20–40 carbon atoms in each molecule, sometimes containing aromatic or aliphatic rings. These lube base stocks also exhibit a small percentage of hetero atoms (such as sulfur, nitrogen, or oxygen) containing compounds. The hetero atoms are a small fraction of the base stock mixture, ranging from 0.5 to 4 wt%, but they have a significant influence on base oil stability and can affect the lubrication performance. The chemical composition, sulfur content, and viscosity of mineral-base lubes are governed by refining processes. The mineral lube oils majorly contain paraffinic (straight or branched chain hydrocarbon), naphthenic (saturated hydrocarbons that contain at least one ring of carbon atoms), and aromatic components (benzene ring-based compounds). Table 1.2 shows chemical structures of mineral oils. Mineral oils are categorized on the basis of relative proportions of naphthenic, paraffinic, and aromatic components.

Table 1.2 Chemical structures of mineral oils

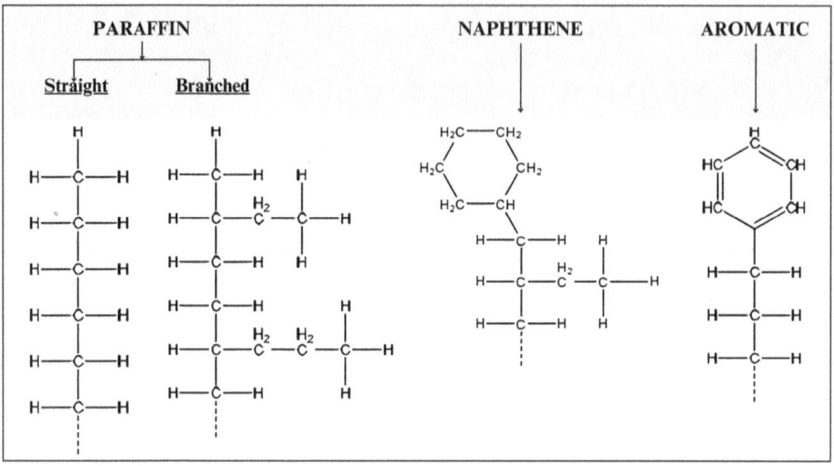

Synthetic oils are made from a more advanced refining process of crude oils. The synthetic oil structure, such as their carbon chain and functional groups, may be designed to possess specific properties and applications. Synthetic oils are utilized for applications that require high performance, such as those needing insulating, thermally resistant, or chemically resistant lubricants [42], and therefore they are more expensive. There are three types of synthetic oils such as synthetic hydrocarbon lubricants, silicon analogues of hydrocarbons, and organohalogens. The α-olefins exhibit a straight carbon chain with an unsaturated site at the end of the chain. The existence of the unsaturated carbons enables oligomerization to form poly-α-olefin (PAO)-based lubricating oils. These oils possess a high viscosity index (120–170), good oxidation stability, and low volatility. However, without double bonds and aromatic structures, the ability of PAOs to dissolve additives is low. The often used synthetic oils for lubrication are PAOs, esters (diesters and polyol esters), polyalkylene glycols (PAGs), silicones, silanes, polyphenyl ethers, etc.

The performance of base oils depends on their physical properties such as pour point, viscosity, viscosity index, and purity. Table 1.3 shows the comparison of properties of different base stocks.

1.1.6 Lubricant additives

Additives are the synthetic chemical substances that are added in lube oils to improve the performance of lubricants. They have the ability to enhance the existing properties, minimize the undesirable characteristics, and introduce novel properties in base oil. Lubricant industries work closely with lube oil and engine manufacturing industries for enhancing the performance and

Table 1.3 Comparison of properties of different base stocks

Properties	*Mineral*	*Vegetable*	*PAOs*	*Diesters*	*Polyol esters*	*PAGs*
Viscosity	Fair	Good	Good	Fair	Very good	Very good
Low-temperature properties	Poor	Poor	Very good	Good	Good	Good
Oxidation stability	Fair	Poor	Very good	Good	Good	Good
Low volatility	Fair	Good	Excellent	Excellent	Excellent	Good
Lubricating properties	Good	Very good	Good	Very good	Very good	Good
Hydrolytic stability	Excellent	Poor	Excellent	Fair	Fair	Very good
Thermal stability	Fair	Fair	Fair	Good	Good	Good
Additive solubility	Excellent	Excellent	Fair	Very good	Very good	Fair
Biodegradability	Poor	Excellent	Poor	Good	Good	Poor

Source: Ref. [43].

durability of engines. Machinery parts that operate in a harsh environment require effective lubricants in order to tolerate higher temperatures and pressures. Reduction of wear and management of lubricant consumption are viewed as crucial factors in the development of innovative anti-wear additives. To modify the properties of a lubricant, various additives such as antioxidant, corrosion inhibitor, viscosity index improver, dispersant, friction modifier, anti-wear, extreme pressure, and pour point depressant (PPD) are added in base oils [44].

The lubricating oils undergo slow oxidation due to either the presence of nascent iron in an oxidative environment or the thermal dissociation of a hydrogen atom from the hydrocarbon chain. This may lead to the formation of a radical carbon chain, which reacts with dissolved oxygen to form peroxides. Peroxides induce further radical formation and the oxidation of hydrocarbons, resulting in higher acidity due to oxide-based products, which further leads to corrosion of tribo surfaces. The corrosive wear is due to the presence of peroxy radicals on the metal surface, and these dissolved metals form salts, which precipitate as sludge. To prevent such events, antioxidants like meal deactivators, radical inhibitors, or peroxide decomposers are added in the lubricants. The sterically hindered mono-, di-, and polynuclear phenol and aromatic amine derivatives are the most efficient antioxidants used in many applications.

Corrosion inhibitors prevent the surfaces from corrosion and rusting in oxidative environments. A corrosion inhibitor is a chemical compound that efficiently minimizes the corrosion rate of the exposed metallic surfaces.

The corrosion inhibitors are usually adsorbed on the metal surface in order to form a protective film or neutralize corrosive acids in the lubricants. The adsorbed thin layer of corrosion inhibitors avoids the exposure of the underlying metal surface to oxygen, water, and other corrodents. Various organic compounds such as zinc dithiophosphates, basic metal sulfonates, metal phenolates, fatty acids, and amines act as corrosion inhibitors in lubricants [45,46].

Viscosity modifiers known as viscosity index improvers are generally polymeric materials that are employed as additives in lubricants to minimize change in viscosity as a function of temperature. The dispersion stability of high molecular weight polymers depends on their chain length, chemical composition, and structure. At a lower temperature, the polymer chains contract and form coils of low hydrodynamic volume without affecting the lubricant viscosity. However, at high temperatures, the molecules expand and increase the viscosity. The viscosity index improvers are primarily used in multigrade engine oils, hydraulic fluids, gear oils, automatic transmission fluids, etc. Viscosity of a lubricant is influenced by factors such as the category of the polymer, its concentration in the lubricant, and molecular weight. Popular viscosity modifiers are olefin copolymer and polymethacrylates.

Wear occurs due to the mechanical failure of heavily stressed interfacial areas. Anti-wear additives are load-bearing additives, which help minimize wear and extend the lifespan of mechanical systems by forming films on interacting surfaces. The extreme-pressure additives prevent failures such as scuffing, galling, and seizure by modifying the metal surface in the boundary surface regime. The extreme-pressure additives interact with the interacting surfaces under severe conditions such as high load, high speed, or high temperature. Since metal dialkyl dithiophosphate compounds (MDTPs) have anti-wear, friction-modifying, extreme pressure, antioxidant, and corrosion-resistant properties, they are used in lubricating oils.

Crude oils and processed products contain a significant quantity of waxes known as paraffins. As the temperature decreases, the paraffins form a crystalline network by trapping molecules of liquid hydrocarbon and leading to stop of the flow. The pour point refers to the lowest temperature where oil undergoes transformation into a semi-solid state, resulting in a loss of its flow properties. It is a crucial property in any lubricant since it sets the lowest limit for the operating temperature. PPDs are the polymeric compounds that preserve lubricant fluidity at a lower temperature by dislocating wax crystals. As a result, they broaden the operating temperature range of base stock and formulated lubricants. There are some specific additives, which prevent the lubricant from forming persistent foam called an antifoam agent. Typically, these chemical substances are silicone polymers or polyacrylates, which accelerate coalescence of air bubbles by minimizing surface tension of the bubble.

Sulfur, halides, and phosphorous-containing additives have attracted significant interest for the use in lubricant formulations. In order to avoid severe wear and seizure, they were designed to interact chemically with the metal surfaces and produce thin layers of sulfides or phosphides. Lubricants contain dialkyl dithiophosphates of various metals such as molybdenum, cadmium, copper, titanium, gadolinium, iron, and antimony [45–48]. Zinc dialkyl dithiophosphate (ZDDP) was invented as an antioxidant and corrosion inhibitor in the 1940s. The anti-wear properties of ZDDP were explored in the 1950s for using it in engine oils. The ZDDP is most widely used for various lubricant applications [49,50]. However, ZDDP-based additives pose risks to human health and environment, are harmful to aquatic life, and can damage the automotive exhaust gas catalyst. As a result, significant measures have been taken to find alternatives to ZDDP-based additives.

In the 1960s, solid lubricant additives were first added to gear oil, and they act as anti-wear and extreme-pressure agents in gear oil. Microparticles of layered structure materials such as graphite and molybdenum disulfide were firstly employed as solid lubricant additives. Due to their intrinsic crystalline structure, these materials have low shear strength and hence exhibit good friction and wear reduction. However, addition of a solid lubricant additive in base oil leads to challenges of dispersion stability. In liquid media, solid particles are generally not stable, especially when they are larger in size. The solid lubricant particles tend to aggregate and flocculate, resulting in their separation from the lubricant through sedimentation.

In the 1980s, nano- and submicron-size particles were firstly used as solid lubricant additives. Compared with the large-size particles, submicron- and nano-size particles have shown intrinsic advantages for lubrication performance. Typically, nano- and submicron-size particles exhibit superior dispersibility in lubricating base media and greatly improve feasibility of solid particles for lubricant applications in order to minimize the wear that may be due to the presence of solid lubricant additives on the surface of the substrate. Solid inorganic lubricant additives were found to be highly effective in order to reduce friction and wear in some applications [51–54]. Layer structural materials such as graphite, molybdenum disulfide (MoS_2), tungsten disulfide (WS_2), and hexagonal boron nitride (*h*-BN) have been extensively used as a solid lubricant additive. A low resistance to shear due to interlamellar weak van der Waals interactions of these layered nanomaterials minimizes the friction between sliding surfaces. The graphite powder is one of the most commonly used solid lubricant additives in high- or low-temperature conditions. Huang et al. investigated the tribological behavior of graphite nanosheets in paraffin oil by using four-ball and a pin-on-disk tribo tester [55]. The layered structure plays an important role in enhancing tribological performance, with 1% identified as an optimum concentration. MoS_2, WS_2, and *h*-BN as solid lubricant additives also enhance the tribological properties by minimizing the friction and wear between the interacting surfaces [56–58].

Recent studies demonstrated that adding nano-size metal particles to a lubricant can help to reduce friction and wear in mechanical systems. A number of metal nanoparticles such as copper (Cu), nickel (Ni), and gold (Ag) were studied for the tribological applications. Yu et al. [59] and Tarasov et al. [60] have reported wear and friction properties of Cu nanoparticles in commercial lubricants under different loads and sliding speeds. Tarasov also reported that the use of motor oil containing copper nanoparticles as a lubricant results in smoother worn surfaces, whereas in the absence of copper nanoparticles in motor oil, the surface appears rough with developed grooves [60]. Chou et al. reported that the presence of nickel nanoparticles in PAO improves the tribological properties [61].

To reduce friction and wear, metal oxide nanoparticles are also used as solid lubricant additives. Different types of metal oxide nanoparticles such as CuO, ZnO, and ZrO have been used as solid lubricant additives in PAO [62]. It is reported that the concentration of nanoparticles at which an effective reduction in friction was achieved does not necessarily reduce wear. The 2% CuO nanoparticles in PAO gave the highest coefficient of friction, but wear was reduced remarkably about 60% in comparison with pure PAO. In some cases, it is reported that the friction coefficient is minimum when the wear rate is high. This might be due to the higher hardness of the metal oxide nanoparticles. Hence, large abrasive wear is caused on the contact surfaces.

1.1.7 Dispersion stability of solid lubricant additives

In the tribology, solid lubricant additives must be thoroughly dispersed in the lube base oil. The lubricants are two-phase (solid–liquid) systems, where the solid phase is dispersed in the liquid phase. Dispersion stability is an important issue in a two-phase system. The lubricating performance of additives is significantly limited by their poor dispersion stability in lube base oil. Hence, improvement of dispersion stability of solid lubricant additives in the lube base oils still remains a big challenge. The dispersion stability of solid lubricant additives is influenced by two main factors such as agglomeration and sedimentation. Small particles have a high surface area, leading to a higher surface energy. Consequently, the tendency of forming agglomerates is relatively high. Minimizing agglomeration plays an important role in enhancing the dispersion stability of solid lubricant additives in lubricants.

Use of surfactants is an easy and cost-effective way to increase the dispersion stability of solid additives in base oil. A minimum quantity of surfactant can greatly influence the inter-surface properties of a system. The surfactant molecule consists of two components: a hydrophobic tail, typically a long-chain hydrocarbon at one end, and a hydrophilic group at the other end. Surfactants have the capability to enhance the degree of contact between two substances by adhering to solid particle surfaces and establishing continuity between the solid particles and the surrounding liquid environment. Modification of the surface by depositing an organic layer onto the particle surface is another effective technique to disperse solid lubricant additives. The functionalization of solid particles is a good strategy to attain long-term dispersion stability of solid particles within the lubricant. Grafting of macromolecules onto carbon nanomaterials has attracted a great interest since macromolecular chains show improved thermal stability and possess stronger steric hindrance to prevent aggregation of carbon nanomaterials. The surface modification of graphene through chemical functionalization by using various long alkyl chains containing organic molecules improves dispersion stability of graphene in organic solvents and lube base oils for better lubrication [63–66]. Our group has reported the extended dispersion stability of octadecylamine-functionalized graphene (ODA-Gr) in hexadecane. The optimum concentration (0.06 mg/mL) of ODA-Gr in hexadecane reduces friction (26%) and wear scar diameter (9%) [67]. The dispersion of alkylated reduced graphene oxide (rGO) in 10W40 engine oil was found to be stable and minimizes the friction and wear by 36% and 35%, respectively [68]. The enhanced tribological properties were attributed to continuous supply of alkylated rGO to the interacting surfaces. This prevents the direct interaction of the interacting surfaces and provides easy shearing. The amino ionic liquids (ILs) attached to the carbon nanomaterials show good dispersion stability in water, dimethylformamide (DMF), and dimethyl sulfoxide (DMSO) [69,70]. The improved dispersion may be due to the electrostatic intersheet repulsion provided by IL units.

Numerous materials have been investigated for their potential as solid lubricant additives, and some of them exhibit promising tribological characteristics. However, there is still some disagreement regarding their lubricating mechanisms. On the basis of previous reports, it is observed that the dispersibility of the solid additives in base oil plays a crucial role in improving the tribological performance. The development of new modification techniques is an important aspect of solid lubricant additive research.

1.2 CONCLUSIONS

Friction and wear are inherent events in the interacting surfaces of mechanical systems, which are responsible for the deformation of surfaces and loss of energy and materials. High friction results in energy loss and requires more power for the functioning of machinery and engines. When engineering components come in contact with each other, especially in the boundary lubrication regime, deformation of surfaces occurs, which causes plastic movement or flow of materials in the form of wear. This reduces the life of engineering components. Lubricants play a crucial role in minimizing friction and wear by forming a protective layer between the surfaces that interact with each other. Lubricants are a mixture of base oil and a variety of additives, tailored for specific uses. Extreme-pressure additives are chemically active substances that interact with surfaces and create a protective layer, ultimately decreasing friction and wear. Nowadays, most of the lubricants contain several critical lubricant additives such as anti-wear, antifriction, antioxidant, corrosion inhibitor, viscosity index improver, pour point depressants, dispersants, antifoam agents, and extreme-pressure additives. The formulation of lubricants for diverse uses is primarily determined by factors like contact pressure, temperature, surrounding conditions, and the nature of engineering surfaces. In recent years, a lot of efforts have been made to develop solid particles and use them as lubricant additives. This is due to the fact that two-phase (liquid–solid) lubricants offer incomparable benefits like remarkable thermal conductivity, significant pressure-standing ability, high resistance to decomposition at temperature, etc. The addition of ZDDP in a wide range of lubricants has long been recognized as an effective method for reducing wear and friction between the engineering surfaces. However, ZDDP has some adverse effects on human health, aquatic wildlife, and environment. Hence, a lot of work has been carried out to replace ZDDP-based additives. The major challenge of using solid lubricant additives is poor dispersion stability in lube base oils. Functionalization by chemical compounds that are compatible with lube base oil and the use of surfactants are easy and economic approaches to improve the dispersion stability of solid additives in base oils.

REFERENCES

1. Williams, J. A. 2005. *Engineering Tribology*. Cambridge: Cambridge University Press.
2. Holmberg, K., Andersson, P., Erdemir, A. 2012. Global energy consumption due to friction in passenger cars. *Tribol. Int.* 47:221–234.
3. Fontaras, G., Samaras, Z. 2010. On the way to 130g CO_2/km—estimating the future characteristics of the average European passenger car. *Energy Policy* 38:1826–1833.
4. Stachowiak, G. W., Batchelor, A. W. 2005. *Engineering Tribology*. Amsterdam: Elsevier Butterworth-Heinemann.
5. Oron, A., Davis, S. H., Bankoff, S. G. 1997. Long-scale evolution of thin liquid films. *Rev. Mod. Phys*. 69:932–976.
6. Khruschov, M. M. 1974. Principles of abrasive wear. *Wear* 28:69–88.
7. Sakamoto, T., Uetz, H., Föhl, J. 1985. Reaction layer formation on S-P extreme pressure additive in boundary lubrication under increasing load. *Wear* 105:307–321.
8. Sarin, R., Tuli, D. K., Verma, A. S., Rai, M. M., Bhatnagar, A. K. 1994. Additive – additive interactions: Search for synergistic FM-EP-AW composition. *Wear* 174:93–102.

9. Williams, J. A. 2005. Wear and wear particles—some fundamentals. *Tribol. Int.* 38:863–870.
10. International Energy Agency. 2010. *Energy Technology Perspectives, Scenarios & Strategies to 2050.* Paris: International Energy Agency.
11. Schwarz, U. D. 2015. Tracking antiwear film formation. *Science* 348:40–41.
12. Wang, Y., Wang, Q. J., Lin, C., Shi, F. 2006. Development of a set of Stribeck curves for conformal contacts of rough surfaces. *Tribol. Trans.* 49:526–535.
13. Stribeck, R. 1902. Die wesentlichen Eigenschaften der Gleit- und Rollenlager (Characteristics of plain and roller bearings). *Zeit. des VDI* 46:1341–1349.
14. Hardy, W., Doubleday, I. 1922. Boundary lubrication – the paraffin series. *R. Soc. Lond. A.* 100:550–574.
15. Zhang, J., Meng, Y. 2015. Boundary lubrication by adsorption film. *Friction* 3:115–147.
16. Spikes, H., Olver, A. V. 2003. Basics of mixed lubrication. *Lubr. Sci.* 16:1–28.
17. Campañá, C. E., Müser, M. H. 2007. In A. Erdemir, J.-M. Martin (eds.), *Theoretical Studies of Superlubricity.* Elsevier, 39–56.
18. Chouhan, A., Mungse, H. P., Khatri, O. P. 2020. Surface chemistry of graphene and graphene oxide: A versatile route for their dispersion and tribological applications. *Adv. Colloid Interface Sci.* 283:102215(1–28).
19. Kanakia, M., Peterson, M. 1987. *Literature Review of Solid Lubrication Mechanisms.* (Southwest Research Inst San Antonio TX Belvoir Fuels and Lubricants Research).
20. Savage, R. H. 1948. Graphite lubrication. *J. Appl. Phys.* 19:1–10.
21. Lancaster, J. A. 1990. Influence of environmental humidity and water on friction, lubrication and wear. *Tribol. Int.* 23:371–389.
22. Pritchard, C., Midgley, J. 1969. The effect of humidity on the friction and life of unbounded molybdenum disulphide films. *Wear* 13:39–50.
23. Scharf, T., Prasad, S. 2013. Solid lubricants: A review. *J. Mater. Sci.* 48:511–531.
24. El-Sherbiny, M., Salem, F. 1986. Tribological properties of PVD silver films. *ASLE Trans.* 29:223–228.
25. Bowden, F. P., Tabor, D. 2001. *The Friction and Lubrication of Solids.* Oxford: Oxford University Press
26. Pooley, C. M., Tabor, D. 1972. Friction and molecular structure: The behaviour of some thermoplastics. *Proc. R. Soc. Lond. A* 329:251–274.
27. John, P., Zabinski, J. 1999. Sulfate based coatings for use as high temperature lubricants. *Tribol. Lett.* 7:31–37.
28. Geim, A. K., Novoselov, K. S. 2007. The rise of graphene. *Nat. Mater.* 6:183–191.
29. Geim, A. K. 2009. Graphene: Status and prospects. *Science* 324:1530–1534.
30. Stoller, M. D., Park, S., Zhu, Y., An, J., Ruoff, R. S. 2008. Graphene-based ultracapacitors. *Nano Lett.* 8:3498–3502.
31. Lee, C., Wei, X., Kysar, J. W. 2012. Measurement of the elastic properties and intrinsic strength of monolayer graphene. *Science* 321:385–388.
32. Balandin, A. A., Ghosh, S., Bao, W., Calizo, I., Teweldebrhan, D., Miao, F., Lau, C. N. 2008. Superior thermal conductivity of single-layer graphene. *Nano Lett.* 8:902–907.

33. Kim, K. -S., Lee, H. -J., Lee, S. K., Jang, H., Ahn, J. H., Kim, J. H., Lee, H. J. 2011. Chemical vapor deposition-grown graphene: The thinnest solid lubricant. *ACS Nano* 5:5107–5114.
34. Berman, D., Deshmukh, S. A., Sankaranarayanan, S. K., Erdemir, A., Sumant, A. V. 2015. Macroscale superlubricity enabled by graphene nanoscroll formation. *Science* 348:1118–1122.
35. Berman, D., Erdemir, A., Sumant, A. V. 2014. Graphene: A new emerging lubricant. *Mater. Today* 17:31–42.
36. Shavanova, K., Bakakina, Y., Burkova, I., Shtepliuk, I., Viter, R., Ubelis, A., Beni, V., Starodub, N., Yakimova, R., Khranovskyy, V. 2016. Application of 2D non-graphene materials and 2D oxide nanostructures for biosensing technology. *Sensors* 16:223(1–23).
37. Bauer, K., Garbe, D., Surburg, H. 1988. Ullmann's encyclopedia of industrial chemistry. *Ullmann's Encyclopedia of Industrial Chemistry*, Vol. 11. Weinheim: Wiley.
38. Bauer, W. H., Finkelstein, A. P., Wiberley, S. E. 1960. Flow properties of lithium stearate-oil model greases as functions of soap concentration and temperature. *ASLE Trans.* 3:215–224.
39. Fan, X., Xia, Y., Wang, L., Li, W. 2014. Multilayer graphene as a lubricating additive in bentone grease. *Tribol. Lett.* 55:455–464.
40. Missala, T., Szewczyk, R., Winiarski, W., Hamela, M., Kamiński, M., Dąbrowski, S., Pogorzelski, D., Jakubowska, M., Tomasik, J. 2015. Study on tribological properties of lubricating grease with additive of graphene. *Adv. Intell. Syst. Comput.* 3:181–187.
41. Jianqiang, H., Huanqin, Z., Li, W., Xianyong, W., Feng, J., Zhiming, Z. 2005. Study on tribological properties and action mechanism of organic cadmium compound in lubricants. *Wear* 259:519–523.
42. Boshui, C., Junxiu, D., Guoxu, C. 1996. Tribochemistry of gadolinium dialkyldithiophosphate. *Wear* 196:16–20.
43. Pirro, D. M., Webster, M., Daschner, E. 2016. *Lubrication Fundamentals, Revised and Expanded.* Boca Raton: CRC Press.
44. Wong, V. W., Tung, S. C. 2016. Overview of automotive engine friction and reduction trends—effects of surface, material, and lubricant-additive technologies. *Friction* 4:1–28.
45. Wan, G. T. Y., Amerongen, E. V., Lankamp, H. 1992. Effect of extreme-pressure additives on rolling bearing fatigue life. *J. Phys. D Appl. Phys.* 25:A147–A153.
46. Minami, I., Mori, S. 2004. *Technische Akademie Esslingen International Tribology Colloquium Proceedings.* Germany: Technische Akademie Esslingen, 14 III, 1593–1598.
47. Minami, I., Hirao, K., Memita, M., Mori, S. 2007. Investigation of anti-wear additives for low viscous synthetic esters: Hydroxylalkyl phosphonates. *Tribol. Int.* 40:626–631.
48. Kar, P., Asthana, P., Liang, H. 2008. Formation and characterization of tribofilms. *J. Tribol.* 130:042306.
49. Fujita, H., Spikes, H. A. 2004. Formation of zinc dithiophosphate antiwear films. *J. Eng. Tribol.* 218:265–277.

50. Palacios, J. M. 1887. Films formed by antiwear additives and their incidence in wear and scuffing. *Wear* 114:41–49.
51. Peng, D. X., Kang, Y., Hwang, R. M. 2009. Tribological properties of diamond and SiO_2 nanoparticles added in paraffin. *Tribol. Int.* 42:911–917.
52. Xue, Q., Liu, W., Zhang, Z. 1997. Friction and wear properties of a surface-modfied TiO_2 nanoparticles as an additive in liquid paraffin. *Wear* 213:29–32.
53. Dong, J., Hu, Z. 1998. A study of the anti-wear and friction-reducing properties of the lubricant additive, nanometer zinc borate. *Tribol. Int.* 31:219–223.
54. Cusano, C., Sliney, H. 1982. Dynamics of solid dispersions in oil during the lubrication of point contacts, Part I - Graphite. *ASLE Trans.* 25:183–189.
55. Huang, H. D., Tu, J. P., Gan, L. P. 2006. An investigation on tribological properties of graphite nanosheets as oil additive. *Wear* 261:140–144.
56. Kimura, Y., Wakabayashi, T., Okada, K., Wada, T., Nishikawa, H. 1999. Boron nitride as a lubricant additive. *Wear* 232:199–206.
57. Gansheimer, J., Holinski, R. 1973. Molybdenum disulfide in oils and greases under boundary conditions. *J. Lubr. Technol.* 95:242–246.
58. Rapoport, L., Fleischer, N., Tenne, R. 2005. Applications of WS_2 (MoS_2) inorganic nanotubes and fullerene-like nanoparticles for solid lubrication and for structural nanocomposites. *J. Mater. Chem.* 15:1782–1788.
59. Yu, H. L., Xu, Y., Shi, P. J., Xu, B. S., Wang, X. L., Qian, L. I. U. 2008. Tribological properties and lubricating mechanisms of Cu nanoparticles in lubricant. *Trans. Nonferrous Met. Soc. China* 18:636–641.
60. Tarasov, S., Kolubaev, A., Belyaev, S., Lerner, M., Tepper, F. 2002. Study of friction reduction by nanocopper additives to motor oil. *Wear* 252:63–69.
61. Chou, R., Battez, A. H., Eras, J. J. C. 2010. Tribological behavior of polyalphaolefin with the addition of nickel nanoparticles. *Tribol. Int.* 43:2327–2332.
62. Battez, A. H., Gonzalez, R., Felgueroso, D., Fernández, J. E., del Rocío Fernández, M., García, M. A., Penuelas, I. 2007. Wear prevention behaviour of nanoparticles suspension under extreme pressure conditions. *Wear* 263:1568–1574.
63. Mungse, H. P., Gupta, K., Singh, R., Sharma, O. P., Sugimura, H., Khatri, O. P. 2019. Alkylated graphene oxide and reduced graphene oxide : Grafting density, dispersion stability to enhancement of lubrication properties. *J. Colloid Interface Sci.* 541:150–162.
64. Mungse, H. P., Kumar, N., Khatri, O. P. 2015. Synthesis, dispersion and lubrication potential of basal plane functionalized alkylated graphene nanosheets. *RSC Adv.* 5:25565–25571
65. Chouhan, A., Mungse, H. P., Sharma, O. P., Singh, R. K., Khatri, O. P. 2018. Chemically functionalized graphene for lubricant applications: Microscopic and spectroscopic studies of contact interfaces to probe the role of graphene for enhanced tribo-performance. *J. Colloid Interface Sci.* 513:666–676.
66. Gusain, R., Mungse, H. P. Kumar, N., Ravindran, T. R., Pandian, R., Sugimura, H., Khatri, O. P. 2016. Covalently attached graphene-ionic liquid hybrid nanomaterials: Synthesis, characterization and tribological application. *J. Mater. Chem. A* 4:926–937

67. Choudhary, S., Mungse, H. P., Khatri, O. P. 2012. Dispersion of alkylated graphene in organic solvents and its potential for lubrication applications. *J. Mater. Chem.* 22:21032–21039
68. Mungse, H. P, Khatri, O. P. 2014. Chemically functionalized reduced graphene oxide as a novel material for reduction of friction and wear. *J. Phys. Chem. C.* 118:14394–14402.
69. Yang, H., Shan, C., Li, F., Han, D., Zhang, Q., Niu, L. 2009. Covalent functionalization of polydisperse chemically converted graphene sheets with amine terminated ionic liquid. *Chem. Commun.* 3880–3882.
70. Kharissova, O. V., Kharisov, B. I., Ortiz, E. G. de C. 2013. Dispersion of carbon nanotubes in water and non-aqueous solvents. *RSC Adv.* 3:24812–24852.

Chapter 2

Type of base oils used in lubricant formulation and their application scenarios

Ajay Kumar and Gananath D. Thakre

2.1 INTRODUCTION

In today's world, obtaining the desired efficiency from a mechanical system is highly impossible without using a proper and efficient lubricant. Over the ages, lubricants have proven their vitality in reducing friction wear and preventing severe damage to mechanical systems. The early lubricants in the form of animal fats and oils have slowly and steadily evolved into the present-day mineral-based lubricants. Depending on the application requirements, chemical processes such as hydrotreating, catalytic hydrocracking, catalytic dewaxing, and modern wax hydroisomerization have significantly improved the lubricant's specific properties, making them suitable for more robust and stringent operating conditions [1,2]. A lubricant is a substance applied between two connecting surfaces in relative motion, reducing friction and wear. In a mechanical system, lubricants work against the frictional heat generated between contacting surfaces, acting as coolants or heat transfer fluids. Solids (including particles and coatings), semi-solids (greases), liquids (mineral and synthetic oils), and inert gases are the four types of lubricants. Among all the lubricants, liquid lubricants are the most ubiquitous in many practical applications. An efficient lubricant forms a thin protective film between the contacting surfaces to provide adequate lubrication. Efficient and effective lubrication can be understood using three regimes: boundary, mixed, elastohydrodynamic lubrication (EHL), and hydrodynamic lubrication [3,4]. The Stribeck curve best explains these lubrication regimes.

2.2 STRIBECK CURVE

The Stribeck curve presents the dependence of the friction coefficient on the operating parameters (load and sliding speed), material properties (surface roughness and elastic modulus), and lubricant properties (viscosity and pressure–viscosity relations). Figure 2.1 depicts the Hersey

DOI: 10.1201/9781003642848-2

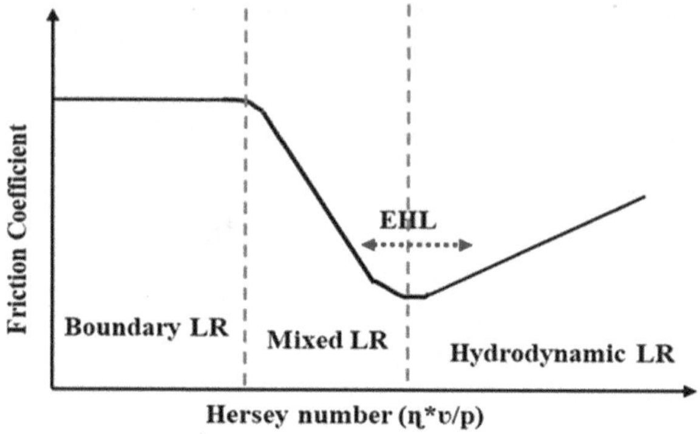

Figure 2.1 Hersey diagram.

diagram, which is plotted between the coefficient of friction and the Hersey number (viscosity*sliding speed/contact pressure) and is a simplified version of the Stribeck curve.

2.2.1 Boundary lubrication (BL) regime

In the BL regime, the lubricant could not form a lubricant film between the two contacting surfaces due to high load or low-speed conditions, resulting in high friction and undesirable wear and tear. The BL regime generally occurs during the start-up and stop-down of any machinery. In the BL regime, the physical properties of lubricants, like viscosity, density, etc., are not as important as the chemical properties of the lubricant and the properties of surfaces in the contact zone. The ratio of the minimum film thickness to the surface roughness (λ_{min} or h/σ) is less than one, which signifies that the asperities of contacting surfaces come into direct contact with no fluid film in between, as shown in Figure 2.2.

2.2.2 Mixed lubrication regime

The BL is dramatically reduced by increasing the sliding speed or decreasing the applied load, forming a lubricating film between the contacting surfaces. The asperities of surfaces in relative motion are separated by increasing the fluid film thickness, resulting in a reduced coefficient of friction (μ). The ratio of minimum film thickness to surface roughness (λ_{min}) ranges from 1 to 3, indicating that the fluid film thickness reduces asperities' contact and protects the surfaces from severe damage, as shown in Figure 2.2.

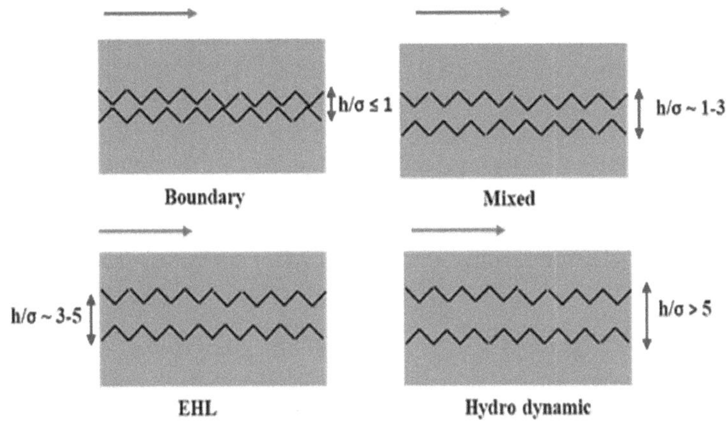

Figure 2.2 Ratio of minimum film thickness to surface roughness observed in the fluid film.

2.2.3 Elastohydrodynamic lubrication (EHL)

The EHL regime occurs when the lubricant is introduced between the rolling and sliding contact. In this regime, the applied load is significantly high, causing the elastic deformation of contacting surfaces in relative motion. Researchers have reported that non-conforming contacts, like those in bearings, gears, etc., are exposed to high contact pressures of the order of 2 GPa or more and come under EHL regimes. In this regime, the effect of the non-Newtonian behavior of the lubricant cannot be neglected because of its impact on the lubricant film thickness.

2.2.4 Hydrodynamic lubrication (HD) regime

In the HD regime, the fluid film forms a stable, thick lubricant film, fully separating the contacting surfaces in relative motion. Based on the numerical value, the ratio of minimum film thickness to the surface roughness (λ_{min}) is $\lambda \geq 5$, which signifies that the fluid forms a relatively thick film that separates the load-carrying surfaces.

2.3 LUBRICANTS AND THEIR TYPES

Lubricants are substances that tend to minimize the friction and wear between the interacting surfaces under relative motion. Lubricants are essential in saving energy, minimizing waste, and developing durable products.

Minimizing the frictional heat using an efficient lubricant can save up to 10% of energy. However, minimizing the wear improves the life of the machine or mechanical system. In recent times, sustainable lubricants have been in demand. Hence, the lubricant manufacturers are inclined to use water-soluble lubricants for metalworking applications. Water is a green lubricant possessing environmental compatibility, biocompatibility, and cost-effectiveness. Water-based lubricants are in demand in the food processing, textile, and pharmaceutical industries. The lubricants are classified as semi-solids (greases), solids (dry lubricants), gases, and liquids (oils).

2.3.1 Semi-solid lubricants (greases)

Grease is an example of a semi-solid lubricant. It is formulated using base oils (mineral/vegetable/synthetic oils) along with thickeners (organic clay, polyurea, lithium, calcium, aluminum, sodium, and other metal soaps) and performance additives for corrosion protection, oxidation stability, and load-bearing properties [5,6]. Typically, depending upon the application conditions such as temperature, pressure, corrosion protection required, and operating conditions, the proper combination of additives and base oils is selected to formulate a grease to meet the performance requirements. Generally, grease composition consists of 70%–80% base oil, 5%–15% thickener, and 0%–7% additives and is classified using National Lubricating Grease Institute (NLGI) grade, as described in Table 2.1.

2.3.2 Solids (dry lubricants)

Solid lubricants are used in acidic environments, high-vacuum applications, nuclear reactors, and other applications where lubrication by lubricating oils and grease cannot be achieved. Solid lubricants are chemical substances used in dynamic contacts to reduce friction and wear between the contacting surfaces and protect them against severe damage. The surface protection

Table 2.1 NLGI classification of greases

S. no.	*NLGI grade*	*Penetration number at 25°C*	*Consistency*
1.	000	445–475	Very fluid
2.	00	400–430	Semi-fluid
3.	0	355–385	Soft
4.	1	310–340	Semi-soft
5.	2	265–295	Multipurpose
6.	3	220–250	Semi-hard
7.	4	175–205	Hard
8.	5	130–160	Very hard
9.	6	85–115	Solid

properties of a solid lubricant depend on the particle shape, size, and application technique (like mechanical, electrochemical, or physical methods) to form a solid lubricating film on the connecting surfaces. Solid lubricants are classified into three categories: structural, mechanical, and chemically active [7].

Structural lubricants possess a layered lattice structure, such as graphite or metal dichalcogenides. Mechanical or mechano-lubricants are lubricants with self-lubricating properties like polymers, soaps, naturally occurring metal oxide films, and chemical coatings. Chemically active lubricants interact strongly with metal surfaces and can form a surface protective layer. Lamellar molybdenum disulfide (MoS_2), graphite, polytetrafluoroethylene (PTFE), and hydrogenated castor oil are examples of chemically active categories of solid lubricants [7].

2.3.3 Gas lubricants

Gases like air, nitrogen, and helium are used in gas-lubricated aerodynamic and aerostatic bearings, which operate in speed ranges of 10000–600000 rpm or temperature ranges of 15°C–800°C or more [7]. Ordinary lubricants would decompose and be incapable of providing adequate lubrication under these operating conditions. The advantage of gas lubricants is that they form an ultra-thin film between the connecting surfaces without dreading vaporization, solidification, or decomposition. Their chemical stability improves the effectiveness of proper lubrication. The disadvantages of gaseous lubricants are that they have a load-carrying capacity limited to 35–70 kPa in the aerodynamic regime and 100 kPa in the aerostatic regime, which causes failure in lubrication [7].

2.3.4 Liquid lubricants (oils)

Generally, liquid lubricants are classified into three groups: mineral-based, synthetic-based, and natural-based.

2.3.4.1 Mineral oils

Mineral base oil, or conventional or petroleum-based oil, is a type of lubricant base stock derived from naturally occurring crude oil through refining. It is the most widely used base oil type in lubricant formulations.

Mineral base oils are complex mixtures of saturated and unsaturated hydrocarbons with wide molecular weight distribution, mainly C_{20}–C_{50} hydrocarbons, including paraffinic, naphthenic, aromatic, and polycyclic aromatics (PCAs) or PCA hydrocarbons (PAHs). It also contains impurities such as sulfur, nitrogen, and oxygen compounds, which can affect its performance properties. The PCAs and PAHs are environmentally harmful and can be removed by hydrotreating and solvent extraction.

Table 2.2 Refining process and their advantage in base oils

Refining process	Advantages
Atmospheric distillation	Separates the hydrocarbons based on volatility into suitable viscosity cuts.
Vacuum distillation	Decrease the amount of aromatic content and remove the undesirable sulfur and nitrogen-containing molecules.
Dewaxing	Remove most wax content and improve the low-temperature flowability of base oil.
Catalytic dewaxing	Convert waxy paraffinic into branch paraffins and improve the low-temperature flowability of base oil.
Deasphalting	Removes asphaltic matters and converts it to suitable feedstock for lube base oils.
Solvent extraction	Removes more high aromatics and improves the oxidation stability.
Hydrotreating	Reduces sulfur and higher aromatic contents.

The production of mineral base oil involves distilling and refining crude oil, during which various hydrocarbon molecules are separated based on their boiling points and molecular weights. This process yields base oils with different viscosities and characteristics suitable for multiple lubrication applications. Refining petroleum crude oils involves various steps, including atmospheric and vacuum distillation, dewaxing, deasphalting, solvent extraction, and hydrotreating [4,8,9]. These refining processes ensure the quality and performance requirements of base oils. The advantages of refining base oils are tabulated in Table 2.2. Mineral oils are often used in lubricants for general-purpose applications where moderate operating conditions are encountered, such as automotive engines, industrial machinery, and hydraulic systems.

2.3.4.1.1 Classification of mineral oils

The mineral oils are most frequently classified using their viscosity index (VI) or how their kinematic viscosity changes with temperature. The formal range of VI ranged from 0 to 100, with the higher VI representing a lower degree of change with temperature and vice versa. Based on base oils' high-temperature sustainability, the hydrotreating method produces the higher VI (above 100) base oils. Based on VIs, the base oils are grouped into low (below 35 VI), medium (35–80 VI), and high (80–120 VI and above) base oils. Additionally, the saturation level of base oil predominantly affects the VI of base oil. The paraffinic oils are more saturated than naphthenic oils and have a higher VI. The VIs of paraffinic and naphthenic oils range between 105 and 120 and 60 and 80, respectively [4,7,10].

However, the American Petroleum Institute (API) and the Technical Association of the European Lubricants Industries (ATIEL) jointly classified the mineral oils into Groups I, II, III, IV, and V based on their sulfur

Table 2.3 API classification of base oils

Base oil group	*Base oil nature*	*Sulfur wt%*	*Saturation wt%*	*Viscosity index (VI)*
Group I	Solvent dewaxed High Viscosity Index oil	>0.03	<90	>80 to <120
Group II	Hydrotreated hydrocracked oil	≤0.03	>90	>80 to 120
Group III	High-VI oil	≤0.03	>90	>120
Group IV	Poly-α-olefins			
Group V	All other base oils (naphthenic base oil, polyalkylene glycol, esters, polyisobutene, etc.)			

Table 2.4 Typical composition of mineral base oils

Typical composition	*Group I base oil*	*Group II base oil*	*Group III base oil*
% Saturation	65–85	93–99	93–99
% Aromatic	15–35	<1–7	<1–5
Sulfur	300–3000	5–300	0–30

content, % saturation, and VI. The paraffinic mineral oils are categorized in Groups I, II, and III. In contrast, the naphthenic mineral oils are classified as Group V. The API classifications and typical compositions of mineral oils are shown in Tables 2.3 and 2.4.

Mineral base oils (>90%) are frequently used in industrial applications. They are cost-effective and provide a reasonable service life but have limited thermal stability for high-temperature applications and a low VI that indicates mineral oils are more likely to thin out at high temperatures and thicken at low temperatures. Thus, there is a need to develop synthetic oils that are more efficient than mineral oils.

2.3.4.2 Synthetic base oil

Synthetic base oils are lubricant base stocks artificially engineered through chemical synthesis rather than derived directly from crude oil [11]. These oils are meticulously designed to exhibit precise performance characteristics and properties tailored for specific applications. Synthesizing these base oils involves engineering the molecular structures to achieve desired attributes such as a higher VI, low volatility, thermal stability, oxidative stability, and enhanced lubricity. Synthetic base oils are commonly used in high-performance lubricants for demanding applications such as automotive engines, aerospace equipment, industrial machinery, and high-speed bearings. They offer superior performance and reliability, particularly when conventional mineral oils may not meet the required standards.

2.3.4.2.1 Types of synthetic oils

There are several synthetic base oils, each with unique properties and characteristics [7,9,10]. Some common types include the following.

2.3.4.2.1.1 POLYALPHAOLEFINS (PAO)

PAO synthetic base oils are derived from the polymerization of alpha-olefins. The polymerization of alpha-olefins results in linear chains of varying lengths. PAOs are characterized by their high purity, uniformity, and lack of impurities such as sulfur, aromatics, and nitrogen. These properties make PAOs ideal for many applications requiring high-performance lubricants and functional fluids. Additionally, they have excellent thermal stability, low volatility, high lubricity, and a high VI, making them suitable for various applications, including automotive and industrial lubricants.

2.3.4.2.1.2 POLYALKYLENE GLYCOLS (PAG)

PAG synthetic base oils are synthesized through the polymerization of alkylene oxide monomers. These monomers typically include ethylene oxide, propylene oxide, butylene oxide, or higher homologs. The most common types of PAGs are polyethylene glycol (PEG) and polypropylene glycol (PPG), but other variations with different alkylene oxides can also be produced. PAGs exhibit good thermal and oxidative stability and high lubricity, making them suitable for applications where extreme temperatures or high loads are encountered, such as gear oils, compressor fluids, metal working fluids, and heat transfer fluids.

2.3.4.2.1.3 POLYOLESTERS (POE)

POE synthetic base oils are esters derived from the esterification of polyols such as glycerol or pentaerythritol with fatty acids. These POE are formed through the reaction of multiple ester linkages, resulting in a chain-like molecular structure. POE exhibit various properties depending on their chemical composition, molecular weight, and structure. They show good oxidative and thermal stability, low volatility, and biodegradability and are compatible with a wide range of materials, including metals, plastics, and elastomers. POE-base oils are often used in refrigeration, compressor oils, metal working fluids, food-grade lubricants, and hydraulic fluids.

2.3.4.2.1.4 DIBASIC ACID ESTERS (DBE)

DBE synthetic base oils are esters produced from the esterification reaction of dibasic acids with alcohols. These esters are typically produced from the reaction between dibasic acids like adipic acid, succinic acid, or sebacic acid and alcohols such as ethanol or methanol. They offer excellent low-temperature properties, high oxidative stability, and compatibility with various materials. DBE base oils are used in metalworking, greases, and hydraulic fluids.

2.3.4.2.1.5 POLYISOBUTYLENES (PIB)

PIB base oils are synthesized from isobutylene monomers. PIBs are characterized by their molecular weight and the presence of repeating isobutylene units in their polymer chains. They offer stable viscosity, low volatility, and high shear stability. PIB base oils are commonly used in two-stroke engine oils and gear lubricants.

2.3.4.2.1.6 ALKYLATED NAPHTHALENES (AN)

AN base oils are produced by the alkylation of naphthalene compounds with various alkylating agents, resulting in the attachment of alkyl groups to the naphthalene ring structure. They offer excellent thermal stability, oxidation resistance, high solvency, and low volatility. AN base oils are often used in high-temperature applications such as metalworking and heat transfer.

2.3.4.2.1.7 PHOSPHATE ESTERS

Phosphate esters are a class of organic compounds derived from the esterification of phosphoric acid (or its derivatives) with alcohols or phenols. These compounds contain one or more phosphate groups ($-PO_4$) covalently bonded to an organic moiety. Phosphate esters are characterized by their diverse properties and applications, including fire-resistant hydraulic fluids, surfactants and emulsifiers, antioxidants, and flame retardants.

2.3.4.2.1.8 POLYPHENYL ETHERS

Polyphenyl ethers, or polyphenyls or polyphenylene ethers (PPE), are a class of high-performance thermoplastic polymers characterized by their repeated phenylene (C_6H_4) units in the polymer backbone. They are synthesized by polymerizing monomers containing phenyl rings, typically biphenyl or diphenyl oxide. They have high-temperature stability, chemical resistance, and electrical insulation properties, making them more suitable for automotive and industrial lubricants.

2.3.4.2.1.9 POLYSILOXANES

Polysiloxanes, also known as silicone polymers or siloxanes, are a class of inorganic-organic polymers characterized by a backbone of alternating silicon and oxygen atoms, with organic groups (such as methyl, phenyl, vinyl, or other functional groups) attached to the silicon atoms. The general chemical structure of polysiloxanes can be represented as $(\text{-Si-O-})_n$, where "n" denotes the number of repeating units in the polymer chain.

Table 2.5 The comparative advantages of mineral, synthetic, and natural oils

S. no.	*Characteristic properties*	*Mineral oils*	*Synthetic oils*	*Natural oils*
1.	Oxidation stability	Low	High	Very low
2.	Pour point	Moderate	Excellent	Moderate
3.	Viscosity index	Low	High	Very high
4.	Volatility	High	Very low	Low
5.	Biodegradability	Poor	Good	Excellent
6.	Toxicity	High	Very low	Low
7.	Lubricity	Poor	Good	Excellent
8.	Additive solvency	Poor	Good	Excellent
9.	Thermal stability	Poor	Excellent	Very poor
10.	Cost	Low	Very high	High

Polysiloxanes exhibit a wide range of properties, depending on the specific chemical structure, molecular weight, and functional groups attached to the silicon atoms. They exhibit excellent thermal stability, chemical inertness, dielectric properties, and biocompatibility. They are generally used to make antifoam agents for metalworking fluids.

The demand for renewable and sustainable lubricants has increased drastically to minimize the impact on the environment globally. In this regard, edible or non-edible vegetable oils are available in the lubricant industry and are referred to as natural oils.

2.3.4.2.1.10 NATURAL OILS

The natural oils are derived from plants, animals, or other natural sources and are composed primarily of fatty acids, glycerides, and other organic compounds. They are extracted from seeds, nuts, fruits, or animal tissues through mechanical pressing, solvent extraction, or other methods. Natural oils have been used for various purposes for centuries due to their diverse properties and applications. The typical natural oils are castor oil, palm oil, rapeseed oil, karanja oil, sunflower safflower oils, etc. They exhibit excellent biodegradability, superior lubricity, a higher flash point, and a high VI. Due to their poor oxidation and thermal stability, they are restricted from being used in high-temperature applications. The comparative advantages of mineral, synthetic, and natural oils are tabulated in Table 2.5.

2.4 CHEMISTRY OF BASE OILS

The finished lubricant contains 80%–95% base oil and 5%–20% performance additive, as required. Generally, base oils are complex mixtures of hydrocarbons ranging in maximum carbon distribution from C20 to C40.

Regarding chemical composition, the base oils are classified as paraffinic and naphthenic, depending on the presence of paraffinic, naphthenic, and aromatic content. A higher paraffinic content implies a paraffinic base oil, and a higher naphthenic content implies a naphthenic base oil [11,12]. The presence of aromatic content determines the color of base oil; the higher the aromatic content, the darker the amber color and the lower the aromatic content, the lighter or/white color of base oil.

The paraffinic base oils have low molecular weight, are wax-bearing, and have low sulfur content, while the naphthenic base oils have high molecular weight, are low wax, and may have sulfur content. Additionally, the presence of types of hydrocarbon chains like straight-chain hydrocarbons, branched-chain hydrocarbons, alkyl cycloalkane hydrocarbons, and alkyl aromatic-based hydrocarbons significantly affects the characteristic properties of base oils [11,13]. For example, higher aromatic and naphthenic content reduces the VI and pour point of the base oil. Lowering the paraffinic and isoparaffinic content increases the VI and pour point of the base oil. The characteristic properties of base oil of the same average molecular weight concerning paraffinic, naphthenic, and aromatic content are shown in Figure 2.3.

The significance of base oil composition in terms of the final lubricant performance is growing, even though additives play an essential role in lubricating oil properties. The importance of base oil composition is especially proven for performance requirements and their specifications. The molecular composition of base oils influences the physicochemical properties of lubricants as they consist of various structural hydrocarbons, and it is not easy to understand their molecular properties without analytical characterization. The following analytical techniques are used to predict

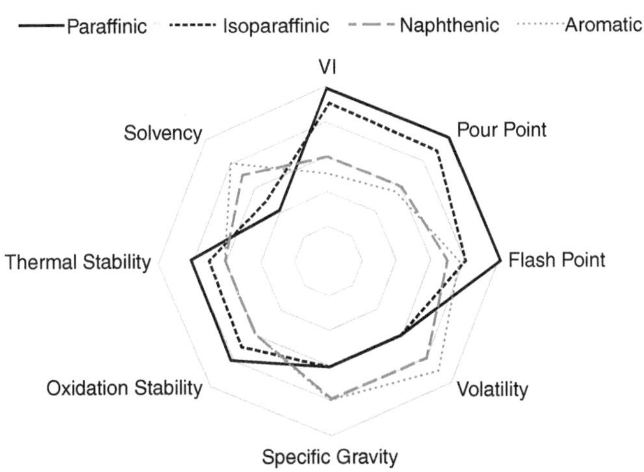

Figure 2.3 Characteristic properties of base oil of the same average molecular weight.

the molecular structure of base oils: nuclear magnetic resonance (NMR), simulated distillation (SIMDIS), Fourier transform infrared (FTIR) spectroscopy, CHNS elemental analysis, high-pressure liquid chromatography, and gel permeation chromatography [14–18].

2.4.1 NMR spectroscopy

NMR is an important technique for determining the molecular characterization of base oils. The characteristic intensity and position of ^1H and ^{13}C peaks provide the types and number of nuclei in the hydrocarbon chain. The chemical shifts in NMR spectra for ^1H NMR and ^{13}C NMR spectra range from δ 0 to 15 ppm and δ 0 to 300 ppm, respectively. The base oils consist of two types of ^1H NMR chemical shifts: peaks at 6.5–8 ppm due to hydrogen atoms bonded to aromatic hydrocarbons and those at 2–0.5 ppm due to hydrogen bonded to aliphatic hydrocarbons. In aliphatic regions, peaks at 0.8 ppm corresponding to hydrogen atoms of the methyl group in paraffinic chains and a sharp peak at about 1.25 ppm representing hydrogen atoms of the methylene groups in paraffinic chains are observed [14,15].

Similarly, in the ^{13}C NMR spectra of base oils, the chemical shifts at 100–150 ppm and 0–50 ppm represent aromatic hydrocarbons and aliphatic hydrocarbons, respectively. In ^{13}C NMR spectra, it is difficult to identify the chemical shift of CH, CH_2, and CH_3 protons. However, this problem can be resolved using the Gated Spin Echo multiphase spectral technique.

2.4.2 SIMDIS

SIMDIS is one of the most widely used analysis techniques for separating the range of petroleum hydrocarbons, crude oils, and petroleum feeds based on their boiling range distribution. This technique is wholly based on gas chromatography. SIMDIS is the only characterization technique capable of analyzing the entire range of volatile hydrocarbons (n-C3 to n-C120) found in petroleum feed. The variety of ASTM test methods that can be used to distill the petroleum feed in their boiling point range is tabulated in Table 2.6.

2.4.3 FTIR spectroscopy

The base oil's paraffinic, naphthenic, and aromatic content can be determined using FTIR analysis as per the BIS 13155 test method. In the FTIR spectra of base oil, the characteristic stretching frequencies at 1650–1500 cm^{-1} and 750–680 cm^{-1} represent the range of aromatic and paraffinic content, respectively, and can be calculated by analyzing the peak area and peak height by utilizing the following equations:

Table 2.6 ASTM SIMDIS test method

ASTM method	*Boiling range*	*Carbon chain distribution*	*Application*
D 3710	−162 to 260°C	nC1 to nC15	Naphthas, gasolines
D 7096	−162 to 280°C	nC1 to nC16	Naphthas, gasolines
D 2887	−162 to 545°C	nC1 to nC44	Naphthas through vacuum gas oil (VGO)
D 7213	100 to 615°C	nC7 to nC60	Heavy naphthas, kerosene, jet, diesel
D 6352	174 to 715°C	nC10 to nC90	Replaces by D7500
D 7500	100 to 735°C	nC7 to nC110	VGO, heavy VGO (HVGO), lube oils
D 5307	−162 to 545°C	nC1 to nC44	Crude oils
D 7169	−162 to 720°C	nC1 to nC100	Crude oil, residues
D 5399	37 to 285°C	nC5 to nC16	Hydrocarbon solvent
D 7398	100 to 615°C	nC7 to nC60	Biodiesels

$$A = \frac{\log\left(\frac{I_0}{I}\right)}{C^* d}$$

$$C_A = 1.2 + 9.8\ A$$

$$C_P = 29.9 + 6.6\ A$$

The naphthenic content is calculated by subtracting the aromatic and paraffinic content from 100 as per the BIS 13155 standard.

$$C_N = 100 - (C_A + C_P)$$

Where A is the absorbance, C is the concentration factor, d is the path length of the cell in mm, and C_A, C_P, and C_N represent the aromatic, paraffinic, and naphthenic content, respectively.

2.4.4 CHNS elemental analyzer

The CHNS elemental analysis provides rapid determination of the carbon, hydrogen, nitrogen, and sulfur constituents present in organic hydrocarbons as per the ASTM D 5291 test method. The CHNS analyzer works on the principle of the Dumas method, which involves the complete and instantaneous oxidation of the tested sample by flash combustion. The combustion gases can be separated by a gas chromatographic column and quantified using a thermal conductivity detector of carbon dioxide, hydrogen nitrogen dioxide, and sulfur dioxide, which can be converted into the percentage of C, H, N, and O elements.

2.4.5 High-pressure liquid chromatography (HPLC)

HPLC is a semiquantitative analysis that can separate the components in an organic mixture by computing the relative peak area of each component. The sample is dissolved in a solvent (mobile phase) such as heptane and pumped through a solid column (stationary phase) composed of an adsorbent at high pressure in the HPLC process of column chromatography. The components of various molecules interact differently with the column, which causes them to separate and then be identified and quantified. HPLC can be used to obtain reliable group-type compositional data on the chemicals in base oils with a low to high boiling range.

2.4.6 Gel permeation chromatography

Gel permeation chromatography separates the mixture of organic compounds based on the molecular size of each component. It is commonly used to determine the relative molecular weight of polymeric samples or mixtures of compounds. The base oils have a complex mixture, and it is handy to determine the molecular weight of the base oils.

2.5 CHARACTERISTIC PROPERTIES OF LUBRICANTS

A good lubricant should have a high flash point, low-temperature stability, a high VI, good lubricity, and high oxidation stability. A good lubricant consists of 90%–95% base fluids and a 5%–10% additive package, depending upon lubricant performance requirements. The selection of an effective lubricant for a particular application depends on its physico-chemical characteristics like viscosity, VI, pour point, flash point, total acid number (TAN), total base number (TBN), sulfated ash, elemental analysis, and tribo-performance analysis like anti-friction, anti-wear, and extreme pressure properties.

2.5.1 Physicochemical analysis and its significance

2.5.1.1 Color

Base oils can come in various colors, depending on their source and refining process. Generally, they range from clear to pale yellow. However, some specialty oils might have a slightly darker hue due to the presence of natural compounds or additives. The basis of the color measurement technique is a visual comparison between the amount of light passing through a specific depth of base oil and the amount of light passing through several colored glasses. A number representing the color of the glass is assigned. Generally, the ASTM D 1500 test method is frequently used for determining the color of base oils and providing a number in the range of 0.5–8, indicating a change in the base oil's color from light yellow or white to dark amber, respectively. However, color does not matter in finished lubricants; only light-colored or water-white finished lubricants are required in some specific applications like pharma-grade, food-grade, and textile lubricants.

2.5.1.2 Density

The density of base oil is a fundamental property that influences its performance, suitability for applications, and overall quality. Base oil manufacturers and lubricant formulators consider density alongside other physical and chemical characteristics when selecting and using base oils. The ASTM D 1298 test method uses standard hydrometers to determine base oil's density and relative density.

2.5.1.3 Viscosity

Viscosity is one of the most important physical properties of a lubricant because it measures the resistance of flow and directly affects the heat dissipation property. Viscosity also determines the lubricant film's strength and efficiency in preventing friction between the machining parts under relative motion. If the oil is too thick under given operating conditions, the machine generates much friction heat and consumes more energy, reducing efficiency. However, if the oil is too thin, the lubricant cannot form a compelling film to prevent machining parts from wear and tear, further reducing the machine's efficiency. The ASTM D 445 standard test method is used to measure the viscosity of lubricants.

2.5.1.4 Viscosity index

VI indicates the temperature-related variation of viscosity. It describes the rate of viscosity change due to temperature change. Higher VI-containing lubricants reduce oil consumption and improve wear resistance at higher temperatures by retaining the viscosity of fluids at operating temperatures. At a lower temperature, a high-VI lubricant consumes less fuel as the energy required for pumping will be lower. VI also plays a critical role in improving the performance of lubricants in specific applications, such as hydraulic and automotive engine oils. The ASTM D 2270 standard test method determines the VI of lubricants.

2.5.1.5 Pour point

The pour point of lubricant indicates the physical condition of oil (solid or flowable) in terms of storage stability in cold environments. Primarily, lubricating oils contain petroleum-derived base oils that contain waxy, long-chain saturated hydrocarbons. These waxy hydrocarbons separate from oil at lower temperatures and result in solidification, which minimizes the pour (flow) properties and causes issues at low or cold operating temperatures lower than the pour point. The pour point of the lubricants is determined using the ASTM D-97 test method.

2.5.1.6 Cloud point

The cloud point is also a crucial factor in determining the low-temperature performance of base oils. In applications where the oil is exposed to cold environments, such as automotive lubricants or hydraulic fluids, the cloud point indicates the temperature at which the oil may start to solidify or form wax crystals. It affects the oil's flow characteristics and lubricating properties, potentially leading to start-up issues. The ASTM D 2500 determines the cloud point of base oils. Due to its waxy content, paraffinic oil has a poorer cloud point than naphthenic oil.

2.5.1.7 Flash point and fire point

A lubricant's flash and fire points determine its operational temperature range and fire safety. The temperature at which lubricating oil produces enough vapor to initiate a flash of fire for a brief period of time when subjected to direct fire is known as the "flash point," and the temperature at which oil vapors catch fire for a period of time is known as the "fire point." The flash point and fire point of lubricants are determined using ASTM D 92 and ASTM D 93 test methods, respectively.

2.5.1.8 Total acid number (TAN)

The TAN value indicates the acidity and the level of acidic content (which can sometimes be from the additive constituents or due to oxidative or thermal degradation of the oil). The TAN value is crucial in maintaining the integrity of the mechanical system and preventing its damage. When oil is exposed to a high temperature, it results in oxidation, which increases the acidity (TAN value). This oxidation process will lead to an increment in the TAN value and affect the internal mechanical components of the system through wear and corrosion. However, in normal operating conditions for an automotive engine oil, the TAN value should not exceed three times the initial value, as a rule of general practice in the industry. TAN is defined as the weight (mg) of standard KOH required to neutralize one gram of the acidity of oil, and it is determined using the ASTM D 664 test method.

2.5.1.9 TBN or reverse alkalinity

TBN values represent the level of reserved alkalinity available for neutralizing the acids formed during the oxidation process. While in operation, the drop in the TBN value of lubricant should not be more than 40% of its initial value, or else it may adversely affect the performance behavior of oil. TBN is the weight (mg) of standard KOH that requires neutralizing one gram of lubricating oil's acidity and is determined using the ASTM D 2896 test method.

2.5.1.10 Sulfated ash

Sulfated ash is the amount of metallic content formed by the combustion of lubricating oil. The metallic content is usually derived from the detergent and anti-wear additives. The high concentration of sulfated ash causes adverse effects on the filter and other machine components, resulting in power loss, wear and tear, and high frictional heat generation. Generally, higher levels of phosphorus and sulfur are responsible for the high concentration of sulfated ash and the formation of corrosive acids. The sulfated ash content is determined using the ASTM D 1266 test method.

2.5.1.11 Aniline point

The aniline point of base oils plays a significant role in determining quality, reliability, performance characteristics, and their suitability for specific applications as lubricants. It is defined as the lowest temperature at which the base oil is entirely miscible with an equal volume of aniline. The ASTM D 611 test method is used to determine the aniline point. Base oils with a higher aniline point exhibit better thermal stability, solubility, and compatibility, resulting in higher performance levels under given operating conditions. Generally, paraffinic oils have a higher aniline point than naphthenic base oils.

2.5.1.12 Demulsibility

Demulsibility of base oils ensures their effectiveness, reliability, and longevity in applications where water contamination is a concern. Base oils with good demulsibility improve equipment performance, reduce maintenance requirements for corrosion, and offer overall cost savings for end-users. The ASTM D 1401 test method determines the demulsification characteristic of base oils. Naphthenic oil has superior demulsibility with water compared to paraffinic oils and is highly recommended for formulating water-based cutting fluids.

2.5.1.13 Elemental analysis

Trace metal analysis generally indicates the presence of elements in the additive package and the wear debris present in lubricating oil. The presence of additive elements improves lubricating oil's performance, while the presence of wear metals harms it. It is necessary to measure the metallic content to maintain the lubricant's effectiveness using an inductively coupled plasma atomic emission spectrometer (ICP-AES) under the ASTM D 5185 test method.

2.6 CONCLUSION

The base oils and additives play a vital role in the performance of the lubricants. The performance of any machine in terms of energy savings, reduced maintenance/downtime, and improved efficiency depends on the selection of a good lubricant. The present chapter discussed the types of base oils, their chemical composition, and their characteristic performance benefits in real-world situations. This chapter delved into the fundamentals of lubricant regimes and their significance in engineering contexts where effective lubrication is essential.

REFERENCES

1. A.G. Abdul Jammel, S. Sarathy, Lube products: molecular characterization of base oils. In: *Encyclopedia of Analytical Chemistry* (2018), John Wiley & Sons, Ltd. King Abdullah University of Science and Technology, Thuwal, Saudi Arabia, 1–14.
2. O.A. Kuti, S.Y. Yang, N. Hourani, N. Naser, W.L. Roberts, S.H. Chung, S.M. Sarathy, A fundamental investigation into the relationship between lubricant composition and fuel ignition quality. *Fuel*, 160, 605–613 (2015).
3. Y. Wang, Q.J. Wang, Lubrication regimes. In: Wang, Q.J., Chung, Y.W. (eds) *Encyclopedia of Tribology* (2013), Springer, Boston, MA. https://doi.org/10.1007/978-0-387-92897-5_149.
4. M.E. Korkmaz, M.K. Gupta, R. Demirsoz, Understanding the lubrication regime phenomena and its influence on tribological characteristics of additively manufactured 316 steel under novel lubrication environment. *Tribol Int*, 173, 107686 (2022).
5. A. Kumar, B. Prasad, B. Sidhiwala, T.C.S.M. Gupta, *Role of Micronized PTFE Powders as Performance Additives in Meeting the Industrial Demands of Greases Proceedings of TRIBOINDIA-2018 An International Conference on Tribology*, Mumbai, India, 2018, Tribology for Reliability and Life.
6. H. Hedayati, A. Mofidi, A. Al-Fadhli, M. Aramesh, Solid lubricants used in extreme conditions experienced in machining: a comprehensive review on recent developments and applications. *Lubricants*, 12(3), 69 (2024).
7. A.R. Lansdown, *Lubricant and Selection: A Practical Guide*, Third Edition (2003), 320, The American Society of Mechanical Engineers New York; Series Editors: M.J. Neale, T.A. Polak, M. Priest.
8. M.J. Neale, *Lubrication and Reliability Handbook* (2000), Elsevier, Newnes (An imprint of Butterworth-Heineann Limited, Woburn, England.
9. R. Shah, M. Woydt, S.C. Tung, Grease. *Lubricants*, 10(3), 45 (2022). https://doi.org/10.3390/lubricants10030045.

10. A. Adhvaryu, J.M. Perez, L.J. Duda, Quantitative NMR spectroscopy for the prediction of base oil properties. *Tribol Trans*, 43(2), 245–250 (2000).
11. K.H. Altgelt, M.M. Boduszynski, *Composition and Analysis of Heavy Petroleum Fractions* (1994), Marcel Dekker, New York.
12. L. Montanari, E. Montani, C. Corno, S. Fattori, NMR molecular characterization of lubricating base oils: correlation with their performance. *Appl Magn Reson*, 14(2–3), 345–356 (1998).
13. B.L. Papke, Mineral oil base fluids. In: Wang, Q.J., Chung, Y.W. (eds) *Encyclopedia of Tribology* (2013), Springer, Boston, MA. https://doi.org/10.1007/978-0-387-92897-5_913.
14. N. Hourani, H. Muller, F.M. Adam, S.K. Panda, M. Witt, A.A. Al-Hajji, S.M. Sarathy, Structural level characterization of base oils using advanced analytical techniques. *Energy Fuel*, 29(5), 2962–2970 (2015).
15. J. Keeler, *Understanding NMR Spectroscopy* (2010), John Wiley & Sons, Inc., Hoboken, New Jersey.
16. U. Maurya, A. Kumar, S. Saini, G.D. Thakre, H.G. Asharaf, Unravelling the effect of chain and branch content on the viscosity of PIB-mineral oil blends by modelling and its tribological properties. *J Mater Eng*, 2(3), 45–56 (2024).
17. T. Mang, W. Dresel, *Lubricants and Lubrication – Base Oils* (2006), Chapter 4, Wiley-VCH Verlag GmbH & Co, KGaA, Germany, 34–62. https://doi.org/10.1002/9783527610341.
18. U. Maurya, V. Vasu, V. Saini, Nano-innovations in lubrication: unveiling the synergistic boundary lubrication potential of zinc titanate with phosphonium ionic liquid. *Tribol Int*, 199, 109987 (2024).

Chapter 3

Surface protectants additives

Vinay Saini and Ajay Pratap Singh Lodhi

3.1 INTRODUCTION

Lubricants have evolved significantly from their early formulations, which primarily aimed to reduce friction between moving parts [1,2]. Today, they are complex chemical mixtures tailored to meet specific performance requirements across diverse applications and formulated according to specific applications (engine, cutting, transmission, etc.) and operating conditions. Conventionally, lubricants can be formulated by using either one or a combination of mineral or synthetic base-stock oils [3]. These base-stock oils are added with chemical additives to impart specific desirable properties or improve existing ones. Typically, the dosage of additives varies between 0.1% and 30% dissolved in oils. Surface protective additives are a key component of this evolution, designed to enhance the functionality of lubricants by providing critical protective features [4]. Figure 3.1 shows the essential properties desired in lubricants, which may be absent in inherent base oils; thus, additives are added to impart certain properties.

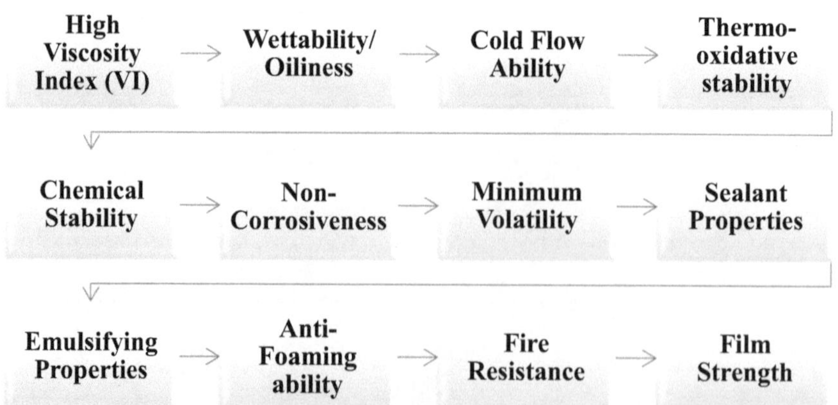

Figure 3.1 Essential lubricant properties [5].

DOI: 10.1201/9781003642848-3

3.1.1 Base oils

Mineral oils (M.O.s) are the most widely used base-stock oils for lubricant formulation worldwide due to their wide availability, cost viability, and better inherent properties compared to vegetable oils [3]. Typically, these base oils are derived from crude mineral oil after a sequence of refining processes. These M.O.s can be further categorized according to the presence of linear branching in their molecular backbone as paraffinic, naphthenic, and aromatic [6], as shown in Figure 3.2.

Based on the purity, viscometrical properties, and quality of lubricants, American Petroleum Institute (API) classifies these mineral base oils as Groups I, II, III, IV, and V [6].

- Group I: Saturates <90% and/or sulfur >0.03%, and (VI) of 80–120.
- Group II: Saturates >90% and sulfur <0.03%, and VI of 80–120.
- Group III: Saturates >90%, sulfur <0.03%, and VI over 120.
- Group IV: Synthetic poly alpha-olefins (PAO) VI over 150.
- Group V: Naphthenic stocks, synthetics, and other non-PAO base stocks.

3.1.1.1 Lubricant additives

Typical lubricants contain approx. 75%–90% as base oil, whereas additives can be in the range from 1 to 30 wt% depending on the applications to enhance the performance without interfering with the functioning of other additives [5,7,8]. Lubricant additives can be in a liquid or particulate form, including nano-size, and can be categorized as chemically active or chemically inert [9,10]. Most of the additives are chemical compounds

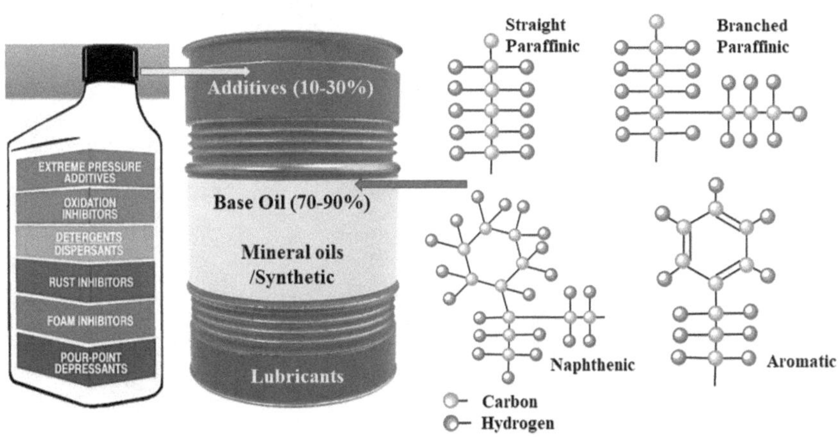

Figure 3.2 Conventional lubricant composition and mineral base-oil structure. (Adopted from Ref. [5].)

that enhance desirable or suppress the undesirable properties of base oil. Thus, overall additives manipulate the key properties of base oils such as viscosity, viscosity index, pour point and oxidation resistance, anti-wear (AW), anti-friction (AF), etc. These additives have to fulfill the following functions:

- It must achieve the function for which it is added.
- It must be soluble in oil. However, particulate additives are the exception to this.
- It must not interfere with the functioning of another additive.

Chemically active additives, such as dispersants, detergents, oxidation inhibitors, rust, corrosion inhibitors, AW, and extreme-pressure additives (EPAs), interact chemically at metal surfaces/interfaces under tribological conditions to generate protective tribofilm on them. Chemically inert additives, such as surfactants, emulsifiers, pour point depressants, foam inhibitors, and viscosity index improvers (VII), improve the physical properties critical to the effective performance of a lubricant. Overall, lubricant additives can be broadly classified into three categories, as shown in Figure 3.3. In a nutshell, lubricant additives fortifying the tribological performance of oils are discussed in the subsections.

3.1.1.2 Surface protectants tribo-additives in oils

AW, AF or friction modifiers (FMs), and EP additives (EPAs) are termed tribo-additives. They are needed to counteract the possible tribo-chemical reaction that occurs at tribo-pair interfaces under certain tribological conditions (contact pressure, temperature, and shear stresses) (Figure 3.4). The possible driving forces for tribo-chemical reactions (tribo-chemistry) under tribological conditions are shown in Figure 3.5.

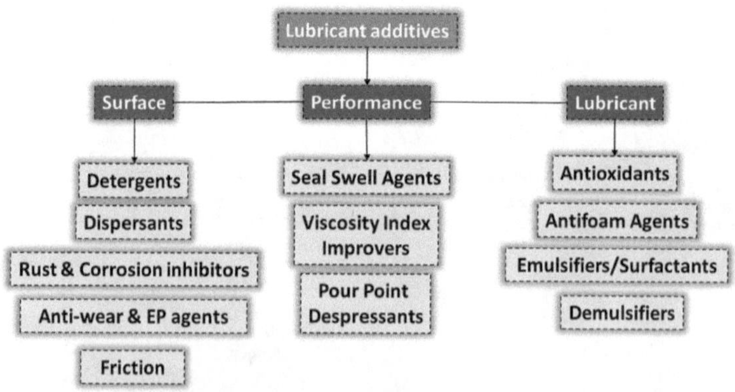

Figure 3.3 Classification of different lubricant additives [5].

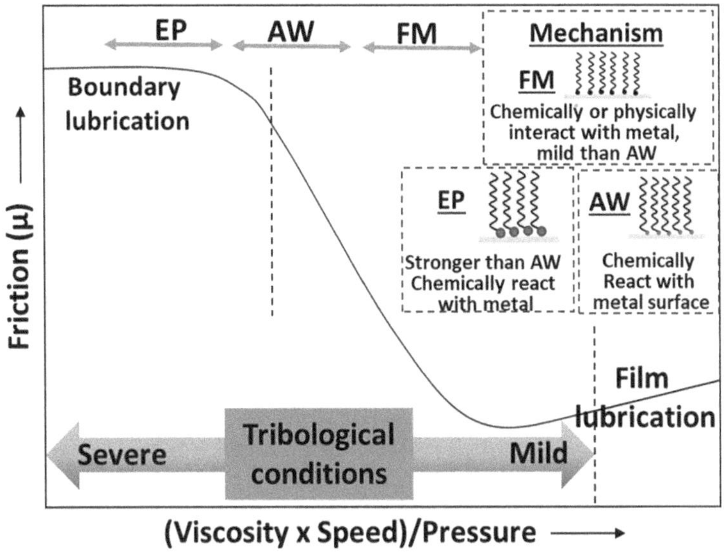

Figure 3.4 Working of range and comparison of oil containing no additives and other with a mixture of FM and EPA [11].

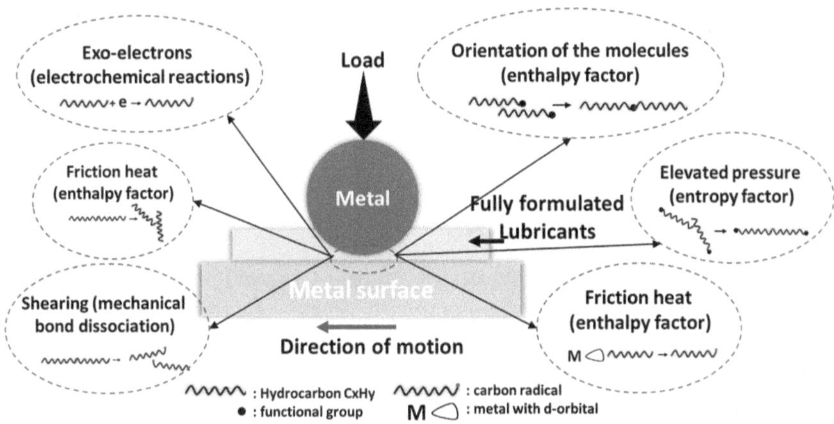

Figure 3.5 Driving forces for chemical reactions at tribo-contact [4].

3.2 ANTI-WEAR ADDITIVES (AWAs)

The AWAs are employed to reduce wear in moderate pressure and high-speed conditions [12–15]. They offer a beneficial tribofilm that reduce asperity–asperity contacts and hence wear. AWAs are used in many lubricating oils to reduce friction, wear, scuffing, and scoring under boundary lubrication conditions. During boundary lubrication, the oil film turns even thinner because of escalating pressure and build-up temperatures leading to increased asperity–asperity contacts which eventually leads to severe wear. AWAs are mostly polar materials such as fatty oils, acids, esters, etc. In particular, long-chain carboxylic acids in combination with sulfur—compounds are widely used as AWA package to prevent wear during mild operating conditions. Some examples of AWA are organophosphate esters, tricresyl phosphate (TCP) and zinc dialkyl dithiophosphate (ZDDP) [3]. ZDDP is the most used AW additive.

3.2.1 Mechanism of AWAs

The core function of AWAs lies in their ability to form a protective tribofilm. This tribofilm is created when the additive reacts chemically with the metal surface under conditions of elevated temperature and contact pressure that arise during mechanical operation. **ZDDP**, for example, breaks down under high temperatures, releasing zinc and phosphorus compounds. These compounds interact with the metal surface, forming a phosphate glass-like layer (long phosphate chain) that adheres to the metal and prevents direct metal-to-metal contact (Figure 3.6). This process is particularly effective in high-stress areas like engine camshafts, valve lifters, and other components subjected to sliding motion. Other commonly used AWAs such as organophosphate esters and TCP follow a similar kind of mechanism (form robust and protective tribofilm) to reduce the wear of mechanical components in relative motion [8,16]. However, they might exhibit different tribo-chemistry and physical properties, offering a range of AW benefits depending on the application.

3.2.2 Applications of AWAs

AWAs are particularly important in automotive and industrial applications where components are subjected to repetitive stress. In automotive engines, for example, the sliding motion between the camshaft and valve lifter requires protection to prevent excessive wear and premature failure.

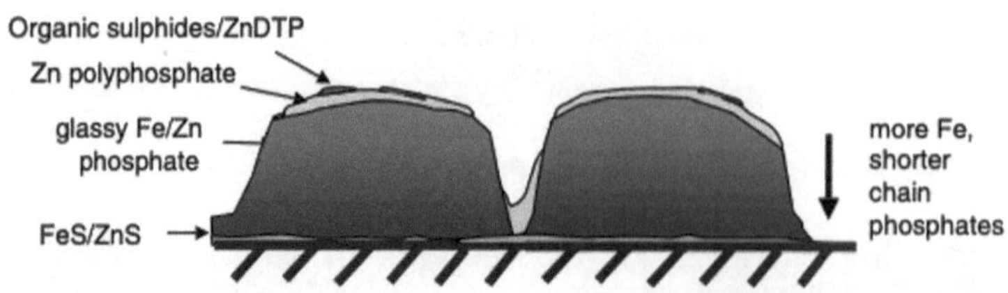

Figure 3.6 Schematic representation of the formation of a ZDDP-derived tribofilm on a metal surface under pressure [7].

In such systems, AWAs provide a protective film that reduces the friction coefficient and prevents surface degradation (wear), even under high-temperature operating conditions. Similarly, in industrial gearboxes, hydraulic systems, and other heavy machinery, AWAs play a critical role in maintaining the integrity of the equipment. These systems often operate under moderate loads and need robust AW protection to avoid equipment breakdowns. AWAs thus contribute to reducing maintenance costs and downtime by extending component life.

3.2.3 Challenges and limitations with AWAs

One challenge facing the continued use of traditional AWAs, particularly ZDDP, is the environmental impact. ZDDP contains phosphorus and sulfur, both of which can lead to the formation of harmful emissions, such as sulfur oxides and phosphates, which contribute to pollution [17,18]. As a result, there has been a push toward developing phosphorus-free or low-phosphorus alternatives, especially in applications subject to stringent environmental regulations.

Another limitation is additive antagonism. ZDDP, for instance, can interfere with other additives in the lubricant, such as FMs or extreme-pressure additives. Balancing these interactions is crucial for optimizing lubricant performance without sacrificing wear protection.

3.3 AF ADDITIVES (AFAS)/FMs

AFAs are also known as FMs since they help reduce the lubricating system's friction coefficient (μ) of the two sliding components. These additives get physio-adsorbed on the surfaces to reduce friction by forming a very low-shear-strength film [7]. AFAs are grouped as organomolybdenum (OM) compounds and organic FMs (OFM). The OMs can be further divided into three forms, sulfur- and phosphorus-containing compounds (e.g., molybdenum dialkyldithiophosphate (MoDTP)), sulfur-containing and phosphorus-free compounds (molybdenum dithiocarbamate (MoDTC)), and sulfur- and phosphorus-free compounds (molybdate ester (ME))] [19]. These additives, such as fatty acids (oleic acid, linoleic acid, etc.) and molybdenum-based compounds, form low-shear films on metal surfaces. The reduction in friction results in enhanced fuel efficiency and less energy loss in automotive and industrial applications. A schematic showing the formation of a low-shear friction-modifying film on a metal surface by oleic acid molecules (Figure 3.7). However, its performance to facilitate continuous film is limited by its concentration and availability at the interface that provides low-shear ability to interacting metallic surfaces of gear and sliding bearings.

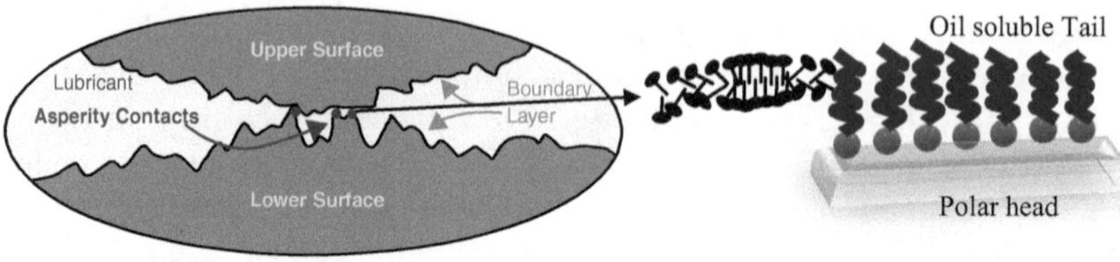

Figure 3.7 Lubrication mechanism of the common friction modifier.

3.3.1 Key applications

FMs are widely used in automotive engine oils and transmission fluids to improve fuel economy and reduce emissions. They are particularly important in low-viscosity oils designed for modern engines.

3.4 EXTREME-PRESSURE ADDITIVE (EPA)

These are a particular class of boundary lubricants that are effective in extreme-pressure conditions. Usually, they remain dormant in oil and get activated only when specific pressure and temperature conditions are attained during sliding [20]. They chemically react with the metal surface to form very hard compounds, having a higher load-bearing capacity and lower shear strength than metal. In particular, these additives are used in high dosage in industrial lubricants pertaining to the metal removing and forming operations, gear oil, etc [21]. EPAs reduce friction, control wear, and prevent severe surface damage. They are also referred to as anti-scuffing additives. The EPAs form a sacrificial layer by chemically thermo-chemical reaction with the sliding metal surfaces. The kinetics of the chemical reaction is a function of the surface temperatures generated by the localized high temperatures due to rubbing at the high-pressure condition that formation and breaking of short-lived micro-welds junctions in between counter asperities. Currently, four main types of EPAs are used worldwide, namely, sulfur, phosphorus, chlorine, and borate. In simple words, EPAs are designed for environments where surfaces are exposed to extreme loads and temperatures, such as in gears and heavy machinery. These additives chemically react with metal surfaces to form a solid protective film, typically composed of sulfur, phosphorus, or chlorine compounds. Under high pressure, sulfurized hydrocarbons react to form iron sulfide layers, which prevent the welding of metal surfaces. Phosphorus-based EPAs function similarly by forming iron phosphate films that reduce surface damage; refer Figure 3.8.

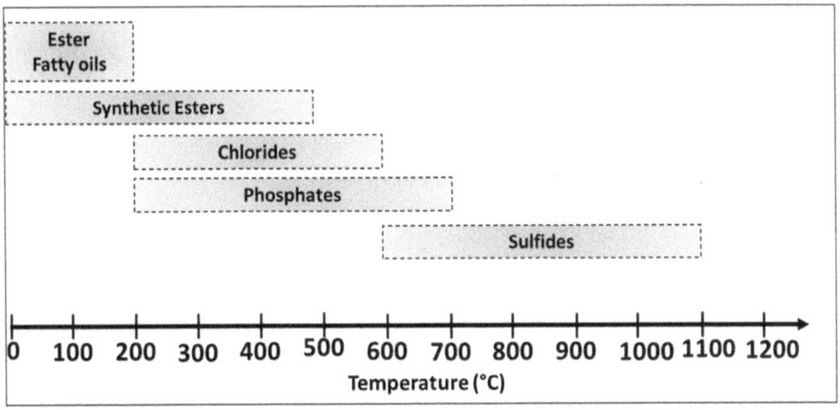

Figure 3.8 A simplified diagram showing the reaction of sulfur-based EPAs under high stress to form an iron sulfide protective layer.

3.4.1 Key applications and challenges

EPAs are critical in gear oils, metalworking fluids, and cutting fluids where high stress and temperature are present. These additives ensure the prevention of wear, scuffing, and pitting under extreme conditions. One of the primary challenges with EPAs is the potential for corrosion, particularly with sulfur-containing compounds. Under certain conditions, sulfur can react with metal surfaces to form corrosive compounds, which can lead to surface damage over time. As a result, the formulation of EPAs must balance the need for extreme-pressure protection with corrosion inhibition.

Another challenge is environmental regulation. Many traditional EPAs, such as those containing chlorine or sulfur, are subject to regulatory scrutiny due to their environmental impact. This has driven the development of alternative EPAs, such as boron or phosphorus-based compounds, which offer comparable performance with reduced environmental concerns.

3.5 CORROSION INHIBITORS

Corrosion inhibitors are critical additives in lubricants designed to protect metal surfaces from chemical attack [3]. Corrosion occurs when metal surfaces react with oxygen, water, or other corrosive agents, leading to the gradual degradation of the metal. This can manifest as rust, pitting, or surface weakening, which ultimately reduces the lifespan of both the lubricant and the equipment it protects. Corrosion inhibitors work by forming a protective layer on the metal surface, avoid corrosive agents' interaction with metal surfaces.

3.5.1 Mechanism of action

Corrosion inhibitors can work through several mechanisms, depending on the chemistry of the additive and the nature of the corrosion threat. One common mechanism is the formation of a passive layer on the metal surface. This layer can either be a physical barrier that prevents moisture and oxygen from reaching the metal or a chemically bonded film that stabilizes the surface, making it less reactive to corrosive agents.

For example, amines and carboxylates are common corrosion inhibitors that work by adsorbing onto metal surfaces and forming a protective film. These compounds are particularly effective in preventing rust formation in steel components, which are prone to oxidation in the presence of water or humid environments.

In other cases, corrosion inhibitors may work by neutralizing acidic compounds in the lubricant or the surrounding environment. For instance, sulfonates are commonly used as corrosion inhibitors in engine oils because they neutralize acidic byproducts that can form during combustion. By preventing these acids from attacking the metal surfaces, the inhibitors extend the life of both the oil and the engine components.

3.5.2 Applications and importance

Corrosion inhibitors are widely used in lubricants for a variety of applications, particularly in systems exposed to moisture, humidity, or other corrosive environments. Marine lubricants, for example, rely heavily on corrosion inhibitors to protect ship engines, gears, and other machinery from the corrosive effects of seawater and salt air. Similarly, hydraulic fluids used in outdoor machinery often contain corrosion inhibitors to protect against rust and oxidation in moist environments. In automotive applications, corrosion inhibitors are used in engine oils, transmission fluids, and coolant systems to protect internal components from oxidation and acid attack. Corrosion inhibitors are also important in metalworking fluids, where they prevent rust formation on both the tools and the workpieces, ensuring smooth operation and longer tool life.

3.5.3 Challenges with corrosion inhibiters

One challenge in formulating corrosion inhibitors is ensuring that they do not interfere with other additives in the lubricant. For example, some corrosion inhibitors may negatively interact with AW or extreme-pressure additives, reducing their effectiveness. Balancing these interactions is crucial for optimizing lubricant performance.

Additionally, corrosion inhibitors must be carefully selected based on the specific operating environment. A corrosion inhibitor that works well in a marine environment may not be suitable for use in high-temperature engine applications. Formulating lubricants with the appropriate corrosion protection for specific conditions remains a key challenge in the lubricant industry.

3.6 ANTIOXIDANT ADDITIVES

Antioxidant additives, such as hindered phenols and aromatic amines, prevent the oxidation of base oils, maintaining their viscosity and preventing the formation of harmful deposits. They act by neutralizing free radicals, which are highly reactive species formed during oxidation. So, this additive does not directly protect the surface but rather limit the formation of free radical that can influence the degradation of interacting metallic surfaces. These are one of important additives that indirectly influence the tribological performance of oils. Since, both synthetic and natural hydrocarbons mineral oils are prone to oxidization during their operation timeframe and produces sludge, acids and varnish. Their oxidation rate is mainly accelerated due to the following:

High temperature
Free radicals
Oxygen-rich environment

The oxidant inhibitors are mainly of three types [22]:

Peroxide decomposers: destroy peroxides which increases the oxidation rate.
Radical inhibitors: removes free radicals as they increase further oxidation.
Metal deactivators: deactivate the metal ion catalyst and inhibit the free radical formation.
Synergistic mixtures: (Homosynergism); usage of a mixture of two different peroxy radical scavengers. Hetrosynergism—usage of two mix oxidant inhibitors such as aminic and phenolic AO for effective suppression of the oxidation rate through synergism

Some of the commonly used different types of organic and organo-metallic antioxidants are shown in Figure 3.9.

Figure 3.9 Types of organic and organo-metallic antioxidants [23,24].

3.7 CONCLUSIONS

Surface protective additives are integral to the advancement of lubrication technology, providing essential protection against wear, corrosion, and oxidation. The ongoing research in this field focuses on developing new additives that offer enhanced performance while minimizing environmental impact. Future studies should explore the synergies between various additives and their long-term effects on lubricant performance in diverse operational conditions. Especially concern the advent of novel nano-additives those have shown high potential for incrementing load carrying capacity of oils and grease, along with in wear preventive properties as single and multi-hybrid additive systems. The implementation of novel antioxidant technologies remains critical for inhibiting oxidative degradation of base oils and improving their storage stability, facilitating the achievement of extended oil drain intervals targeted by the lubricant industry.

REFERENCES

1. Maurya U, Vasu V. Ionic liquids and its potential in lubricants: A review. *SSRN Electron J* 2018. https://doi.org/10.2139/ssrn.3320367.
2. Maurya U, Vasu V. Ionic liquids as a potential sustainable green lubricant for machining in the era of industry 4.0. In *Advances in Sustainable Machining and Manufacturing Processes*. Boca Raton: CRC Press; 2022. pp. 135–55. https://doi.org/10.1201/9781003284574-9.
3. Rizvi SQA. *A Comprehensive Review of Lubricant Chemistry, Technology, Selection, and Design*. West Conshohocken, PA: ASTM International; 2009.
4. Maurya U, Vasu V. Boehmite nanoparticles for potential enhancement of tribological performance of lubricants. *Wear* 2022;498–499:204311. https://doi.org/10.1016/J.WEAR.2022.204311.
5. Mang T, Dresel W. eds. *Lubricants and Lubrication*. 2nd ed. KGaA, Weinheim: WILEY-VCH Verlag GmbH & Co.; 2007.
6. Mang T, Lingg G. Base oils. In Dresel W, Mang T. (eds.). *Lubricants and Lubrication*. KGaA: Wiley Online Library (Wiley-VCH Verlag GmbH & Co.); 2017. pp. 51–82. https://doi.org/10.1002/9783527645565.ch4.
7. Spikes H. Friction modifier additives. *Tribol Lett* 2015;60:5. https://doi.org/10.1007/s11249-015-0589-z.
8. Spikes H. The history and mechanisms of ZDDP. *Tribol Lett* 2004;17:469–89. https://doi.org/10.1023/B:TRIL.0000044495.26882.b5.
9. Upendra M, Vasu V. Synergistic effect between phosphonium-based ionic liquid and three oxide nanoparticles as hybrid lubricant additives. *J Tribol* 2020. https://doi.org/10.1115/1.4045769.
10. Saini V, Bijwe J, Seth S, Ramakumar S. Augmenting the lubrication performance of nano-oils through synergistic co-functioning of nanoparticles. *Tribol Int* 2023;182:108332. https://doi.org/10.1016/j.triboint.2023.108332.
11. Minami I. Molecular science of lubricant additives. *Appl Sci* 2017;7:445. https://doi.org/10.3390/app7050445.
12. Maurya U, Vasu V, Saini V. Nano-innovations in lubrication: Unveiling the synergistic boundary lubrication potential of zinc titanate with phosphonium ionic liquid. *Tribol Int* 2024;199:109987. https://doi.org/10.1016/J.TRIBOINT.2024.109987.
13. Saini V, Bijwe J, Seth S, Ramakumar SSV. Interfacial interaction of PTFE sub-micron particles in oil with steel surfaces as excellent extreme-pressure additive. *J Mol Liq* 2020:115238. https://doi.org/10.1016/j.molliq.2020.115238.
14. Saini V, Bijwe J, Seth S, Ramakumar SSV. Unexplored solid lubricity of titanium nanoparticles in oil to modify the metallic interfaces. *Appl Surf Sci* 2022;580:152127. https://doi.org/10.1016/J.APSUSC.2021.152127.
15. Saini V, Bijwe J, Seth S, Ramakumar SSV. Potential exploration of nano-talc particles for enhancing the anti-wear and extreme pressure performance of oil. *Tribol Int* 2020;151:106452. https://doi.org/10.1016/j.triboint.2020.106452.

16. Spikes HA. Triboelectrochemistry: Influence of applied electrical potentials on friction and wear of lubricated contacts. *Tribol Lett* 2020;68:1–27. https://doi.org/10.1007/s11249-020-01328-3.
17. Maurya U, Vasu V, Kashinath D. Ionic liquid-nanoparticle-based hybrid-nano-lubricant additives for potential enhancement of tribological properties of lubricants and their comparative study with ZDDP. *Tribol Lett* 2022;70:11. https://doi.org/10.1007/s11249-021-01551-6.
18. Maurya U, Vasu V, Kashinath D. Three-way compatibility study among nanoparticles, ionic liquid, and dispersant for potential in lubricant formulation. *Mater Today Proc* 2022. https://doi.org/10.1016/J.MATPR.2022.03.329.
19. Tang Z, Li S. A review of recent developments of friction modifiers for liquid lubricants (2007 – present). *Curr Opin Solid State Mater Sci* 2014. https://doi.org/10.1016/j.cossms.2014.02.002.
20. Alves SM, Schroeter RB, Bossardi JC dos S, Andrade CLF de. Influence of EP additive on tool wear in drilling of compacted graphite iron. *J Brazil Soc Mech Sci Eng* 2011;33:197–202. https://doi.org/10.1590/S1678-58782011000200011.
21. Srinivas V, Thakur RN, Jain AK. Antiwear, antifriction, and extreme pressure properties of motor bike engine oil dispersed with molybdenum disulfide nanoparticles. *Tribol Trans* 2017;60:12–9. https://doi.org/10.1080/10402004.2016.1142034.
22. Mortier RM, Fox MF, Orszulik ST. *Chemistry and Technology of Lubricants.* Dordrecht: Springer; 2010.
23. Soleimani M, Dehabadi L, Wilson LD, Tabil LG. Antioxidants classification and applications in lubricants. In Johnson D. (ed.). *Lubrication Tribology, Lubricants and Additives.* London, UK: IntechOpen Limited; 2018. p. 23. https://doi.org/10.5772/intechopen.72621.
24. Rudnick LR. Additives for industrial lubricant applications. In Rudnick LR. (ed.). *Lubricant Additives* London: CRC Press (Taylor & Francis Group); 2017. pp. 473–84.

Chapter 4

Lubricant additives

Protectant agents

Manoj Kumar Gupta

4.1 INTRODUCTION

In the real life, we encounter a number of assemblies and sub-assemblies. There is relative motion between different components of these assemblies and sub-assemblies. During the relative motion, the coefficient of friction between different material combinations comes into the picture. During asperity–asperity contact, this friction leads to heat generation due to the contact between surfaces of the tribo-pairs. Consequently, there are other detrimental effects, and hence, it is necessary to reduce the friction coefficient. For this purpose, certain lubricating materials are added. The primary purpose of oil is to decrease friction and wear through the lubricating film between pairs. A lubricant smoothens the instrument by taking off and suspending sludge, carbon, and varnish deposits. Lubricants also provide surface protection due to oxidation and corrosion. In addition to this, lubricants dissipate heat away from the parts.

Based on the origin source, there are four kinds of oil, namely, vegetable, animal, mineral, and synthetic oils. Vegetable oils (VOs) are triglycerides extracted from plants. These oils are less stable at higher temperatures due to rapid oxidation. Major VOs are extracted from rapeseed, sesame, safflower, canola, sunflower, etc. Animal oils are lipid materials produced from the animals. Hard fats (stearin) and soft fats (lard) are two main animal fats. Examples of animal oils are fish oil, lard oil, sperm oil, whale oil, and tallow oil. Most common mineral oils are petroleum-based lubricants. Only approximately 1% of crude oil products are used as lubricants, including insulating oils [1]. Mineral oils are derived from petroleum distillation. Synthetic oils contain esters, polyethers, silicones, and halogenated fluids and are used to develop lubrication properties as required. Polyalphaolefins, polyisobutenes, alkylated aromatics, and cycloaliphatics are four categories of synthesized hydrocarbons. Organic esters are classified into dibasic acid esters and polyol esters. Water-soluble and hydrocarbon-soluble polyglycols are two types of polyglycols. Other synthetic lubricants include phosphate esters, silicon-containing fluids, halogenated fluids, and polyphynol

DOI: 10.1201/9781003642848-4

esters. Oils are also classified according to Society of Automotive Engineers (SAE) and API. API has classified base oils into five groups, from Group I to Group V.

During the refining of crude oil, a majority of mineral base oils are produced [2]. The main differences between crudes are the relative quantities of respective hydrocarbon, which is present in each source of raw oil. Since base oil is the maximum part of a lubricant, base oil characteristics greatly affect lubricant properties. For effective lubrication, it must possess good cleansing and dispersing ability, appropriate viscosity, higher film strength, slipperiness, nontoxicity, lower corrosivity, lower pour point, and low flammability and volatility. Criteria for base oil selection include its density and viscosity at lower and higher temperatures. Selection can also be done by considering foaming characteristics and seal compatibility. In addition to this, oxidation resistance, viscosity index, cloud point, pour point, volatility, and flash point are also criteria for base oil selection.

Generally, commercial oils are mixtures where the composition is selected as per the application [3]. Paraffinics (%Cp), naphthenics (%C_N), and aromatics (%C_A) are three main components of the base oil [1]. Paraffinic mineral oils are formed by the solvent extraction process or hydrocracking. These are non-ring long-chained structures and have a high pour and flash point. These oils are used in engine oils, textile, and rubber industries. Naphthenic oils are extracted by crude oil distillates to obtain a saturated ring structure. Low viscosity and low flash point are the properties of naphthenic oils. Aromatic oil is obtained as a product of paraffinic oil refining. It includes non-saturated ring structures and have a high flash point. Chemical forms of mineral oils are shown in Figure 4.1. Molecular weight, refractive index, and density are necessary to determine the amount of each compound in a mineral oil. This is calculated through an equation as per ASTM D3238. To achieve the desirable properties as per the different applications, chemicals are added to the base oil [4].

Modern history of additives in oils started in the early 20th century. During those days, fatty oils and sulfur were added to base oils to improve lubrication, especially at higher loads. After the 1930s, additives were used commonly due to the development of compact and faster engines. Current lubricants are a mixture of chemicals that ensure lower friction, which provides other functionalities, for example, reduced wear, sealing, and soot control. A formulated oil comprises a base fluid [6] and a performance package. The performance package comprises additives that help improve the lubricating ability of base oil. Most additives except some viscosity index improvers (VII) and pour point depressants (PPDs) contain an oleophilic-lipophilic hydrocarbon group and a polar functional group. The polar functional group consists of oxygen, nitrogen, sulfur, and/or phosphorus [4,7].

Additives may be categorized in a number of distinct ways, i.e., they protect base oil, improving performance of the base lubricant, and protect

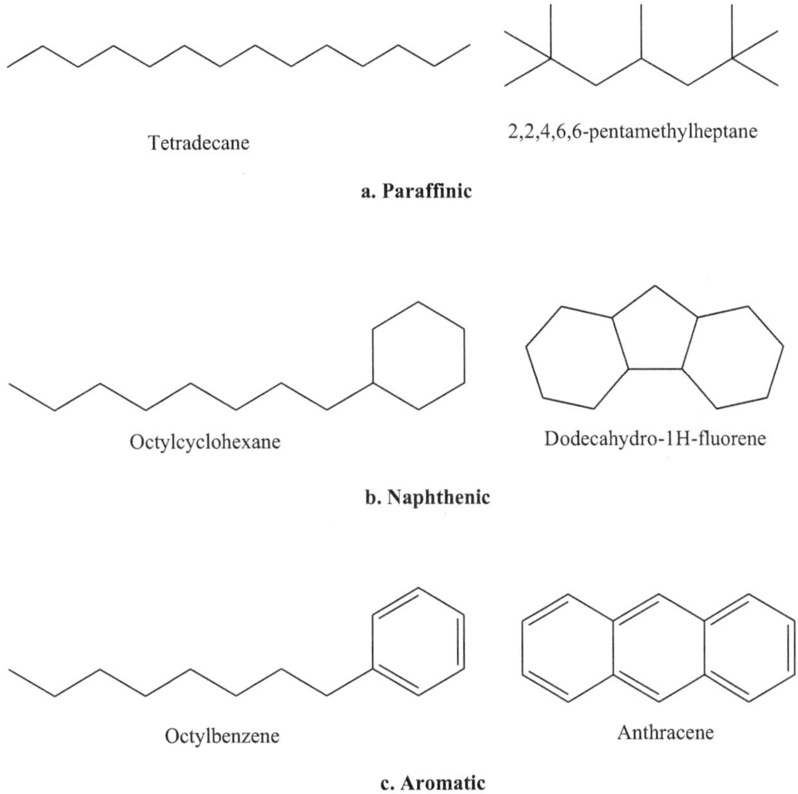

Figure 4.1 Chemical forms of mineral oils [5].

the application area. Minami [8] segregated additives on the basis of their role, i.e., tribo-improvers, rheo-improvers, mainteners, and auxiliaries. Tribo-improvers enhance the tribological performance of the formulated oil. VIIs and PPDs are classified as rheology improving additives. Maintainers keep contact surfaces and oil in a fair state. Other additives are categorized as auxiliaries. Figure 4.2 represents a schematic for base oil to commercial oil formation.

Due to the chemical reactivity of additives, they can intermingle with other additives of the package, either synergistically or antagonistically [9]. Misra et al. [10] used different polymers and reported synergistic and antagonistic effects in wear-preventive characteristics. They summarized that the total response of all additives resulted synergistically or antagonistically depending on the interaction of individual components. Hence, it is the duty of the oil formulator to optimize the antagonistic synergistic effects of additives in formulated oil, which is ready for commercial use.

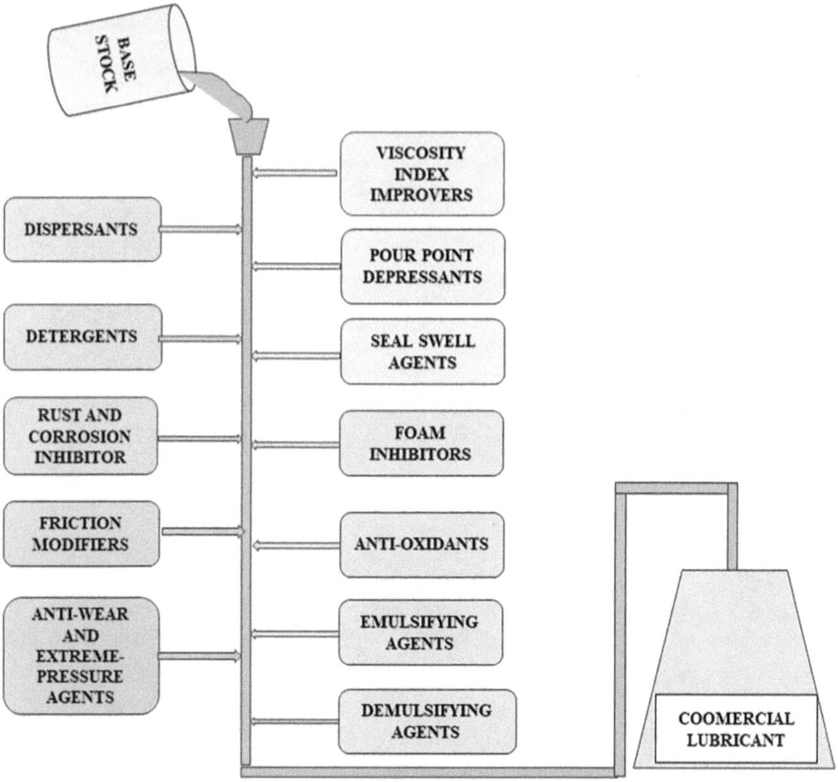

Figure 4.2 Schematic for base oil to commercial oil formation.

4.2 SEAL SWELL AGENTS

A standard oil seal is having an outer circular metal disc along with inner flexible rubber. Inner flexible rubber is attached to the metal through the vulcanization process. There are a number of uses of seals in machinery and other industrial applications. Oil seals are used to protect bearing and shafts from foreign materials and simultaneously do not allow oil or grease to leak from the system. Rotary seal is an oil seal that seals the rotating member. Oil seals and "O" rings are used in pumps, pipe connections, and pistons. These are also used in all hydraulic systems, automotive power train systems, compressors, etc.

Seals in automotive serve a number of purposes. Among the different uses of the seal, the most prominent is easy access to the malfunctioning parts during the repair. Minimization of contaminants and loss of the lubricant are reduced by the use of seals in the automotive components. A number of polymeric materials are used for making the seals. These polymeric materials comprise fluoroelastomers, polyacrylates, nitrile rubber, and polysiloxane silicones. The critical challenge with the seal is to maintain the integrity. If the lubricant is lost from the system due to the integrity issue of the seal, then wear of the instrument will happen and it may lead up to equipment failure. In addition to the above problem of seal, it may fail due to shrinkage or elongation. Seals may deteriorate after becoming brittle [11]. The destruction of elastomer used in the seals is evaluated by soaking it in the oil for a long duration and calculating the tensile strength, change in volume, tendency to elongate, hardness, and rupture.

There are two mechanisms which are used to explain the seal damage. First is the abrasion of the seal due to oil debris, and the second mechanism includes the attack of different oil components on the seals. Due to the installed oil filtration system in most of the instruments, abrasive damage is not common. However, damage due to oil is of primary concern. As we know, modern oils are oil blends with additive packages. Specific base stocks, for example, those consisting of ester type, are having more susceptibility to extract the plasticizer because they are having a higher polarity. Since additives have the capability to diffuse into the seal material, this can change the seal properties and can remove the plasticizer. For example, dispersants cause seal damage, especially to fluoroelastomer seals [12]. If oil components diffuse into the seals, then there are fair chances of lubricant-related damage. Due to the diffusion of oil components into the seal, the seal's hardness changes consequently, leading to swelling. Generally, the higher the nitrogen, the more will be the seal problems. Problems occur due to the lower molecular weight molecules present in the dispersant. They include free amine or low molecular weight succinimides and succinamides.

To overcome this problem, an additive is added to the polymer of the seals, which gives flexibility along with strength. Polyesters, a few phosphorus derivatives can be used as seal swell agent additives in the oils. The use of these materials is very common in transmission and hydraulic oils. Seal swell agents (SSAs) are used to address the issue of seal shrinkage because this seal shrinkage causes oil leakage [13]. Functions of SSAs in lubrication systems are as follows:

- To segregate from harmful particles
- Helping in maintaining pressure
- To minimize contamination

Elastomeric sealing elements are generally used for transmissions and hydraulic instruments. The presence of an over-based detergent in the lubricating oil often greatly extends the life of the seals due to the suppression of acid build-up [14]. Figure 4.3 shows the structure of elastomers.

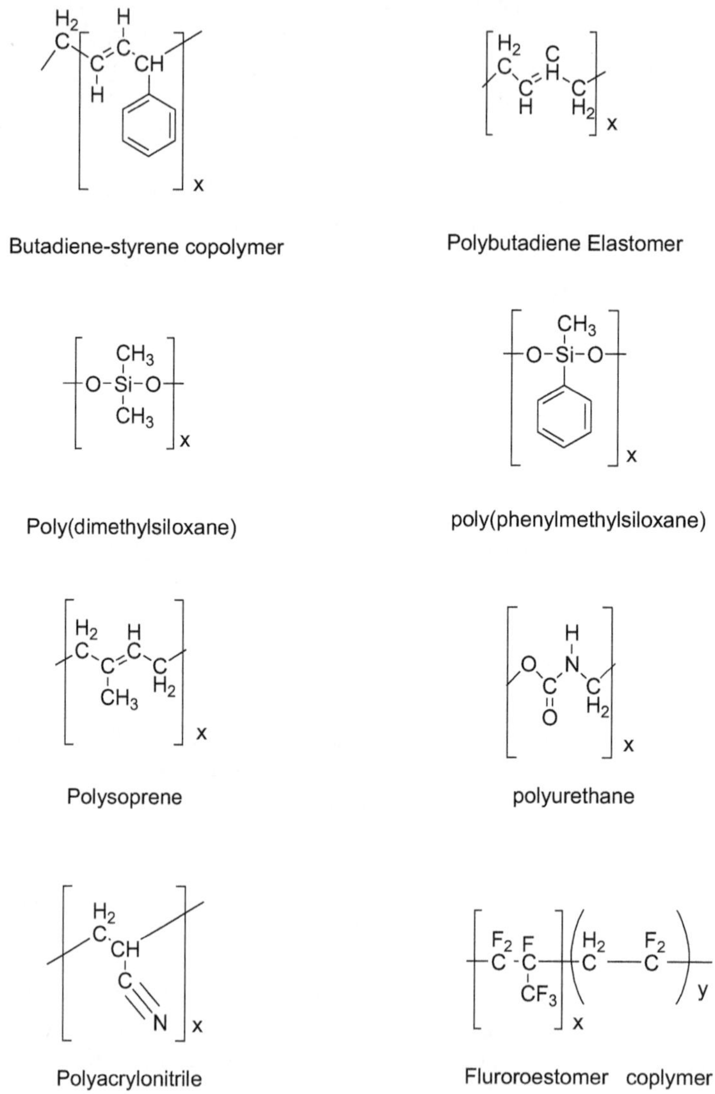

Figure 4.3 Molecular structure of elastomers used in seals [4].

4.3 VISCOSITY INDEX IMPROVERS

Viscosity is a significant property of lubricants. It is classified into dynamic and kinematic viscosity. Dynamic viscosity is based on the resistance against the shear, while kinematic viscosity is based on the resistance against the gravity. In the practical applications, there are a number of variables on which viscosity depends. Among these variables, temperature is one of the variables in which viscosity changes appreciably. For industrial practitioners, the unfortunate truth is that the mineral base oil loses its effectiveness at higher temperatures due to reduced viscosity because of heat. This also results in diminishing lubricant film on the application surfaces. Hence, it necessitates the study of the viscosity–temperature relation for finding the solution to problems related to viscosity–temperature phenomenon. There is a term called VI, which relates the variations in lubricant's kinematic viscosity to changes in temperature. This is an empirical number. The higher the VI value of a lubricant, the greater the resistance to thin out at higher temperatures or to thicken out at lower temperatures. Oils, which have a smaller viscosity–temperature dependency, will be a suitable candidate for use in varying temperature conditions [15].

VI is evaluated as per the ASTM D2270 [16]. Here, first of all, the kinematic viscosities of the given oil sample are calculated at 40°C and 100°C in accordance with ASTM D445. For measuring the kinematic viscosities, capillaries are used as per the specifications given in the standard. L, Y, H, and U are defined as per ASTM D2270 to evaluate VI in mm^2/s [17]. Now, if the kinematic viscosity of the sample at 100°C is more than 2.0 mm^2/s and less than or equal to 70 mm^2/s, then "L" and "H" should be extracted from Table 4.1 of ASTM D2270. In case the measured value of kinematic viscosity is not present in the table, then it can be interpolated from the range (of the evaluated value of kinematic viscosity) of the table. If the kinematic viscosity of the sample at 100°C is greater than 70 mm^2, then the values of "L" and "H" are calculated using ASTM D2270. After evaluating and extracting the required parameters to calculate the value of VI, check $U>H$ or $U<H$. Further, ASTM D2270 is used to determine the viscosity index.

A higher value of VI represents a relatively small change of viscosity with respect to temperature and vice versa. Commercial oils for internal combustion engines (ICE) are typically multi-grade engine oils, with VI ranging from 150 to 200. Since we need to enhance the performance of commercial oils for temperature-varying applications, it necessitates a VII. The required purpose is met by using additives called VIIs; these are added to base oils for enhancing viscometric and rheological properties.

VIIs are also called viscosity modifiers or viscosity improvers. VIIs are used in power steering oils, multi-grade engine lubricants, automatic transmission oils, gear oils, and hydraulic oils. At lower temperatures, VII molecules remain in the coiled form. At high temperatures, two key characteristics of VIIs are thickening efficiency and shear stability. Thickening efficiency is the amount of polymeric additive that must be employed in a lubricant formulation to reach the desired kinematic viscosity at a particular temperature, generally at 40°C [18].

Selby [19] proposed the mechanism of VII. Flory [20] reported that the radius of gyration of polymer is the function of the interactions between polymer chain segments and solvent molecules. The performance of a dispersant is measured in relation to its efficiency for dispersancy and thermal and oxidative stability [21,22]. Researchers examined the effect of VIIs on viscosity of engine oils and proposed a model which was a linear equation with an intercept. The proposed model relates changes of final oil viscosity with respect to base oil viscosity [23]. VIIs [24–26] are mainly categorized into five classes of polymers used in modem lubricants [27]. Some of the VIIs are shown in Figure 4.4.

4.4 POUR POINT DEPRESSANT

Whole dewaxing of the base oils in petroleum refining is practically impossible due to process constraints and economics. In addition to this, since wax offers a high VI property [29], the presence of wax is desirable. The pour point is defined on the basis of the viscosity–temperature relationship. Generally, straight-chain paraffin (amount of wax) in oils is represented in terms of the pour point. At lower temperatures, wax is likely to disintegrate as crystals having lattice-type structures. Separated crystals can trap a considerable amount of oil through association and inhibit oil flow that leads to improper lubrication of the instrument. Under specified conditions, the pour point is the lowest temperature at which a lubricant will flow. At the pour point, flow of the oil sample takes place due to gravity alone. The pour point is significant in diesel fuels and base oils. Typically, lubricant suppliers are providing the pour point of supplied oils to determine oil's pumping ability and low-temperature applications suitability. With the increase in the viscosity, the pour point increases. For example, an ISOVC 46 has a pour point of –39°C, while ISOVC 460 has a pour point of –15°C [30]. To overcome the problem of wax formation during processing and transportation, especially at low temperatures, certain chemicals are added in selected amounts. These added chemicals are known as PPD. Removal of the residue waxes from the crude oil is costly as compared to the addition of PPDs to lower the pour point of oil. Additives used for VI improvement are commonly used as PPDs. PPDs work as wax-crystal modifiers and alter the crystal size. High-viscosity lubricants may stop flowing at lower temperatures because of very high wax formation [31]. PPDs have fundamentally no effect on the cloud point and quantity of separated wax. The performance evaluation of PPDs can be carried out through ASTM D97, ASTM 3829, and ASTM D2602 standards.

The mechanism of pour point reduction using PPDs is still part of research, and investigations are being carried out in this direction. Soni et al. [32] synthesized comb-shaped polymeric diesters with different

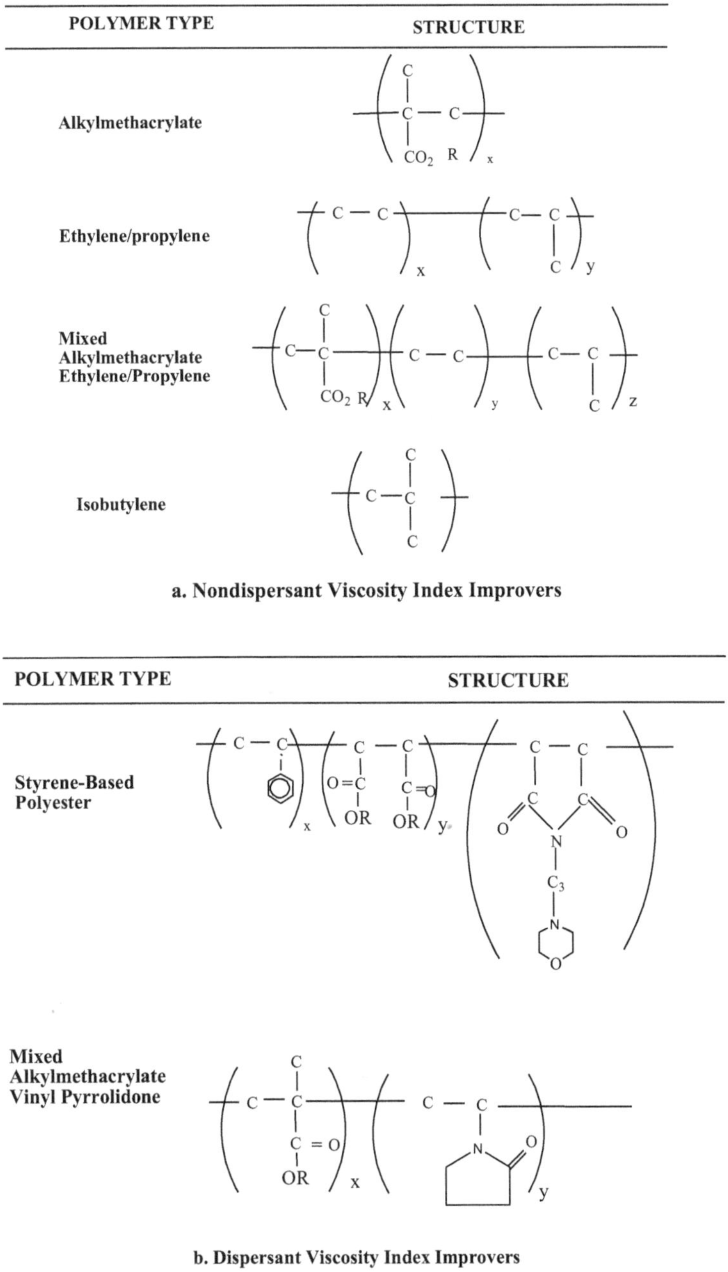

Figure 4.4 Viscosity index improvers: (a) non-dispersant VII and (b) dispersant VII [28].

unsaturated C_{22} esters for crude oil pour point reduction by providing various nucleation sites for wax precipitation. The mechanism justified by the adsorption of molecules onto the wax crystals. However, reduction of the pour point by PPDs from about −6.67°C to −1.1°C is more common [33] PPDs act via surface adsorption onto the emergent, growing, wax crystals. The subsequent surface layer of PPD hinders the growth of the wax crystals and the capability to adsorb oil and form gels. The restrictions of PPDs are controlled by the nature of oil and the concentration and nature of PPD. Different PPDs have different efficiencies, and the efficiency is maximum at an optimum concentration. Of the commercial PPDs, alkyl aromatics are believed to function through the absorption mechanism and aliphatic polymers through co-crystallization [34].

A good PPD can reduce the pour point of a lubricant by 40°C at the maximum. PPDs are used in mineral oil-based lubricants. Two general types of PPDs are used:

1. Alkylaromatic polymers
2. Polymethacrylate co-crystals

Most commercial PPDs are organic polymers. However, a few non-polymeric materials are shown to be effective as PPDs. Commercial PPDs incorporate polyalkyl methacrylates, polyalkyl fumarates, oligomerized alkylphenols, and other mixed hydrocarbons. In Figure 4.5, the structure of PPDs is shown.

In addition to the above PPDs, a number of other PPDs are listed below [24,31]:

- N-alkyl-2-aminopyridinum
- Polymeric additives based on vinyl acetate copolymers
- Oleic acid-based or maleic anhydride-based polymeric additives
- Oil-based and water-based viscosity reducer
- Aromatic polyisobutylene succinimides
- Styrene and vinyl acetate (SVA) copolymers
- Bio-based flow improver
- Hexatriethanolamine mono oleate
- Aminated copolymer and ethylene-vinyl acetate copolymers (EVA)
- Modified maleic anhydride copolymers
- Paraffin inhibitor from VO
- Anhydride and esters of n-alkyl alcohol-based polymeric additives
- Poly-hydrazide
- Triethanolamine (TEA)
- Ethylene acrylic alkyl ester copolymer wax dispersants
- Hetero atom compounds
- Multifunction copolymer

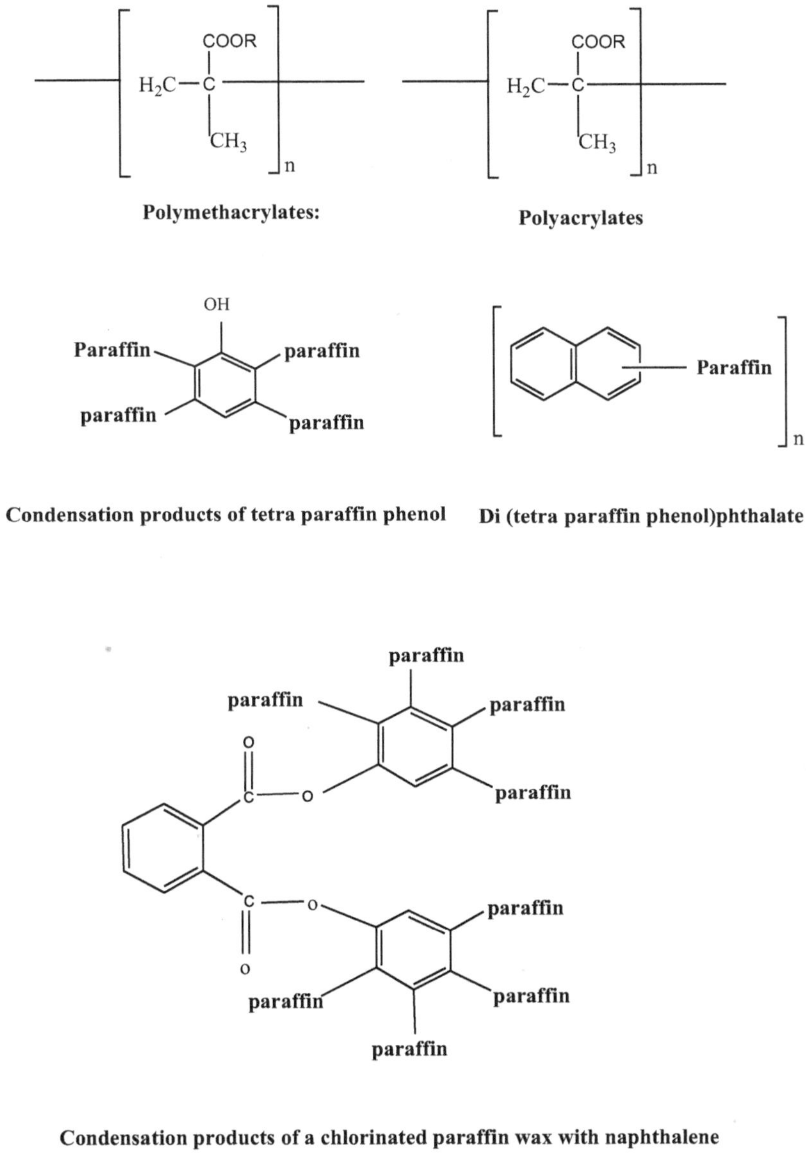

Figure 4.5 Structure of PPDs [35].

- Polymeric PPDs
- Emulsified acrylate polymer
- Copolymer solubilized in organic solvent with an aromatic hydrocarbon
- Oleic acid-based polymeric flow improvers

- Comb-type polymeric acid alkylamide-co-α-olefin
- Spirocompound as clean viscosity reducer
- Low molecular weight poly biphenyl acrylate
- Methacrylate-based polymeric additives
- Polymeric structures as cold flow improvers
- Phthalic anhydride-polyethylene glycol ester (PAP)
- PE-b-PEP de-block copolymers
- Modulating long-chain paraffin crystallization with comb polymers
- Polymeric additives based on vinyl acetate copolymers
- Polymethyl methacrylate-graphene oxide nano-hybrid polymers
- Hydrophobically modified polyampholyte
- Four sets of copolymeric additives
- Synthesized polymeric PPD from olive oil
- Comb-type copolymers of poly(maleic anhydride-co-α-olefin-co-styrene) (MASC)
- Maleic anhydride copolymers (MAC) and poly(ethylene butene) polymers
- Vinyl acetate-biphenyl fumarate copolymer
- *Madhuca longifolia*
- Esters of styrene-maleic anhydride copolymers
- Poly(n-alkyl itaconate-co-vinyl acetate)
- Poly(styrene-co-octadecyl maleimide)
- Biosurfactants from *Pseudomonas putida* and *Halomous xianhensis* (bacteria)
- Long-chain acrylic ester polymers with polar monomers
- Graphene-based nanocomposite PPDs

4.5 ANTIOXIDANTS

Oxidation is a chemical process that occurs in oils when exposed to an oxidizing agent, for example, oxygen. The oxidation process can be catalyzed by iron and copper. Reactions of oxygen with hydrocarbon polymers and oils, with varying temperatures and pressures of oxygen, are not desirable, and these lead to damage to application materials [36,37].

Oxidation of oils is the most unwanted phenomenon during the application. Because of the prevailing working environment, the temperature, metals, and air in a given application lead to enhanced oil oxidation. As we know, an atom of oxygen has reactivity with a carbon atom because of unsatisfactory valences. In addition to this, due to thermal, mechanical, and chemical stresses in the oil, bonding between successive carbon atoms weakens, which leads to oxidation of oil. The rate of oxidation depends on the following:

- Chemical stability of mineral oil, which is governed by the degree of refining
- Temperature
- Catalytic act of dissolved metals
- Aeration
- Water, for example, automobile crank cases and gear cases have water present in cooler parts

Oxygen atoms may form CO, CO_2, or a chain of reactions, leading to aldehydes, ketones, acids, etc. Sludge, gum, varnish, lacquer, carbon deposits, acidic (organic) compounds, etc. are the main oxidation products. Many acids, e.g., H_2SO_4, HNO_3, and HCl are also produced due to the presence of sulfur, nitrogen, or chlorine in the mineral oils or diesel fuels. These acids lead to corrosion. Moreover, oxidation products get deposited on various parts of the engine and can block oil supply pipelines, filters, etc. apart from leading to deterioration in the lubricating efficiency of oils. Oil oxidation creates the following problems [38]:

- As oxidation products are sticky and due to the inclusion of these sticky oxidized products, viscosity of oil increases.
- Increase in acidity leads to reduced alkalinity.
- Deterioration in friction and wear performance.

Since oxidation is the most undesirable process for oils, it is essential to enhance the oxidation stability of lubricants to the optimum extent. Addition of antioxidants (AOs) to lubricating oils controls the oxidation process since they preferentially undergo oxidation and save oil molecules. Once these are consumed, oil is vulnerable to oxidation.

4.5.1 Autoxidation of hydrocarbons

Thermally induced hydrocarbon oxidation is called autoxidation. It is categorized in accordance with two temperature phases, i.e., "low,"30°C to 120°C, and "high," >120°C.

4.5.1.1 Oxidation of hydrocarbons at low temperatures (30°C–120°C)

Initiation of radical chain reactions, its spread, branching, and termination are four stages of the mechanism of oxidation of hydrocarbons at low temperatures [36,37]. At moderate temperatures and partial pressure of oxygen, a radical chain reaction begins [39]. The rate of commencement is very slow [40]. In the next step, chain propagation takes place. Peroxyl radicals are present in relatively high concentrations, as compared to other radicals, due to their low reactivity [41]. In the third step, different types of hydroperoxides are generated. At low concentrations, they may yield an alkoxy and a hydroxy radical [42]. The last stage is the termination of the radical chain reaction. Auto termination is called the "chain termination reaction." This phase may be affected by the combination of radical species and alcohols.

4.5.1.2 Oxidation of hydrocarbons at high temperatures (>120°C)

Above 120°C, the oxidation can be separated into primary and secondary phases. The first stage is similar to low-temperature oxidation. However, in this case, the first stage has less selectivity and enhanced reactivity [43–45]. The secondary phase includes polycondensation and polymerization reactions of high molecular weight that produce intermediates form products which are not soluble [46].

4.5.2 Oxidation stability of base oils

All modern lubricants are prone to oxidation [47,48]. Every base oil has a threshold, outside which oxidation inhibitors or AOs are required. The model for high-temperature oxidation is described by researchers [49].

There are different classifications for AOs [50]. Based on the source, AOs are categorized as natural and synthetic. Based on the solubility, AOs are classified into oil-soluble and water-soluble. Oil-soluble organic AOs are important for lubricating oils. Based on the action mechanism, AOs are segregated as primary AOs (radical scavengers), secondary AOs (peroxide decomposers), and metal deactivators [51]. Some of the AO examples are hindered phenols, ethers, arsenic organo Zn and Cu compounds, etc. Figure 4.6 shows the mechanism of oxidation.

Some of the oil-soluble AOs are compounds of sulfur, phosphorus, sulfur-phosphorus, aromatic amine, hindered phenolic, organo-alkaline earth salt, organo-zinc, organo-copper, and organo-molybdenum [47,48]. Due to the enhanced rate of oxidation [52], if this process is not controlled, the lubricant decomposition will lead to oil thickening, sludge formation, and the formation of varnish, resin, and corrosive acids [53,54].

Lubricating oils are susceptible to degradation by oxygen. The oil oxidation [55] process is the major cause of oil thickening. AOs are additives added in lubricants to minimize and delay the oxidative degradation of oil. The hydroperoxide decomposers transform the chain-propagating hydroperoxides into alcohols. The radical scavengers remove the radicals from the oxidation process by transferring a hydrogen atom. Figure 4.7 shows a few AO structures.

Intitaion:

$RH + O_2 \longrightarrow R' + HO_2$

$RH + O_2 + RH \longrightarrow R' + H_2O + R'$

Propagation:

$R' + O_2 \longrightarrow ROO'$

$ROO' + RH \longrightarrow ROOH + R'$

Peroxide Decompostion

$ROOH \longrightarrow RO' + O'H$

$2ROOH \longrightarrow RO' + ROO' + H_2O$

$RO' + ROOH \longrightarrow$ Various Products

$\left[\begin{array}{l} ROO' \longrightarrow \\ ROOH \longrightarrow \end{array}\right.$ Non-radical products

$X + ROOH \longrightarrow$ Free radicals

$Y + ROOH \longrightarrow ROH + YO$

$Z + ROOH \longrightarrow$ Inactive products + Z

$M + ROOH \longrightarrow$ Free radicals

Termination

$ROO' + ROO' \longrightarrow$ Inactive products

$ROO' + IH \longrightarrow ROOH + I'$

RH	=	Organic substrate
ROO'	=	Peroxy Radical
ROOH	=	Hydroperoxide
X,Y,Z	=	Decomposers of organic peroxides
M	=	Metal
I^*	=	Stable or unreactive radical
IH	=	chain breaking Inhibitor

Figure 4.6 Mechanism of oxidation [47].

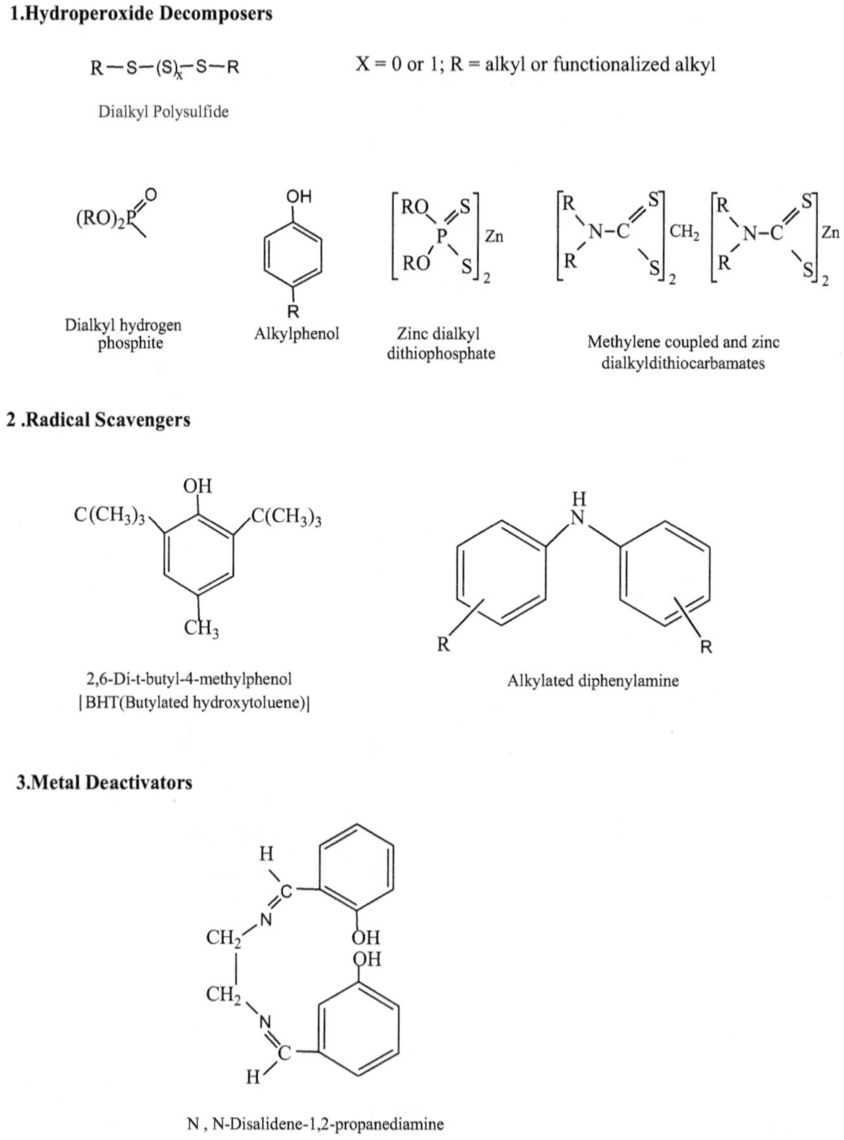

Figure 4.7 AO structures [4].

The two most common test methods for oxidation resistance are ASTM D 943 and ASTM D 2272 [56]. The CRC L-60 test is used for gear oils. Huang et al. [57] synthesized phenolic AO (3,5-di-tert-butyl-4-hydroxy-phenyl) propionic acid (DBHP). It was functionalized with ZnO nanoparticles (NPs). It was reported that DBHP-ZnO NPs exhibited better

AO performance. The harmful effect of dispersants on the oxidation stability of the ZnDTP-added oil was increased due to calcium detergents [58] and, to a great extent, magnesium detergents. Hence, the additive combinations are synergistic and/or antagonistic, as reported by a number of researchers. Interaction patterns also depend on the composition of base oil and additive concentration in a formulated oil [59].

4.6 ANTI-FOAM AGENTS

Foaming facilitates the construction and coalescence of gas bubbles on the surface of a lubricant. Foaming is required in certain applications, for example, flotation, washing, and cleaning. However, in a number of cases, foaming is undesirable in lubricant-related applications, distillation, and pumping. Over filling of sumps, air leaks, and contamination may result in foaming. Loss of oil out of a vent may be because of foaming. Additionally, too much foam in the liquid lubricant can starve bearings and pumps, resulting in pump cavitation, performance degradation of systems, and even failure of components and/or systems. Gases are soluble in mineral-based oils up to a certain limit. For example, at room temperature, in mineral oils, eight to nine times of air (by volume) is soluble and is invisible. Bubble production is expected during lubrication. This yields an air lock, which interrupts oil pump supply and leads to oil shortage [60].

There are a number of issues because of the foam during the process. The destructiveness of foam is due to the following:

- Degradation of lubrication and enhanced wear: Crossing the saturation limit of gas in the oil, small bubble formation will take place. The formed bubbles may be suspended and make the appearance of oil hazy. The bubbles gradually upsurge to the surface. Due to this, solid-to-solid contact increases due to poor film continuity of oil with entrained air [61].
- Cooling and heat dissipation performance reduction: We know that one of the responsibilities of lubricating oil is to dissipate heat under circulation. However, due to the presence of entrained air in oil, the heat dissipation function of lubricating oil is affected adversely [62].
- Deterioration of dispersing and cleaning effect: At high temperatures, oxidative metamorphism of oil increases due to the increased contact area between oil and air because of foam. This generates more carbides and sludge, resulting in weakened cleaning properties [61].
- Degradation of the anticorrosion and antirust effect, air locking, and flow interruption [63], aggravating oxidation and deterioration of lubricating oil [64], and safety hazards occur because of the harmful effects of the foam [65].

This enhances oil's oxidation and reduces the oil change cycle. It also decreases fluidity and lubricity of the lubricant, while intensifying wear. The amount of soluble gases depends on the gas type and the temperature of the oil [66]. Dissolved gases influence fluid properties like foaming, viscosity, cavitation, heat transfer, boundary lubrication, oil oxidation, and metal oxidation. Foaming in lubricants is undesirable, which might lead to an insufficient oil supply of the lubricant in a circulation system. Foaming is affected by the surface tension property of base oil, the presence of detergents, corrosion inhibitors, acidic contamination, etc. Among the physical properties of lubricating oil, surface tension, density, and viscoelasticity of the surface film have a great influence on foaming [67,68]. The greater the difference between the surface tensions of lubricating oil and defoaming agent, the earlier the defoaming agent diffuses into the foam film [69].

4.6.1 Defoaming methods

Defoaming methods can be divided into physical and chemical defoaming methods. Under physical defoaming, two sub-classifications are physical bubble suppression and physical bubble bursting. Under chemical defoaming methods, the two sub-classifications are chemical bubble suppression and chemical bubble bursting. The main feature of the defoaming process is the small surface tension of a defoaming agent. Additionally, the defoaming process characteristics are the insolubility and the highly dispersed colloidal state of a defoaming agent.

To overcome the harmful effects of foaming, certain chemicals are added to the lubricating oil in appropriate amounts. These additives are called anti-foam agents (AF), defoaming agents, or foam inhibitors. Silicones are generally used as anti-foam additives in most of the lubricants. They have low surface tension (in the range of 21–25 dynes/cm at 35°C) over a wide viscosity range. The addition of foam inhibitors reduces foaming, and most commercial lubricants contain foam inhibitors. A few ppm of silicone is commonly used [69]. Figure 4.8 shows the structure of some common foam inhibitors.

There are a number of literature studies that describe the solubility of different gases in mineral oils and also provide mathematical models for evaluating solubility. One of the standard is ASTM D2780. In this standard, the amount of fixed soluble gas is estimated in liquid. Here, using the extraction process, gas is segregated physically and the quantity is measured (volumetrically). Successive qualitative analysis for extracted gas can be carried out through appropriate techniques. ASTM D 3427 is used to measure the relative tendency to release entrained gas from the oil. This standard used the gas bubble separation method for the measurements. The ASTM D892 test is used to assess a lubricant's foaming tendency.

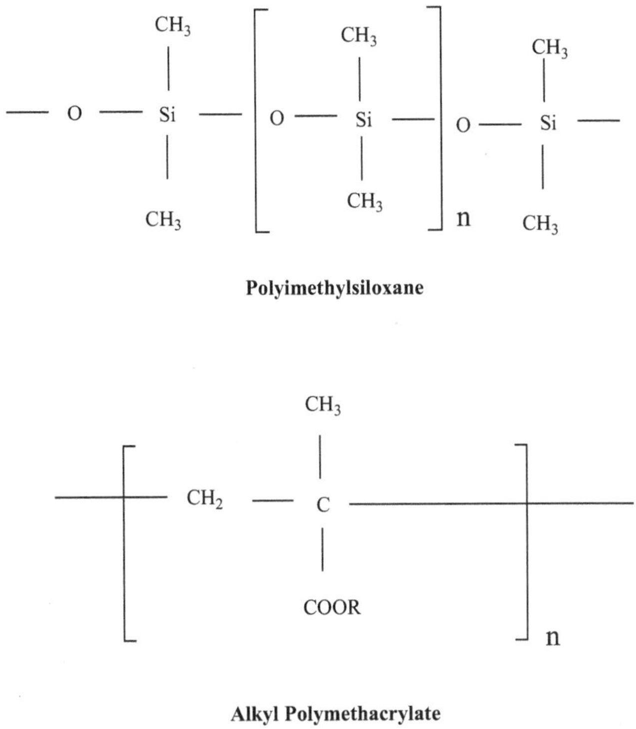

Figure 4.8 Structure of some common foam inhibitors [4].

4.7 EMULSIFYING AGENTS AND DEMULSIFYING AGENTS

4.7.1 Emulsifying agents

Corrosion in pumps occurs due to the water generated from crude oil. It also poisons the downstream refinery catalysts [70,71]. Henceforth, there is a need for a variety of properties for casting off water emulsification from crude oil. Thus, in the oil industry, the stability of crude oil emulsion should be improved. According to Abdulredha et al. [70], in the petroleum industry, from the beginning till the refinery process, emulsion plays a very critical role. Inclusion of resins and asphaltenes in oil-in-water emulsions through different surface-active materials makes them difficult to handle [72].

We can find emulsions in everyday life, for example, in mayonnaise, yogurt, milk, butter, pharmaceuticals, paints, cream, and others. Likewise, lubricants, emulsion water, surfactants, and energy-related processes are also familiar areas where emulsions can be found. Here, focus is given to emulsions with respect to the oil industry. Emulsions are also called

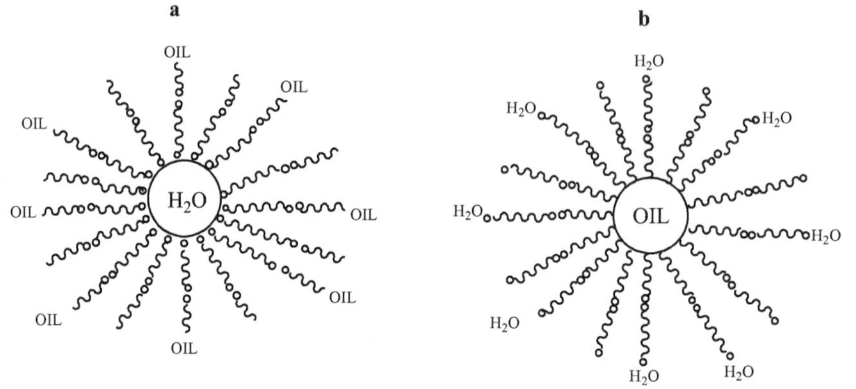

Figure 4.9 A representation of (a) water-in-oil emulsion and (b) oil-in-water emulsion [4].

emulgents or emulsifiers. Emulsion is a colloid in which one state is distributed into another. Distribution state is identified as interior, while the other as continuous or exterior [73]. Emulsion is generated through distributing one liquid state into another in the form of small droplets. Because of separation resistance and coalescence, dispersed state's droplet is wrapped into the continuous state. The existence of an agent at the interface leads to stability in relation to separation resistance. In general, in the oil industry, water-in-oil (W/O) and oil-in-water (O/W) emulsions are two basic types of emulsions, as shown in Figure 4.9.

The W/O emulsion is created by dispersing water globules in a continuous state (Figure 4.9a), while O/W emulsion is produced by dispersing oil globules in a continuous state (Figure 4.9b). The W/O emulsion is mostly observed in the production of water, whereas the O/W is regularly encountered during the production of oil [74]. Additionally, due to their fire-retardant property and lower cost and because they are easier to dispose of, O/W emulsions have a variety of applications in many industries. Multiple emulsions are developed using O/W or W/O emulsions, and these are distributed into other immiscible phases over time [75].

Emulsions act by decreasing the interfacial tension between the two phases and establishing an interfacial film. Formation of a stable emulsion depends on the selection of emulsifying agents. According to prior studies [70], contact between two immiscible liquids, the presence of surface-active components for an emulsifying agent (through the resins and asphaltenes), and adequate mixing strength for distributing a liquid into another are three important features for forming crude oil emulsion.

Emulsions should have a good lubricating property and be easy to demulsify for disposal. The emulsion should not harm metals and seals. Due to the existence of detergents, some lubricants have affinity to form emulsions in the presence of water. The primary use of water and mineral oil emulsion is in hydraulic and metalworking applications.

4.7.1.1 Emulsifying agent's working mechanism

The mechanism of emulsifying agents includes different chemical and physical processes. According to surface tension theory, emulsification takes place due to the reduction of interfacial tension between two phases. As per the repulsion theory, an emulsifying agent forms a film over one globule phase, which repels both phases. Because of this repulsive force, one phase remains suspended in the dispersing medium. According to the viscosity modification mechanism, hydrocolloid emulgents, polyethylene glycol, and polymers increase the viscosity of the medium.

4.7.1.2 Classification of emulsifying agents

Emulsifying agents can be categorized on the basis of chemical structure and mechanism of action. They are classified into synthetic, natural, finely dispersed solids, and auxiliary based on the chemical structure, while based on the mechanism of action, they are classified into mono-molecular, multi-molecular, and solid particle films.

In addition to the above classification of emulsifying agents, they are also classified on the basis of type of emulsifier. This classification includes those stabilized by an electric charge, stabilized by colloids, and stabilized by powders. In the stabilization by electric charge emulsion, oil (with a negative charge) is present in a very small amount. Hydrophilic solutions are stabilized by hydration and an electric charge. Emulsions stabilized by colloids are most important in food preparation, for example, eggs, flour, milk, etc. In the category of stabilization, fine particles—lampblack, mustard, and paprika—are used for emulsifying agents. This type of emulsion keeps them from coalescing [76].

4.7.2 Demulsifying agents

Stable water with a higher content of asphaltenes is formed in oil emulsions when producing crude oil. However, in the oil industry, the allowed water content value in crude oil is less than 0.5%. Hence, prior to transportation, breaking of emulsion into two phases is required. The process of breaking emulsion into water and oil in two different phases is known as demulsification, and the property by virtue of which oil separates from water is called demulsibility [77]. During demulsifier addition to the formulation, foam formation reduces, while water separation increases. Physical, biological, and chemical treatments are used for demulsifying emulsions [78]. Emulsion performance of the lubricant is enhanced with the use of a proper demulsifier. Chemical demulsification is most commonly used for treating emulsions using demulsifiers or surfactants [79].

Structurally, most demulsifiers are oligomers or polymers with up to 100000 g/mol molecular weight. Demulsifiers consist of 5%–50% polyethylene oxide in a combined form. Emulsifiers and demulsifiers are fundamentally surface-active agents called surfactants. They have the capability to adsorb onto surfaces to alter them (film-forming agents) or onto interfaces to alter the boundary between any two immiscible phases (interfacial free energies). They comprise hydrophilic (water-liking) and hydrophobic (water-disliking) moieties [80].

Non-ionic and ionic are emulsifying and demulsifying agent types based on the charging status of the polar part. The ionic emulsifying and demulsifying agents are further categorized as cationic (positive charge) and anionic if the charge is negative [81]. Applications of demulsifiers include fluids for automatic transmission oils.

Destabilization of the emulsifying film can be achieved through the addition of a chemical demulsifier. It is also achieved by combinations of the above listed techniques. A good candidate for a demulsifier should be able to separate into the oil and water phases [82]. Some of the emulsifiers and demulsifiers are shown in Figure 4.10.

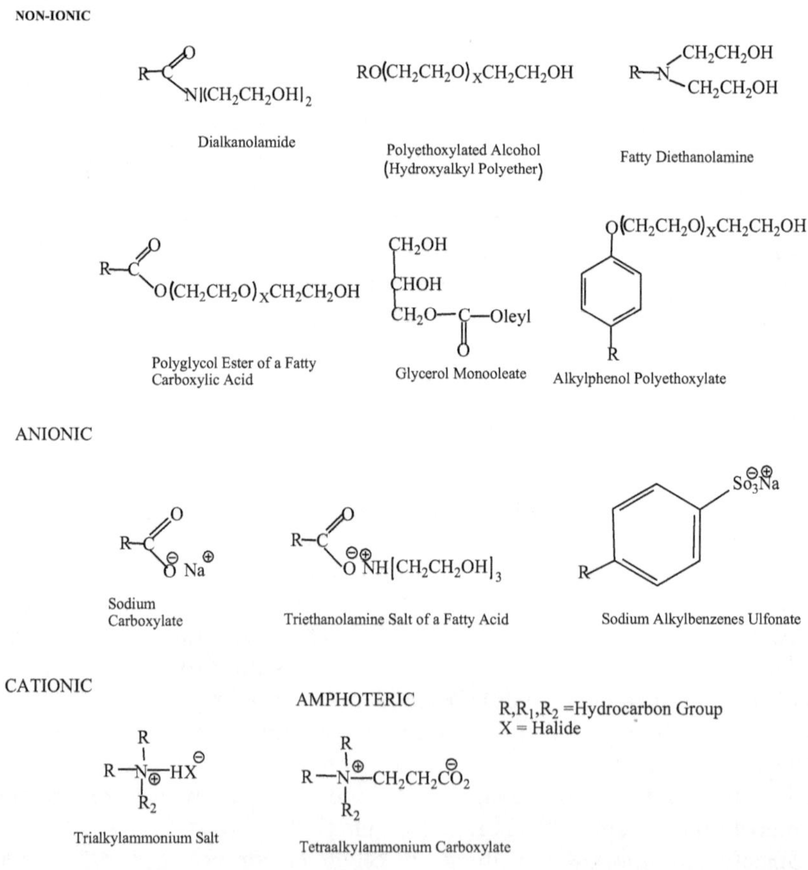

Figure 4.10 **Emulsifiers and demulsifiers [83].**

4.8 CONCLUSION

From an application point of view, lubricants can be roughly categorized into two types: engine oil and non-engine oils. Engine oils (EOs) are used in ICE, which operate on different fuels. A major concern in ICEs is the contact between oil and combustion products resulting in deposits on the components and wear of different parts. Performance evaluations for EOs are developed by the combined efforts of the API, ASTM, SAE, and other country-specific organizations. Original equipment manufacturer (OEM) recommendations are over and above other standards. In terms of application, EOs can be subcategorized as:

- Automotive, stationary, railroad, and marine diesel engine oils
- Gasoline EO
- Aviation EO
- Natural gas EO
- Alternate fuel EO
- Two-stroke EO

Non-engine lubricants can be categorized as follows:

- Automotive and industrial gear lubricants
- Automatic, manual, and power transmission oils
- Metalworking fluids
- Power steering lubricants
- Tractor and industrial hydraulic oils
- Shock absorber fluids
- Miscellaneous industrial oils
- Greases

The roles of different lubricant additives for engine and non-engine applications are summarized in Table 4.1 for reference.

New additives are mixed with others in the base oil as per the requirement of the customer. Further, the developed multi-additive added oil is screened in a number of registered bench tests. Developed oils have to be run for a long time and in severe conditions as compared to lab conditions. This creates the necessity of conducting accelerated tests using formulated oil. Hence, bench tests are made to simulate the real field condition closely to experience the actual service. Such tests are highly required, because they permit the inexpensive assessment of additives. After passing the bench tests, a full-fledged testing with the formulated oil is carried out. Actual equipment is used during full-fledged testing. For automotive applications, field trials may be essential as per the requirement in specific uses.

Table 4.1 Common additive types used for different applications

Application	*Subcategory*	*Dispersant*	*Detergent*	*AW/EP*	*AO/OI*	*CI/MD*	*FM*	*PPD*	*AF/FI*	*VII/VM*	*Other*
Engine lubricants	Gasoline	√	√	√	√	√	√	√	√	√	
	Diesel	√	√	√	√	√		√	√	√	
	Stationary gas	√	√	√	√	√		√			
	Aviation	√	√	√	√	√		√			
	Two-stroke cycle	√	√	√	√	√					√
Non-engine lubricants	Automatic transmission fluid	√	√	√	√	√	√	√	√	√	√
	Automotive gear oil (GO)			√	√	√	√	√	√	√	
	Industrial GO			√	√	√	√	√	√	√	
	Tractor hydraulic oil		√	√	√	√	√	√	√	√	√
	Industrial hydraulic foil		√	√	√	√		√	√	√	√
	Metal working fluids		√	√	√	√	√		√		√
	Greases		√	√	√	√				√	√

AW, anti-wear; EP, extreme pressure; AO, antioxidants; OI, oxidation inhibitor; CI, corrosion inhibitor; MD, metal deactivator; FM, friction modifier; PPD, pour point depressant; AF, anti-foam; FI, foam inhibitor; VII, viscosity index improver; VM, viscosity modifier; Other, other includes couplers, dyes, diluents, and emulsifier.

REFERENCES

1. Hsu, S. M. 2004. Molecular basis of lubrication. *Tribol. Int.* 37:553–559. https://doi.org/10.1016/j.triboint.2003.12.004.
2. Durant, W. D., L. M. Teintze. 1991. World-wide supply and demand of lubricants, AM911. *NPRA Annual Meeting,* Denver, Colorado.
3. Neale, M. J. 1993. *Lubrication: A Tribology Handbook.* London: SAE, Warrendale, PA as R-130.
4. Rizvi, S. Q. A. 2009. *A Comprehensive Review of Lubricant Chemistry, Technology, Selection, and Design.* Baltimore: ASTM International.
5. Prince, R. J. 2010. Base oil from petroleum. In: Mortier, R., Fox, M., Orszulik, S. (eds) *Chemistry and Technology of Lubricants*, 3–34. Dordrecht: Springer. https://doi.org/10.1023/b105569_6.
6. Ripple, D. E., J. F. Fuhrmann. 1984. *Perlbrmance Comparisons of Synthetic and Mineral Oil Crankcase Lubricant Base Stocks.* Germany: Presented at the Fourth International Colloquium Esslingen.
7. Leslie, R.R. 2003. *Lubricant Additives: Chemistry and Applications.* Marcel Dekker.
8. Minami, I. 2017. Molecular science of lubricant additives. *Appl. Sci.* 7(5):445.
9. Hsu, S. M., A. L. Cummings. 1983. Interactions of additives and lubricating base stocks, lubricant and additive effects on engine wear. *SAE* SP558:61.
10. Misra, A. K., A. K. Mehrotra, R. D. Srivastava. 1975. Antiwear characteristics of additives: Synergistic and adverse effects. *Wear* 31:345–357. https://doi.org/10.1016/0043-1648(75)90168-4.
11. Mortier, R. M., M. F. Fox, S. T. Orszulik. 2010. *Chemistry and Technology of Lubricants.* 3rd ed. Heidelberg: Springer. https://doi.org/10.1023/b105569.
12. Bellaire, D., S. Thielen, C. Burkhart et al. 2022. Investigation of radial shaft seal swelling using a special tribometer and magnetic resonance imaging. *ACS Omega* 7(14):11671–11677. https://doi.org/10.1021/acsomega.1c06521.
13. Chen, X., C. Donaghy, A. N. Kurchan. 2013. *Seal Swell Additive.* Patent No. WO 2013/109568 A1.
14. Keller, R. W. 1987. Effects of lubricants on seal elastomers. *SAE* Paper No. 871626.
15. Covitch, M. J., K. J. Trickett. 2015. How polymers behave as viscosity index improvers in lubricating oils. *Adv. Chem. Eng. Sci.* 5(2):134–151. https://doi.org/10.4236/aces.2015.52015.
16. ASTM, D2270. 1991. *Standard Practice for Calculating Viscosity Index from Kinematic Viscosity at 40 and 100 °C.* West Conshohocken, PA: ASTM International. https://www.astm.org/Standards/D2270.htm.
17. Dean, E. W., G. H. B. Davis. 1929. Viscosity variations of oils with temperature. *Chem. Metall. Eng.* 36:618–619.
18. Sparnacci, K., T. Frison, E. Podda et al. 2022. Core-crosslinked star copolymers as viscosity index improvers for lubricants. *ACS Appl. Polym. Mater.* 4(12):8722–8730. https://doi.org/10.1021/acsapm.2c01039.
19. Selby, T. W. 1958. The non-Newtonian characteristics of lubricating oils. *ASLE Trans.* 1:68–81. https://dx.doi.org/10.1080/05698195808972315.

20. Flory, P. J. 1952. *Principle of Polymer Chemistry*. Ithaca, NY: Cornell University Press.
21. Kapuscinski, M. M., R. E. Jones. 1987. *Dispersant-Antioxidant Multifunction Viscosity Index Improver*, U S Patent Number 4699723.
22. Kapuscinski, M. M., T. E. Nalesnik, R. T. Biggs et al. 1990. *Dispersant Anti-Oxidant VI Improver and Lubricating Oil Composition Containing Same*, U S Patent Number 4948524.
23. Otto, M., F. L. Miller, A. J. Blackwood et al. 1934. Motor oils having viscosity index of 120 predicted as definite need. *Oil Gas J.* 33(26):98–106.
24. Maurya, U., V. Vasu, V. Saini. 2024. Nano-innovations in lubrication: Unveiling the synergistic boundary lubrication potential of zinc titanate with phosphonium ionic liquid. *Tribol. Int.* 199:109987. https://doi.org/10.1016/J.TRIBOINT.2024.109987.
25. Stambaugh, R. L. 1994. Viscosity index improvers and thickeners. In: Mortier, R. M., Orszulik, S. T. (eds) *Chemistry and Technology of Lubricants*. Dordrecht: Springer. https://doi.org/10.1007/978-1-4615-3554-6_5.
26. Stambaugh, R. L. 1992. Viscosity index improvers and thickeners. In: Mortier, R. M., Orszulik, S. T. (eds) *Chemistry and Technology of Lubricants*. Dordrecht: Springer. https://doi.org/10.1007/978-1-4615-3272-9_5.
27. Otto, M., M. C. Martin. 1938. *High Molecular Weight Iso-Olefine Polymers and Process of Producing the Same*. US-2130507-A.
28. Liston, T. V. 1992. *Engine Lubricant Additives What They Are and How They Function©*, 389–397. Park Ridge: STLE.
29. Schilling, A. 1968. *Motor Oils and Engine Lubricants*. Great Britain: Scientific Publications.
30. Baker, A. E. 1984. Lubricant properties and test methods. In: Booser, E. R. (ed) *Handbook of Lubrication, Theory and Practice of Tribology*. 1st ed, 481–515. Boca Raton: CRC Press.
31. Khaklari, G. H., P. Talukdar. 2021. A review of various pour point depressants used for flow assurance in oil industries. *IJEAST* 6(1):335–352.
32. Soni, H. P., K. S. Kiranbala, A. Agrawal et al. 2010. Designing maleic anhydride-α-olifin copolymeric combs as wax crystal growth nucleators. *Fuel Process. Technol.* 91:997–1004.
33. Chen, B., Y. Sun, J. Fang et al. 2010. Effect of cold flow improvers on flow properties of soybean biodiesel. *Biomass Bioenergy* 34:1309–1313.
34. Baczek, S. K., W. B. Chamberlin. 1988. *Petroleum Additives- Encyclopaedia of Polymer Science and Engineering*. 2nd ed. Hoboken, New Jersey: John Wiley and Sons.
35. Crawford, J., A. Psaila, S. Orszulik. 2010. Miscellaneous additives and vegetable oils. In: Mortier, R., Fox, M., Orszulik, S. (eds) *Chemistry and Technology of Lubricants*, 189–211. Dordrecht: Springer. https://doi.org/10.1023/b105569_6.
36. Lazar, M., J. Rychly, V. Klimo et al. 1989. *Free Radicals in Chemistry and Biology*. Boca Raton: CRC Press.
37. Denisov, E. T., I. V. Khudyakor. 1987. Mechanisms of action and reactivities of the free radicals of inhibitors. *Chem. Rev.* 87:1313–1357.
38. Gaw, W. J., E. D. Black, B. J. Hardy. 1997. Finished lubricants benefits from higher quality base oils. In *Presented at Hart's World Conference on Refining Technology and Reformulated Fuels*, Texas.

39. Veprek-Siska, J. 1985. Does oxidation by molecular oxygen proceed without catalyst. *Oxid. Commun.* 8(3–4):301–307.
40. Emanuel, N. M., E. T. Denisov, Z. K. Maizus. 1967. *Liquid Phase Oxidation of Hydrocarbons.* New York: Plenum.
41. Ingold, J. A. 1983. Chapter 7: Analytical and safety aspects of organic peroxides and related functional groups. In: Patei, S. (ed) *The Chemistry of Functional Groups: Peroxides.* New York: Wiley & Sons.
42. Ingold, K. U. 1969. Peroxy radicals. *Acc. Chem. Res.* 2(1):1–9.
43. Jensen, R. K., S. Korcek, L. R. Mahoney et al. 1979. Liquid-phase autoxidation of organic compounds at elevated temperatures. 1. The stirred flow reactor technique and analysis of primary products from n-hexadecane autoxidation at 120–180°C. *J. Am. Chem. Soc.* 101:7574.
44. Maurya, U., V. Vasu, D. Kashinath. 2022. Ionic liquid-nanoparticle-based hybrid-nanolubricant additives for potential enhancement of tribological properties of lubricants and their comparative study with ZDDP. *Tribol. Lett.* 70:11. https://doi.org/10.1007/s11249-021-01551-6.
45. Korcek, S., R. K. Jensen. 1975. Relation between base oil composition and oxidation stability at increased temperatures. *ASLE Trans.* 19:83–94.
46. Perez, J. M., F. A. Kelley, E. E. Klaus et al. 1987. *SAE* Paper 872028.
47. Ingold, K. U. 1961. Inhibition of autoxidation of organic substances in liquid phase. *Chem. Rev.* 61:563–589.
48. Johnson, M. D., S. Korcek, M. Zinbo. 1983. Inhibition of oxidation by ZDTP and ashless antioxidants in the presence of hydroperoxidcs at 160°C, lubricant and additive effects on engine wear. *SAE* SP.55:71–81.
49. Hsu, S. M., A. L. Cummings. 1983. *SAE* Paper 831682:51–60.
50. Soleimani, M., L. Dehabadi, L. D. Wilson et al. 2018. Antioxidants classification and applications in lubricants. In: Johnson, D. W. (ed) *Tribology, Lubricants and Additives*, 23–42. Rijeka, Croatia: IntechOpen. https://doi.org/10.5772/intechopen.72621.
51. Howard, J. 1980. Chapter 12, Inhibition of hydrocarbon autoxidation by some sulphur containing transition metal complexes. In: Prior, W. A. (ed) *Frontiers of Free Radical Chemistry*, 237–282. New York: Academic Press.
52. Abou El Naga, H. H., A. E. M. Salem. 1984. Effect of worn metals on the oxidation of lubricating oils. *Wear* 96:267–283.
53. Kreuz, K. L. 1969. Gasoline engine chemistry as applied to lubricant problems. *Lubrication* 55:53–64.
54. Kreuz, K. L. 1970. Diesel engine chemistry as applied to lubricant problems. *Lubrication* 56:77–88.
55. Ludema, K. C. 1996. *Friction, Wear, Lubrication- A Textbook in Tribology.* Boca Raton: CRC Press.
56. Kauffman, R. E. 1994. Rapid determination of remaining useful lubricant life. In: Booser, E. R. (ed) *Handbook of Lubrication and Tribology*, 89–100. Boca Raton: CRC Press.
57. Huang, L., C. Zhou, Y. Zhang et al. 2019. DBHP-functionalized ZnO nanoparticles with improved antioxidant properties as lubricant additives. *Langmuir* 35(12):4342–4352. https://doi.org/10.1016/j.triboint.2018.01.042.
58. Yamada, Y., J. Igarashi, K. Inoue. 1992. Effects of metallic detergents on the antioxidant and antiwear properties of zinc dialkyldithiophosphates. *Lub. Eng.* 48(6):511–518.

59. Hsu, S. M., P. Pei, C. S. Ku. 1989. Mechanisms of additive effectiveness. *Lub. Sci.* 1:165–184.
60. Ren, C., X. Zhang, M. Jia et al. 2023. Antifoaming agent for lubricating oil: Preparation, mechanism and application. *Molecules* 28:3152. https://doi.org/10.3390/molecules28073152.
61. Cen, Z. Y., B. Y. Chen, S. J. Yang. 2019. Talking about antifoaming agents in lubricating oils. *J. Shandong Ind. Technol.* 6:38.
62. Zhao, R. P. 2020. Causes, hazards and treatment measures of lubricating oil foam in gear box of wind turbine. *Constr. Mach. Equip.* 51:108–112.
63. Zou, X. R. 2001. The cause, harm and treatment of foaming of lubricating oil. *Mech. Eng. Autom.* S1:164–165.
64. Wu, Y. 2015. *Study on the Foaming Problems of the Cold Heading Process Lubricating Oil.* Shanghai: East China University of Science and Technology.
65. Chang, J. H. 2015. *Study on the Problems of Gear Oil Bubble Properties.* Shanghai: East China University of Science and Technology.
66. Klaus, E. E., E. J. Tewksbury. 1984. Liquid lubricants. In: Booser, E. R. (ed) *Handbook of Lubrication, Theory and Practice of Tribology.* Vol. 11, 229–254. CRC Press.
67. Faujdar, E., H. Negi, R. K. Singh et al. 2020. Study on biodegradable poly(olefins-co-pinene) architectures as pour point depressant and viscosity index improver additive for lubricating oils. *J. Polym. Environ.* 28:3019–3027
68. Zhan, C., A. Saint-Jalmes, M. Receveur et al. 2022. Detailed characterization of aeration in lubricating oils by an ultrasonic approach. *Tribol. Int.* 175:107782.
69. Li, W. D., D. Wang, H. Y. Liu. 1996. Silicone oil's dispersancy and anti-foaming characteristics in oil. *Lubr. Oil* 11:33–35:
70. Abdulredha, M. M., S. A. Hussain, L. C. Abdullah. 2020. Overview on petroleum emulsions, formation, infuence and demulsifcation treatment techniques. *Arab. J. Chem.* 13:3403–3428.
71. Song, X., P. Shi, M. Duan et al. 2015. Investigation of demulsifcation efficiency in water-in-crude oil emulsions using dissipative particle dynamics. *RSC Adv.* 5:62971–62981.
72. Hajivand, P., A. Vazin. 2015. Optimization of demulsifer formulation for separation of water from crude oil emulsions. *Brazilian J. Chem. Eng.* 32:107–118.
73. Lim, J. S., S. F. Wong, M. C. Law et al. 2015. A review on the effects of emulsions on flow behaviours and common factors affecting the stability of emulsions. *J. Appl. Sci.* 15:167–172.
74. Saad, M. A., M. Kamil, N. H. Abdurahman et al. 2019. An overview of recent advances in state-of-the-art techniques in the demulsification of crude oil emulsions. *Processes* 7:1–26.
75. Khan, B. A., N. Akhtar, H. M. S. Khan et al. 2011. Basics of pharmaceutical emulsions: A review. *Afr. J. Pharm. Pharmacol.* 5:2715–2725.
76. Abdulredha, M. M., S. A. Hussain, L. C. Abdullah et al. 2022. Water-in-oil emulsion stability and demulsification via surface-active compounds: A review. *J. Petrol. Sci. Eng.* 209:109848. https://doi.org/10.1016/j.petrol.2021.109848.
77. Langevin, D., J. -F. Argillier. 2016. Interfacial behavior of asphaltenes. *Adv. Colloid Interface Sci.* 233:83–93.

78. Zolfaghari, R., A. Fakhru'l-Razi, L. C. Abdullah et al. 2016. Demulsification techniques of water-in-oil and oil-in-water emulsions in petroleum industry. *Sep. Purif. Technol.* 170:377–407.
79. Issaka, S. A., A. H. Nour, R. M. Yunus. 2015. Review on the fundamental aspects of petroleum oil emulsions and techniques of demulsification. *J. Petrol. Environ. Biotechnol.* 6(2):1.
80. Alao, K. T., O. R. Alara, N. H. Abdurahman. 2021. Trending approaches on demulsification of crude oil in the petroleum industry. *Appl. Petrochem. Res.* 11(3):281–293. https://doi.org/10.1007/s13203-021-00280-0.
81. Becher, P. 1957. *Emulsions: Theory and Practice*, 209–231. Reinhold: American Chemical Society Monograph Series.
82. Raya, S. A., I. M. Saaid, A. A. Ahmed et al. 2020. A critical review of development and demulsification mechanisms of crude oil emulsion in the petroleum industry. *Prod. Eng.* 10:1711–1728.
83. Karsa, D. R. 1987. *Industrial Applications of Surfactants – An Overview.* Cambridge, England: Royal Society of Chemistry.

Chapter 5

Lubrication regimes and mechanisms

Neha Sharma

5.1 INTRODUCTION TO LUBRICATION

Lubrication is the process of reducing friction between two surfaces in contact with each other. It is an essential part of many industrial processes, and it plays a key role in preventing wear and tear on machinery. Lubrication can be used to reduce friction between moving parts in machines, to protect surfaces from corrosion, and hence increase equipment life, reduce energy consumption, and improve safety. Lubrication is used in a wide variety of industries, including automotive industry, manufacturing industry, power generation industry, aerospace industry, and food processing industry. Therefore, engineers can choose the ideal lubricant for a given application and guarantee that the machine's components function smoothly and effectively by being aware of the various lubrication regimes and their mechanics. Detailed studies of the variation in frictional characteristics of fluid lubricated contacts with variation in contact load, speed, and viscosity were first carried out by Richard Stribeck and Mayo D. Hersey [1], who identified and defined various regimes of lubrication through generation of a two-dimensional plot known as 'Stribeck curve'. A basic and well-known idea in the fields of lubrication and tribology is the Stribeck curve. The concept shows how friction behaves in lubricated contacts in relation to entrainment speed, roughness, and viscosity of the lubricant (or sometimes simply roughness alone). Even though the study's conclusions were in line with those of previous investigations, Richard Stribeck and Mayo Hersey are acknowledged as the research pioneers of the 20th century for their extensive research on friction in the railroad sector, which resulted in the development of the Stribeck curve. The major function of any lubricant is to provide a thin protective film. However, variations in load and speed of contacting elements generate different frictional characteristics, categorized into different regimes of lubrication, viz., boundary, mixed, and hydrodynamic regime of lubrication.

 DOI: 10.1201/9781003642848-5

5.2 LUBRICATION REGIMES

Four common regimes of lubrication comprising of boundary, mixed, elastohydrodynamic, and hydrodynamic lubrication were identified by Stribeck, who studied the fluid film lubricated journal bearing system [2,3] as shown in Figure 5.1.

The observed coefficient of friction (μ) in the y-axis is plotted against a composite parameter of dynamic viscosity (η), speed (v), and applied load (p). This parameter at the x-axis is also known as Hersey's number (H). As displayed in the graph, the dependence of the coefficient of friction on the Hersey's number varies according to the lubrication regime. The highest values of coefficient of friction are encountered under dry contact when full contact occurs at the asperity tips of contacting surfaces. As the relative velocity and the viscosity increases, the coefficient of friction decreases, a phenomenon which is mostly encountered in several metal forming processes and in flat rolling. As the viscosity and the speed increase further, the coefficient drops to a minimum and the hydrodynamic regime of full separation of the metal surfaces by the lubricant layer begins. This regime is also called full-film lubrication with a relatively thick film (>0.25 µm) at the interface of the contact. Metal-to-metal contact is completely avoided, making the coefficient of friction only dependent on viscosity of the lubricant; hence frictional resistance is mainly caused by the need to overcome shear strength of the entrained lubricant film. Elastohydrodynamic lubrication falls under the subcategory of hydrodynamic lubrication, wherein the asperities are elastically deformed by the hydrodynamic pressure of a very thin fluid film (0.025–5 µm) [4].

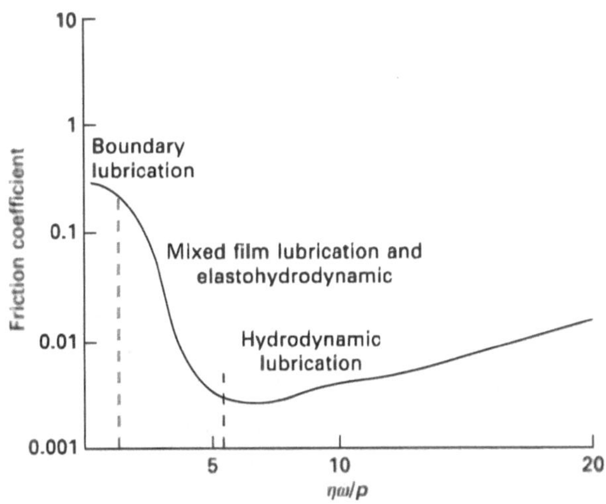

Figure 5.1 Stribeck curve.

In mixed lubrication (ML) regimes, the asperities of metal surfaces occasionally come in contact in some areas; this may be observed for components working at high loads and low speed.

5.2.1 Boundary lubrication

The contact between the solids (or respective boundary lubricant films) determines the amount of friction created in the contact if one starts with very severe conditions or inadequate lubrication; where the parameter λ is quite small, the solid-to-solid connections support the entire load. When compared to the friction created by the viscous shear of the lubricant layers, the friction created in the solid-to-solid contact is typically very high.

The friction coefficient for this type of lubrication is usually between 0.08 and 0.15, and it is known as the boundary lubrication (BL) regime (Figure 5.2).

In general, BL should be avoided. Experts in lubrication generally agree that friction might be at its maximum during the BL regime. This happens during low speed, high load, shutdown, and startup.

Any situation where there is a chance of abrasion and/or adhesion and the asperities of two lubricated surfaces in relative motion coming into physical contact defines a BL regime. Tribologists and lubrication experts have proposed that up to 70% of wear happens when machinery is starting up and shutting down.

A primary technique for mitigating border lubrication involves using the appropriate viscosity of the lubricant. Metal-to-metal contact happens when a lubricant with insufficient viscosity is unable to maintain the separation of the metal surfaces. The molecular friction of the oil will rise in a lubricant with an excessive viscosity. Oil layers glide past one another as a result of this internal oil shearing motion, raising operating temperatures and causing energy loss [4–6].

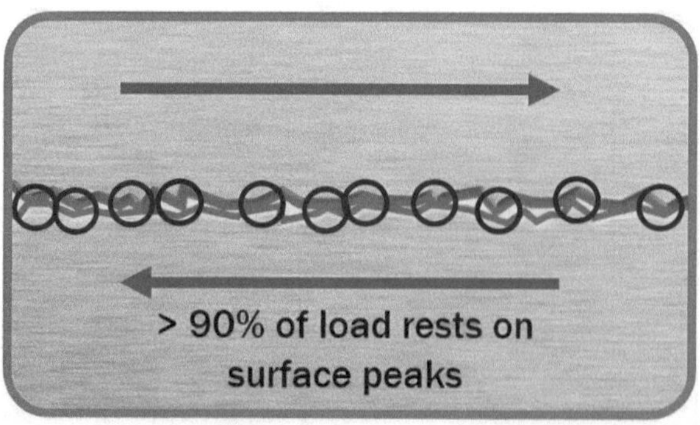

Figure 5.2 Boundary lubrication regime.

One potential solution to mitigate the BL regime issue is to employ a fully formulated lubricant that includes additives for severe pressure or antiwear. These additives react to the high temperature and pressure by interacting with the metal asperities that have come into contact.

5.2.2 Mixed lubrication

When the lubricant layer thickness increases and the working conditions become less severe, this regime is recognized. Only a portion of the particles come into direct contact with the lubricant when the thickness of the lubricant begins to divide the surfaces and carry part of the load. As seen in Figure 5.3, the total friction in this instance is the result of the friction of the lubricant layers as well as the friction of the solid-to-solid contact friction.

This is the gray area that lies between the hydrodynamic/elastohydrodynamic and BL regimes. The system now enters complete film lubrication, either hydrodynamic or elastohydrodynamic lubrication, as the oil film thickness develops even more. Overall friction lowers because lubricant layer friction is usually lower than solid-to-solid contact friction. The ML regime is applicable in this instance. The friction coefficient μ_{ML} in ML can be explained as follows:

$$\mu_{ML} = \alpha\mu_{BL} + (1 - \alpha\mu_{FF}) \tag{5.1}$$

Where μ_{BL} is the friction coefficient in BL regime, μ_{FF} is the hydrodynamic friction and is the ratio of the load carried by the solid-to-solid contact to the load carried by the lubricant. From this relation, it is clear that with the increase of the solid-to-solid contact, friction drops, since up to a certain point, $\mu_{ML} \gg \mu_{FF}$.

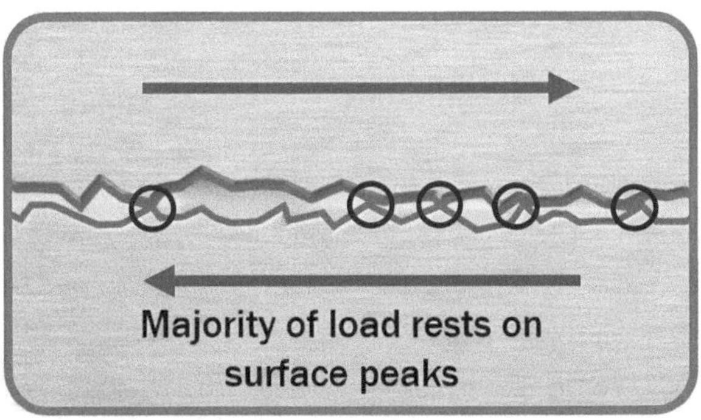

Figure 5.3 ML regime.

5.2.3 Hydrodynamic lubrication

This situation of the lubrication regime arises when a machine starts to rotate and the loads and speeds are such that an oil wedge forms between the bearing surfaces and the shaft. There is less chance of asperity contact since this oil wedge elevates the shaft off the bearing surface. In order to minimize wear and friction, this is the ideal state. When working conditions improve, film thickness increases. Solid-to-solid contact finally ceases to occur upon the achievement of full-film (FF) lubricating conditions. Here, the lubricant layers' viscous shear creates friction. Figure 5.4 illustrates how friction increases throughout the whole film lubrication regime, despite the disappearance of the solid-to-solid contact.

The rise in hydrodynamic friction is the cause of this behavior. One may be concerned that increasing viscosity, speed, or decreasing load is necessary to get a higher film thickness. With the exception of a decrease in load, when friction force reduces more slowly than with a normal load, all result in an increase in shear stresses and friction coefficient [5,7–9].

5.2.4 Lambda ratio

In the fields of lubrication and tribology, the lambda ratio is a well-known metric that is used to differentiate between lubrication regimes [10,11]. The definition of lambda ratio is the ratio

$$\lambda = \frac{h}{\sigma}$$

$$\sigma = \sqrt{Rq1^2 + Rq2^2}$$

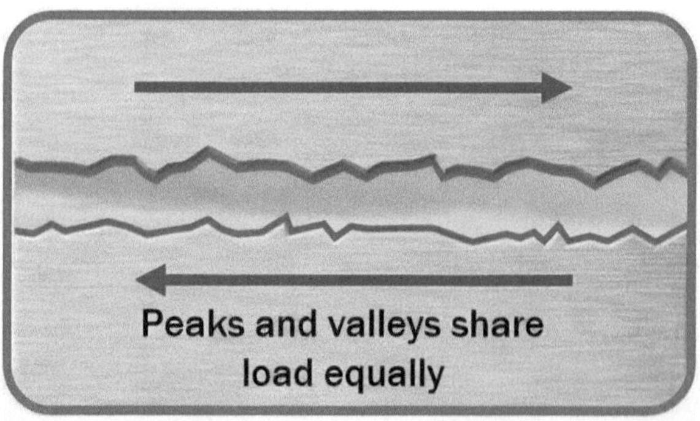

Figure 5.4 Hydrodynamic lubrication.

Here, $Rq\ i$ is the standard deviation of the surface heights of the i-th surface in contact

Typically, the lubrication regimes are identified based on as follows:

1. $\lambda \geq 3$, FF lubrication regime;
2. $0.1 < \lambda < 3$, ML regime; both the characteristics of the lubricant and the "solid" contacts have an impact on friction. The solid contacts and the lubricant each contribute a portion of the overall load.
3. $\lambda \leq 0.1$, BL regime; friction is primarily described by a boundary friction coefficient, which is a reasonably constant characteristic value of a given system. Friction is totally determined by the shear strength of the surface boundary layers. It is important to note that the lambda ratio range provided above is only an indication and could vary significantly in actual situations. At lambda values as low as $\lambda = 0.6$, complete film lubrication can occasionally be produced.

5.3 MECHANISMS IN LUBRICATION

5.3.1 Boundary lubrication mechanisms

Boundary lubrication occurs when there is direct contact between two surfaces in relative motion, with only a thin layer of lubricant separating them. The mechanisms involved in BL are crucial for minimizing friction and wear in these conditions:

5.3.1.1 Mechanisms in BL

5.3.1.1.1 Adsorption

5.3.1.1.1.1 PHYSISORPTION

Molecules of adsorbate may attach or detach from a surface without any irreversible changes to the surface or the adsorbate. Most liquids and gases physisorb to most solid surfaces, but there is almost always an upper temperature limit to this process. In physisorption, van der Waals or dispersion forces provide the bonding between substrate and adsorbate, as illustrated in Figure 5.5.

5.3.1.1.1.2 CHEMISORPTION

Molecules from the lubricant chemically adhere to the metal surface due to attractive forces, forming a protective film. Chemisorption or 'chemical adsorption' is an irreversible or partially irreversible form of adsorption, which involves some degree of chemical bonding between adsorbate and substrate, as illustrated schematically in Figure 5.6.

This film reduces direct metal-to-metal contact and minimizes friction [12–15].

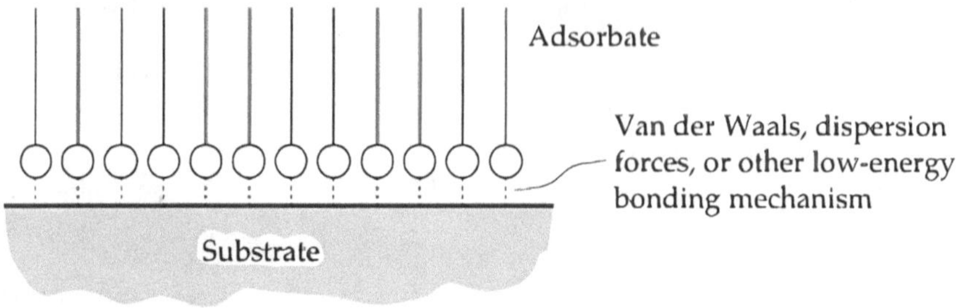

Figure 5.5 Schematic illustration of physisorption.

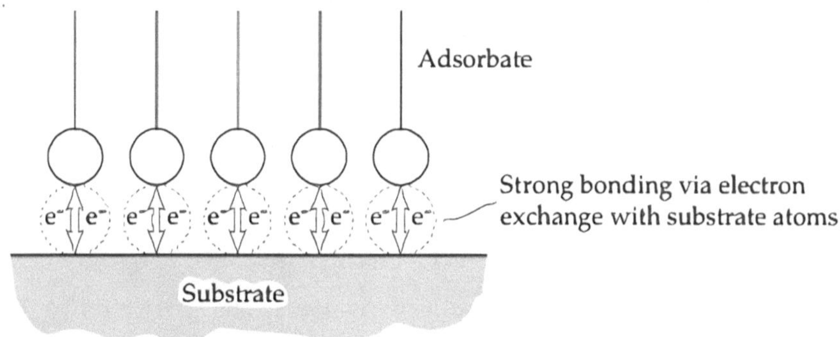

Figure 5.6 Mechanism of chemisorption.

5.3.1.1.2 Formation of tribofilms

- Under BL conditions, chemical reactions between the lubricant additives and surface materials can form tribofilms. These films are a result of reactions between the lubricant additives and surface asperities, forming a protective layer that reduces friction and wear. The adsorption or boundary additives control the adsorption type of lubrication and are also sometimes known as 'friction modifiers'. In such cases, the lubricant interacts with the asperities and forms a reverse micelle monolayer formation with the polar groups at the end of each molecule. The molecules, with their polar end, attach to the surface, as shown in Figure 5.7.

5.3.1.1.3 Antiwear additives to modify surfaces

Additives are made especially for lubricating boundaries. They contain substances that, when subjected to high pressure and heat, react to generate protective layers that lessen wear and shield metal from metal.

In order to protect contacting surfaces at higher temperatures above the range of effectiveness of adsorption or boundary agents, antiwear additives were designed and manufactured. There are several different types of

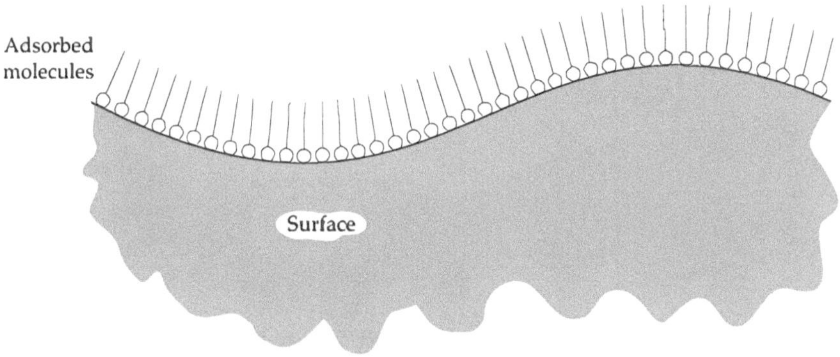

Figure 5.7 Adsorption of the lubricants by boundary additives.

antiwear additives that are currently used in oil formulations. For example, in engine oils, the most commonly used antiwear additive is zinc dialkyldithiophosphate (ZnDDP), and in gas turbine oils, tricresylphosphate or other phosphate esters are used. Phosphorous additives are used where antiwear protection at relatively low loads is required. These additives react with the surfaces through the mechanism of chemisorption, and the protective surface layer produced is much more durable than that generated by adsorption or boundary agents. Common examples of these additives are ZnDDP, tricresylphosphate, dilaurylphosphate, diethylphosphate, dibutylphosphate, tributylphosphate, and triparacresylphosphate. These additives are used in concentrations of 1%–3% by weight.

ZnDDP is a very important additive commonly used in engine oil formulations. It was originally developed as an anti-oxidant and detergent, but it was found later that this compound also acted as an antiwear and mild extreme-pressure (EP) additive. The term 'antiwear' usually refers to wear reduction at moderate loads and temperatures, whereas the term 'EP' is reserved for high loads and temperatures. Although some authors recognize ZnDDP as a mild EP additive, it is generally classified as an antiwear additive in the literature. The chemical structure of ZnDDP is shown in Figure 5.8.

Tricresylphosphate (TCP) has been used as an antiwear additive for more than 50 years. Like ZnDDP, it functions by chemisorption to the operating surfaces, which is explained in detail in the chapter on 'Boundary and Extreme Pressure Lubrication'. TCP is very effective in reducing wear and friction at temperatures up to about 200°C. Beyond this temperature, there is sufficient energy input to the surface for the chemisorbed films to desorb, and it is believed that the compound will then form less effective, much weaker, thick phosphate films with limited load capacity [16–18]. TCP has also been found to delay the onset of rolling contact fatigue in bearings when used as an additive of 1% concentration in a synthetic ester lubricant [19].

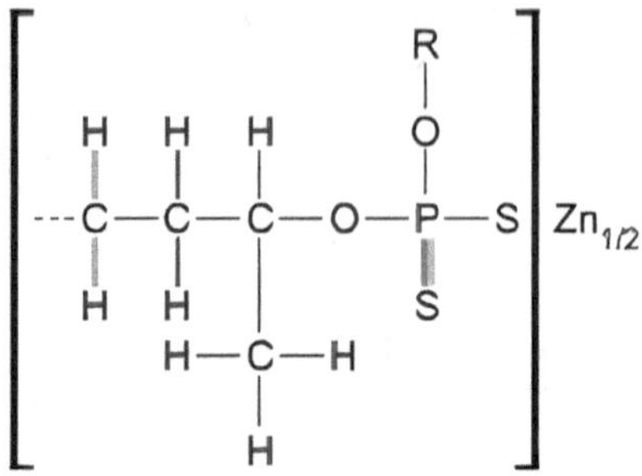

Figure 5.8 Chemical structure of the zinc dialkyldithiophosphate (ZDDP) additive.

Other antiwear additives such as dilauryl phosphate, dibutylphosphate, diethylphosphate, tributylphosphate, and triparacresylphosphate are also being used in lubricant formulation. They function in the same manner as ZnDDP or TCP by producing chemisorbed surface films.

5.3.1.1.4 Shear-thinning behavior

- Some lubricants exhibit shear-thinning behavior, meaning their viscosity decreases under high shear forces. This property helps the lubricant flow more easily between surfaces, reducing friction even under BL conditions.

5.3.1.1.5 Inhibition of oxidation and corrosion

- Boundary lubrication additives often contain compounds that inhibit oxidation and corrosion, protecting the metal surfaces from degradation. Mineral oils inevitably oxidizes during service and these causes significant increases in friction and wear, which affects the performance of the machinery. The main effect of oxidation is a gradual rise in the viscosity and acidity of oil. This effect is demonstrated in Figure 5.9, which shows the variation of viscosity and acidity of a mineral oil as a function of oxidation time [20].

Most lubricating oils in present use contain anti-oxidant additives to delay the onset of severe oxidation of the oil. These are either natural anti-oxidants or artificially introduced additives that are able to suppress oxidation, and any differences in the oxidation resistance of oils largely depend on the presence of these inhibitors. Natural sulfur- or nitrogen-containing compounds, which are present in mineral oils, act as oxidation inhibitors by scavenging the radicals produced by the oxidation process.

Widely used anti-oxidant additives are ZnDDP, metal deactivators, and simple hydrocarbons such as phenol derivatives, amines, and organic

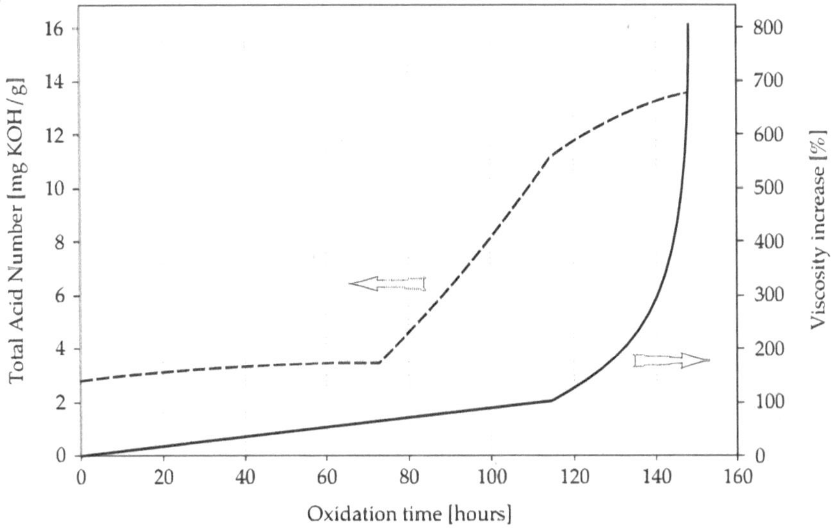

Figure 5.9 Effect of oxidation on viscosity and acidity of mineral oil.

phosphates. Sulfur and phosphorus in the elemental form or incorporated into organic compounds are also effective as anti-oxidants and antiwear additives

5.3.1.2 Mechanisms of ML regime

In a ML regime, both boundary and hydrodynamic lubrication mechanisms are concurrently active within the contact interface between moving surfaces. This interface is dynamic, with fluid film thickness varying across the contact area due to surface roughness and operating conditions. If the lubricating film separating the surfaces is such that it allows for some contact between the deformed asperities, then this type of lubrication is considered in the literature as 'mixed' or 'partial lubrication'. The contact load is shared between the contacting asperities and the film when mixed or partial lubrication prevails; cavities are generally filled with the lubricant along with the asperity–asperity interaction occurring in parallel, as shown in Figure 5.10.

Understanding the mechanisms within this regime involves recognizing how these lubrication modes coexist and influence friction and wear: Understanding these mechanisms is crucial in formulating effective lubricants and additives specifically tailored to withstand BL conditions. They aim to create protective layers, minimize direct metal-to-metal contact, and reduce friction and wear even when the lubrication conditions are challenging.

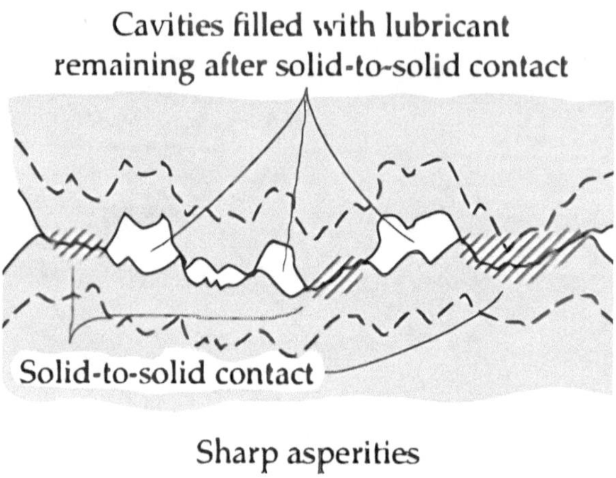

Figure 5.10 ML snapshot.

- Adsorption and tribofilm formation: Even in a predominantly hydrodynamic regime, BL mechanisms like adsorption of lubricant molecules and formation of protective tribofilms on asperities continue to occur. These help reduce direct metal-to-metal contact and mitigate wear.
- Extreme pressure additives: Additives within the lubricant react under high pressures, forming protective layers that further minimize contact between asperities.

This ML regime is quite common in many engineering applications, such as in machine components like gears, bearings, and cams, where operating conditions often lead to a combination of fluid film lubrication and direct contact between surfaces.

In the mixed regime, a thin fluid film is present, generated by the lubricant's viscosity and the relative motion between surfaces. This film, although not as substantial as in full hydrodynamic lubrication, contributes to reducing friction by separating surfaces and bearing part of the load. Some lubricants exhibit shear thinning, where their viscosity decreases under high shear rates. This property aids in maintaining a more robust fluid film even in regions prone to BL.

5.3.1.2.1 Asperity contact and deformation

Intermittent contact: Asperities on the surfaces intermittently come into contact, experiencing local BL, while the bulk of the surfaces are separated by the fluid film, exhibiting hydrodynamic behavior.

Surface deformation and smoothing: The high pressure in areas of contact can lead to plastic deformation of asperities, which can over time smoothen surface roughness, potentially reducing the need for extensive BL.

5.3.1.2.2 Additive and surface interaction

Additive redistribution: Additives present in the lubricant can redistribute themselves between regions of higher and lower pressure, aiding in reducing wear in areas experiencing BL.

Surface chemistry effects: Surface interactions and chemistry play a role in determining the effectiveness of BL and the formation of protective layers.

5.3.1.3 Mechanisms of hydrodynamic lubrication

Hydrodynamic lubrication involves the formation of a continuous fluid film between moving surfaces, which separates them to minimize direct contact and reduce friction. Hydrodynamic lubrication occurs when two surfaces are in relative motion. This relative motion initiates the formation of a fluid film between the surfaces. As the surfaces move relative to each other, the lubricant is drawn into the converging gap between them. This entrainment of the lubricant into the contact zone is facilitated by the converging geometry of the surfaces. The relative motion between the surfaces creates a pressure gradient within the lubricant. This pressure gradient leads to the generation of hydrodynamic pressure within the fluid film. The hydrodynamic pressure within the fluid film supports the load and prevents direct metal-to-metal contact between the surfaces. This results in the formation of a continuous and stable fluid film, which acts as a lubricating layer. The hydrodynamic pressure generated within the fluid film balances the external load and separates the surfaces by pushing them apart. This separation minimizes friction and wear, providing smooth operation

Several mechanisms play crucial roles in hydrodynamic lubrication.

5.3.1.3.1 Fluid film formation

- Pressurization of lubricant: The relative motion between surfaces leads to the pressurization of the lubricant in the converging zone ahead of the contact. This pressure increase forces the lubricant into a wedge-shaped film between the surfaces. There are generally four mechanisms of the pressure, which are governed by the Reynolds equation, as shown in Figure 5.11.

Figure 5.11a shows the hydrodynamics effect. When the lower surface moves at velocity U, the clearance along the motion direction gradually reduces to force the lubricant to flow from the large inlet to the small outlet.

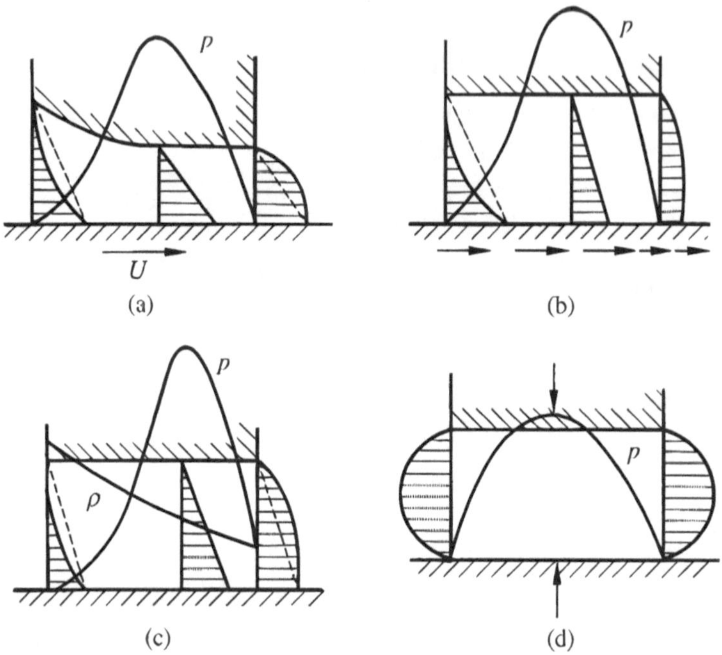

Figure 5.11 Four mechanisms of pressure. (a) physical wedge, (b) stretch, (c) density wedge and (d) squeeze.

The flow along the motion direction gradually reduces (the triangle diagram in the figure). Based on the continuity condition, the pressure as shown in Figure 5.11a is inevitably produced. The flow induced by pressure will reduce the flow into the inlet, but increase the flow out of the outlet in order to maintain the flow balance of each section. Thus, the fluid flow along a convergent clearance will have a positive pressure, while the flow along a divergent clearance generally produces a negative pressure.

Figure 5.11b shows the stretching effect. When the solid surface changes its length due to elastic deformation or other reasons, the relative surface velocity will cause the surfaces to move relatively to generate pressure. In order to generate positive pressure, the relative velocity should be forward, as shown in the figure.

Figure 5.11c is the effect of density variation. When the density of a lubricant gradually decreases along the direction of the motion, it will cause a different mass flow to generate pressure although the volume is the same. The temperature variation may cause the variation of density. Although the pressure generated by the effect of density variation is not very high, it also allows a surface to possess a certain load-carrying capacity.

Figure 5.11d shows the effect of squeezing. When two parallel surfaces are under the action of a normal force, the thickness decreases and pressure is produced. However, when the two surfaces are separated, it will produce negative pressure.

5.3.1.3.2 Surface geometry and roughness effects

- Surface contouring: Surfaces with specific contours (like grooves or textures) or designed roughness patterns aid in generating the required pressure distribution to sustain the fluid film.
- Asperity deformation: Under hydrodynamic conditions, asperities on the surfaces deform elastically due to the pressure exerted by the lubricant film, reducing direct contact and minimizing wear.

5.3.1.3.3 Entrapment and convergence

- Fluid entrainment: Lubricants are entrained into the converging gap between surfaces due to their relative motion, which supports load and maintains separation.
- Fluid convergence: The fluid converges toward the high-pressure region between the surfaces, creating a wedge-shaped film that separates the moving parts, as shown in Figure 5.12.

5.3.1.3.4 Temperature and viscosity effects

- Viscosity variation: Changes in temperature affect the viscosity of the lubricant, altering its ability to sustain the fluid film. Temperature control is important to maintain optimal viscosity for effective hydrodynamic lubrication.

Understanding these mechanisms aids in designing lubrication systems, selecting appropriate lubricants, and optimizing operating conditions to

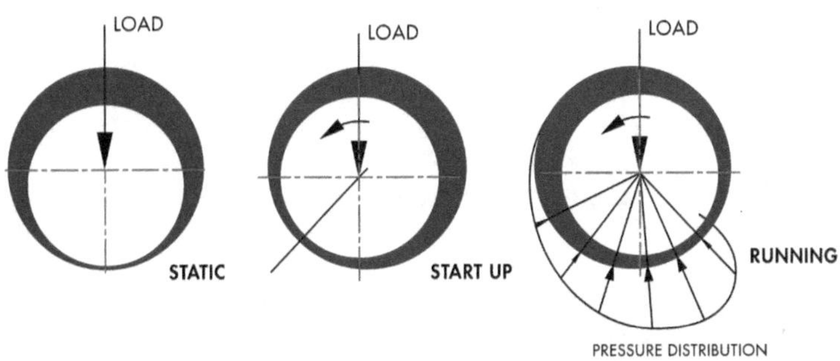

Figure 5.12 The fluid entrainment before and after running to form a wedge-shaped film.

ensure effective hydrodynamic lubrication. Proper management of these mechanisms is crucial in various applications, including machinery, engines, and other mechanical systems to minimize wear and friction [16–19,21].

5.4 CONCLUSION

Understanding lubrication regimes and mechanisms is pivotal for various industries and engineering disciplines. Comprehensive knowledge of this subject enables the formulation of effective lubricants and additives that minimize friction and wear between moving parts, leading to longer-lasting machinery and reduced maintenance costs. Proper lubrication strategies increase the reliability and durability of machinery by preventing premature wear and component failures, thus prolonging equipment lifespan. Furthermore, efficient lubrication reduces energy consumption and material waste, contributing to sustainability efforts by minimizing resource depletion and environmental impact.

In conclusion, comprehending lubrication regimes and mechanisms is fundamental for improving efficiency, reliability, and sustainability across various industries. It plays a crucial role in enhancing performance, reducing maintenance costs, and contributing to overall operational excellence and innovation. This knowledge forms the foundation for designing and maintaining systems that operate efficiently and reliably over extended periods.

REFERENCES

1. Bhushan, B. (2000). *Modern Tribology Handbook*. Boca Raton, USA: CRC Press.
2. Gwidon, W. S., & Andrew, W. B. (2014). *Engineering Tribology*. 2nd ed. Oxford, UK: Butterworth-Heinemann.
3. Meurant, G. (2009). *Tribology: A Systems Approach to the Science and Technology of Friction, Lubrication, and Wear*. Vol. 1. Amsterdam, Netherlands: Elsevier.
4. Hsu, S. M., & Gates, R. S. (2005). Boundary lubricating films: Formation and lubrication mechanism. *Tribology International*, 38(3), 305–312.
5. Zuidema, H. H. (1952). *The Performance of Lubricating Oils*. New York: Reinhold Publishing. pp. 26–31.
6. "The Four Lubrication Regimes Explained." Machinery Lubrication. www.machinerylubrication.com/Read/30741/lubrication-regimes. Accessed 7 June 2025.
7. Holmberg, K., & Erdemir, A. (2017) Influence of tribology on global energy consumption, costs and emissions. *Friction*, 5, 263–284. https://doi.org/10.1007/s40544-017-0183-5

8. Mourier, L., Mazuyer, D., Ninove, F. P., & Lubrecht, A. A. (2010). Lubrication mechanisms with laser-surface-textured surfaces in elastohydrodynamic regime. *Proceedings of the Institution of Mechanical Engineers, Part J: Journal of Engineering Tribology*, 224(8), 697–711.
9. Hsu, S. M. (2004). Molecular basis of lubrication. *Tribology International*, 37(7), 553–559.
10. Cann, P. M. E. B., Ioannides, E., Jacobson, B., & Lubrecht, A. A. (1994). The lambda ratio—A critical re-examination. *Wear*, 175(1–2), 177–188.
11. Morina, A., Green, J. H., Neville, A., & Priest, M. (2003). Surface and tribological characteristics of tribofilms formed in the boundary lubrication regime with application to internal combustion engines. *Tribology Letters*, 15, 443–452.
12. Yang, L., Neville, A., Brown, A., Ransom, P., & Morina, A. (2014). Friction reduction mechanisms in boundary lubricated W-doped DLC coatings. *Tribology International*, 70, 26–33. ISSN 0301-679X. https://doi.org/10.1016/j.triboint.2013.09.020. https://www.sciencedirect.com/science/article/pii/S0301679X13003307.
13. Clark, D. B., Klaus, E. E., & Hsu, S. M. (1985). The role of iron and copper in the oxidation degradation of lubricating oils. *Lubrication Engineering*, 41, 280–287.
14. Braun, O. M., & Naumovets, A. G. (2006). Nanotribology: Microscopic mechanisms of friction. *Surface Science Reports*, 60(6–7), 79–158.
15. Zhou, Z. R., & Vincent, L. (1999). Lubrication in fretting—a review. *Wear*, 225, 962–967.
16. Maurya, U, Vasu, V., & Saini, V. (2024). Nano-innovations in lubrication: Unveiling the synergistic boundary lubrication potential of zinc titanate with phosphonium ionic liquid. *Tribology Letters*, 199, 109987. https://doi.org/10.1016/J.TRIBOINT.2024.109987.
17. Maurya, U., & Vasu, V. (2022). Boehmite nanoparticles for potential enhancement of tribological performance of lubricants. *Wear*, 498–499, 204311. https://doi.org/10.1016/J.WEAR.2022.204311.
18. Faut, O. D., & Wheeler, D. R. (1983). On the mechanism of lubrication by tricresylphosphate – the coefficient of friction as a function of temperature for TCP on M-50 steel. *ASLE Transactions*, 26, 334–350.
19. Trivedi, H. K., Forster, N. H., & Rosado, L. (2011). Rolling contact fatigue evaluation of advanced bearing steels with and without the oil anti-wear additive tricresyl phosphate. *Tribology Letters*, 41, 597–605.
20. Maurya, U., Vasu, V., & Kashinath. D. (2022). Ionic liquid-nanoparticle-based hybrid-nanolubricant additives for potential enhancement of tribological properties of lubricants and their comparative study with ZDDP. *Tribology Letters*, 70, 11. https://doi.org/10.1007/s11249-021-01551-6.
21. Hsu, S. M., Ku, C. S., & Pei, P. T. (1983). Oxidative degradation mechanisms of lubricants, aspects of lubricant oxidation. Editors, W. H. Stadtmiller & A. N. Smith, *ASTM Committee D-2 Symposium on Petroleum Products and Lubricants*, Miami, Florida. West Conshohocken, PA: ASTM Technical Publication. Vol. 916. pp. 27–48.

Chapter 6

Lubricating greases

Sooraj Singh Rawat and
Harsha Arakere Puttaswamy

6.1 INTRODUCTION

The primary goal of greases is to reduce friction and wear caused by the relative movement of two interacting surfaces. Before studying the function of greases and their lubrication mechanism, certain questions concerning greases arise. What are the chemical composition and attributes of greases? How do greases differ from lubricants, and which performs better? Can grease be used instead of a lubricant? Is there any effect of temperature on the lubricating performance of grease? Is the content of grease affected by its application? This chapter makes every effort to answer all these questions.

6.2 ADVANTAGES AND DISADVANTAGES OF LUBRICATING GREASES OVER LUBRICATING LUBES

The primary purpose of lubricants or greases is to provide a lubricating coating between pairs of surfaces in contact with each other, hence reducing friction and minimizing wear. The benefits and drawbacks of lubricating greases compared to lubricating oils principally stem from the differences in their flow characteristics.

6.2.1 Advantages

- Grease has intrinsic sealing characteristics against dirt and foreign particles.
- Greases exhibit an apparent viscosity as the temperature decreases.
- The grease flow occurs solely due to the application of external force.
- Greases are not susceptible to stop-start issues.
- Greases exhibit a higher degree of resistance to water.
- Greases can function well across a broad spectrum of temperatures.
- Greases facilitate the dispersion of solid ingredients.

 DOI: 10.1201/9781003642848-6

6.2.2 Disadvantages

- Greases display inferior thermal conductivity.
- Because of their strong ionic character, greases are more prone to oxidation.
- Due to their high viscosity, greases have a speed limit.

6.3 GREASE COMPOSITION

A grease comprises three essential components: a lubricating base oil, a thickening agent, and additives. The distribution of constituent percentages in grease may vary depending on their intended purpose. In the composition of a traditional grease, it is typically seen that the additive content ranges from 0% to 10%, the thickening agent content ranges from 5% to 20%, and the base lube oil content ranges from 80% to 90% [1]. Greases are typically categorized based on the type of thickener included in their formulation. Greases exhibit a semi-solid or gel-like physical state, occasionally manifesting as a solid substance. Grease is a composite substance consisting of a lubricating base oil and a thickening agent. The physical interaction between the lubricating base oil and the thickening agent facilitates the achievement of a stable colloidal structure in grease. Greases are often favored for the lubrication of bearings and gears due to their ability to exhibit semi-permanent lubrication properties, reducing maintenance expenses.

6.3.1 Base oil

The composition of the base oil primarily consists of hydrocarbon chains of varying lengths. The primary components of grease consist of lubricating base oil, which may be derived from mineral, synthetic, or vegetable sources. The length of the hydrocarbon chain exerts a substantial impact on both the tribological and rheological characteristics. The viscosity of lubricating base oil is directly related to the length of the hydrocarbon chain. The consistency of the grease synthesized is influenced by the viscosity of the lubricant base oil utilized. However, the consistency of grease is governed by the volume and variety of thickeners used [2].

6.3.1.1 Mineral oil

Crude oil is a complex mixture of hydrocarbons with varying chemical compositions. Crude oil is composed of a diverse array of hydrocarbon compounds, including alkanes, alkenes, alkynes, alicyclics, and aromatics. Additionally, it contains substances such as waxes, petroleum, asphalt, and non-hydrocarbons, among others. Mineral oil is obtained by the process of fractional distillation of crude oil. Mineral oil exhibits exceptional fluid film lubricating capabilities across various temperatures. Mineral oil is characterized by its affordability and extensive availability in a diverse array of viscosities, making it easily accessible. Furthermore, mineral oil is abundant in nature, further contributing to its widespread availability. Hence, mineral oil is considered a favorable choice as a foundation stock to produce greases. Mineral oil exhibits good oxidation and hydrolytic stability when subjected to temperatures below 100°C. Mineral oils show compatibility with a diverse range of grease thickeners and additives [3–5]. Mineral oils can be classified into three main types based on their chemical composition:

- Paraffinic
- Naphthenic
- Aromatic

Paraffinic oils exhibit non-polar characteristics, with their hydrocarbon chains being either linear or branched in structure. The viscosity index (VI) of straight-chained paraffinic oils is superior to that of branched-chained paraffinic oils. Paraffinic oils are widely favored in synthesizing lubricants and greases within the mineral oil category. Olefins, also known as alkenes, are hydrocarbon chains containing a double carbon bond. The compounds are classified as unsaturated hydrocarbons. Paraffinic oils demonstrate superior oxidation stability in comparison to olefinic or naphthenic oils [6]. The solvency capacity of paraffinic oils is lower compared to olefins.

Naphthenic oils, also known as cycloalkanes, are characterized by one or more saturated cyclic rings. Each cyclic ring consists of six carbon atoms. Naphthenic oils exhibit a greater polarity and have superior solvency capabilities than paraffinic oils [7]. Paraffinic oils have an elevated pour point, flash point, and VI compared to olefinic or naphthenic oils. Naphthenic oils exhibit a diminished VI and reduced flowability at low temperatures. The naphthenic oils exhibit inferior oxidation stability and viscosity–temperature properties at elevated temperatures.

The hydrocarbon chain of aromatic oils possesses a cyclic structure consisting of six carbon atoms, which are organized in a conjugated or alternating pattern of single and double bonds. The aromaticity of these oils arises from the presence of conjugated double bonds. Aromatic oils exhibit notable characteristics such as high density, favorable low-temperature fluidity, low melting point, and poor viscosity–temperature properties. The thermal stability of aromatic oil is comparatively superior to that of paraffinic or naphthenic oils [7]. This oil is primarily used in heat transfer fluids due to its exceptional thermal stability. Aromatic oils possess a significant degree of polarity, making them readily soluble in water.

6.3.1.2 Synthetic oil

Technological advancement necessitates using high-performance lubricants to achieve optimal performance and ensure seamless operations even in challenging environments. Consequently, producing high-performance oils involves synthesizing one or more basic stocks to achieve the requisite performance characteristics in challenging environmental conditions. Synthetic oils, which are artificially produced oils, are sometimes referred to as synthetic oils and tend to be comparatively more costly than mineral oils. Synthetic oils are produced by synthesizing separate and pure ingredients, resulting in a well-defined molecular structure and weight distribution. Consequently, synthetic oils exhibit more uniform physical and chemical properties than mineral oils. According to research findings, synthetic oils demonstrate exceptional fluidity at low temperatures, show superior oxidation stability, have a higher VI, and experience less evaporation losses than mineral oils [8]. Polyalphaolefins (PAO), perfluoropolyethers (PEPE), organic esters, synthetic hydrocarbons, and silicones are the predominant synthetic oils employed in the manufacturing of grease [9–11]. Synthetic lubricating greases formulated using oil as their foundation exhibit more utilization in missiles, airplanes, and space-related endeavors.

6.3.1.3 Vegetable oil

Vegetable oils can be categorized into two main groups, edible oils and non-edible oils, derived from various plant seeds or kernels. Vegetable oils are considered an environmentally friendly product because of their excellent biodegradability. The biodegradability of mineral oil is relatively low, ranging from 20% to 40%. Still, synthetic oils exhibit a higher biodegradability rate of 65% to 100%, and vegetable oils demonstrate even better biodegradability, ranging from 90% to 98% [12]. According to previous research [13,14], vegetable oils have several advantageous properties compared to mineral oils. These include a higher flash point, improved metal adhesion, greater lubricity, increased load-bearing capacity, higher VI, enhanced shear stability, reduced evaporation losses, and decreased volatility. However, vegetable oils possess certain limitations, including reduced oxidation stability and diminished fluidity at lower temperatures. The insufficiency of vegetable oils can be improved through various chemical interventions.

6.3.2 Thickener

A standard grease formulation typically has a thickening content ranging from 5% to 20%. The thickening is responsible for converting the lubricating base oil into a uniform gel-like structure, altering the characteristics of the lubricating base oil. The primary constituent of grease plays a pivotal role in establishing the fundamental attributes of greases and classifying them into discrete categories. A wide range of thickening agents is available for the formulation of grease. The use of thickening significantly influences several physicochemical features of grease, including its consistency, bleeding rate, water washout characteristics, and other related factors. Generally, thickeners can be categorized into two groups: soap-based and non-soap-based.

6.3.2.1 Soap-based thickener

Soap-based thickeners are prepared through the saponification reaction between a fatty acid and an alkali, forming water as a by-product. Soap is a polymorphous thickener that does not interact with lubricating base oil at normal room temperature conditions. However, soap can form a colloidal dispersion when exposed to higher temperatures. The classification of soaps is based on four categories: alkali metals (such as lithium and sodium), aluminum metals, alkaline-earth metals (such as calcium and barium), and heavy metals (such as lead and zinc) [15].

Adding a complexing agent to plain metallic soap results in the formation of a substance known as complex soap. The complex agents can be classified into two categories: organic and inorganic. Benzoic acid is widely regarded as the primary complexing agent for aluminum complex grease, and it has demonstrated efficacy in various other grease formulations. The prevalent complexing agents include benzoate, chloride, carbonate, and metallic acetate. The primary purpose of a complexing agent is to enhance the dropping point of ordinary grease. The addition of a complexing agent [16] has been found to result in a 50°C–100°C increase in the dropping point of typical soap-thickened grease. Complex greases are extensively utilized in applications that need elevated temperatures. In general, it can be observed that complicated grease with a higher concentration of thickener tends to exhibit a reduction in oil separation. Hence, complex greases are unsuitable for low-temperature applications.

The length of the hydrocarbon chain has a notable influence on the thickening effect, and the highest level of thickening is achieved with carboxylic acids containing 18 carbon atoms [16]. The solubility of soap in lubricant base oil increases or decreases as the carbon chain length increases. Mineral oils have more excellent solubility in unsaturated carboxylic acids, resulting in a reduction in the dropping point and a degradation of the thickening effect. The utilization of the aforementioned substance is limited due to its inadequate oxidation stability [16].

6.3.2.2 Non-soap-based thickener

There are two primary categories of non-soap-based thickeners: inorganic and organic. The absence of phase change in inorganic thickeners at high temperatures necessitates reliance on the oxidation stability of the utilized lubricating base oil to determine their maximum operating temperature. The morphology of the inorganic thickener resembles that of a finely divided powder, exhibiting a notable degree of porosity and surface area, which enables it to absorb lubricating base oil effectively. Inorganic thickeners, such as bentonite clays and silica, are commonly utilized in many applications. Silica gel greases are a type of lubricating greases that are created through the thickening of lubricant base oil with highly dispersion silica. Highly scattered silica is commonly generated by pyrogenic and precipitation techniques. The term 'bentone greases' describes greases produced through the thickening of lubricating base oil using bentonite clays. Additional clays employed in grease manufacturing include attapulgite (palygorskite), vermiculite, and hectorite. Bentonite clay is composed of montmorillonite or beidellite, sometimes accompanied by zeolite-group minerals. Montmorillonite is composed of stacked aluminosilicates.

Polytetrafluoroethylene (PTFE), carbon black, amides, anilides, urea derivatives, and polymer thickeners often employ organic thickeners characterized by plate-like or spherical small particles. Polyurea grease has enhanced oxidative stability due to its independence from the saponification process of fatty acids, resulting in the absence of acids and metals. Carbon black, a cost-effective and thermally stable thickening agent, is generated by thermal degradation and the incomplete combustion of hydrocarbons. Greases that are produced through the thickening of lubricating base oil with carbon black are commonly known as carbon black grease. This type of grease demonstrates effective performance in high-temperature applications.

Polymers are commonly employed as thickeners and can be classified into two main categories: high molecular weight compounds and fluoropolymers. Polyfluorohydrocarbons and polysiloxanes are frequently employed as polymer materials in the manufacture of grease. In addition to those mentioned above, many high-performance polymers, both natural and synthetic, such as polyisobutylene, polymethacrylates, and rubbers, are utilized as additives in the formulation of greases [1,17]. The solubility of polymeric substances with high molecular weight has significant limitations regarding hydrocarbons and other lubricating oils.

PTFE is a polymer composed of carbon chains and is extensively utilized in the production of a diverse range of electrical and chemical industry products. PTFE is commonly employed as a thickening agent during grease production [10]. PTFE-thickened greases exhibit exceptional chemical stability and possess notable resistance to elevated temperatures. The utilization of PTFE as an ingredient in titanium complex grease has been shown to exhibit exceptional tribological performance, as evidenced by previous research [18]. Polyurea is a chemical compound that is formed by the reaction between amines and isocyanates. Polyurea has a notable level of oxidation stability due to the absence of acids and metal residues, making it highly suitable for application as a thickening in grease compositions [19,20].

Soap-based thickeners exhibit deficiencies when subjected to exceedingly low or high temperatures. An instance of concern pertains to the behavior of aircraft greases when exposed to low temperatures at high altitudes. The conveyor chains employed in boundary furnaces or bread ovens are greased with greases that are subjected to elevated temperatures. Soap-based greases are susceptible to temperature variations, tending to liquefy at high temperatures and solidify at low temperatures.

6.3.2.3 Mixed thickener

The mixed thickener is composed of several soap cations. The formulation can be achieved through the saponification reaction involving a specific acid and two separate bases. The characteristics of a mixed thickener exhibit variations from those of its constituent components. Lithium–calcium grease exhibits superior water resistance and shear stability compared to standard lithium grease. The addition of a 20% w/w concentration of calcium thickener to raw lithium soap has been found to enhance the dropping point and anti-friction (AF) and anti-wear (AW) performance of standard lithium grease [16]. Adding a minor concentration of clay thickener [21] has enhanced the extreme-pressure (EP) property and heat resistance qualities of aluminum complex grease.

6.3.3 Additives

Additives are substances added to greases or lubricants to enhance their physicochemical and tribological properties. These substances can include chemicals, fillers, or finely powdered solids. The primary purpose of additives is to augment the favorable characteristics and mitigate the unfavorable attributes of greases or lubricants. The service life of greases can be negatively affected by various circumstances, including but not limited to high speed, excessive temperature, harsh loading conditions, thermal and mechanical stress, and ambient environmental conditions. Greases need to meet the intended service requirements for which they are produced. Additives are incorporated to fulfill service specifications and enhance grease longevity. A conventional lubricant is commonly formulated with a mixture of 0%–10% additives, with a specific proportion contingent upon their intended use. In the formulation of greases, several additives such as anticorrosive agents, EP additives, tackiness agents, antioxidants, AF agents, anti-foaming agents, and AW agents are commonly employed. The introduction of some additives has the potential to alter the microstructure of grease [22], or certain additions can improve the inherent properties of the underlying lube oil. According to the study, the average diameter of lithium soap's fiber has been reported as 75 nm. However, when graphene-based nanoadditives were introduced, the average diameter of the lithium soap's fiber was seen to be 65 nm [23].

Antioxidants are incorporated into the grease formulation to impede the process of oxidation throughout the storage period, extending the grease's overall lifespan. In high-temperature applications, it is generally necessary to incorporate antioxidant compounds into mineral oil-based greases. Common examples of antioxidant additives include zinc dithiocarbamates, zinc dithiophosphate, aromatic amines, metal dialkyl dithiophosphate, hindered phenols, sulfurized phenols, and diphenylamine, among others [24].

Anticorrosive additives are included in grease formulations to protect against electrochemical oxidation and dissolving of metallic surfaces. Anticorrosive additives can be roughly categorized into two types: cathodic inhibitors and anodic inhibitors. The use of cathodic inhibitors effectively prevents the cathodic dissolving process, reducing the rate of oxygen and hydrogen ions (H+) reduction rate. Conversely, anodic inhibitors impede the process of anodic dissolution in metals, particularly in the case of steel. The incorporation of anticorrosive compounds in grease serves to diminish the rate of hydroxide and peroxide generation. Common examples of anticorrosive additives include 2,6-di-tert-butyl-4-methyl phenol, diphenylamine, imidazoline, sarcosine derivatives, metal naphthenates, metal sulfonates, phosphate esters, and amides, among others [25].

Tackiness additives are incorporated into grease formulations to impart a dense texture and enhance the cohesive and adhesive properties of the grease while interacting with the surfaces of the tribological pairs. Polybutene, polyisobutylene, copolymers, ethylene-propylene olefin copolymers, and latex compounds are widely recognized as prominent instances of tackiness additives [24].

EP additives are used to improve the load-carrying capability of greases. In gear lubrication, EP additives are typically utilized to minimize scuffing or scoring under extreme loading conditions. Sulfur and phosphorus are used as EP additives in greases to improve load-carrying ability. Sulfides and disulfides, sulfurized lipids, phosphate and thiophosphate esters, zinc dialkyl dithiophosphate, graphite, and other EP additives are common examples [9,26].

AF and AW additives are added to greases to reduce friction and protect tribo-pairs from wear. AF and AW additives react with the contact surfaces to generate a protective beneficial tribofilm that inhibits direct metal-to-metal interaction. Examples of AF and AW additives include zinc dialkyl dithiophosphate, molybdenum disulfide, carbon-based compounds, and metal oxides [5,27,28].

6.4 OVERVIEW OF THE GREASE MARKET

In 2018, the global aggregate production of various types of greases amounted to approximately 1174 thousand metric tonnes (TMT), with India contributing around 82 TMT [21]. Table 6.1 summarizes the percentage distribution of various types of greases produced, categorized according to the type of thickener used. The dominant portion of the overall demand is attributed to conventional and complex lithium soap-produced greases, accounting for around 72% of the total market share. Table 6.2 summarizes the percentage distribution of generation of different greases, categorized according to the basic lube oil. Mineral oil is commonly preferred as the primary basic stock for the manufacture of lubricating grease. In 2018,

Table 6.1 Percentage distribution of distinctive grease production on the basis of a thickener [29]

Type of thickener	*World (%)*	*India (%)*
Aluminum soap	0.29	2.31
Aluminum complex soap	3.95	0.13
Calcium hydrated soap	3.02	3.09
Calcium anhydrous soap	4.81	2.72
Calcium sulfonate soap	3.68	1.06
Calcium-complex soap	0.80	0.11
Lithium soap	50.94	73.58
Lithium complex soap	21.31	9.44
Sodium soap	0.42	2.64
Other metallic soap	0.88	3.69
Polyurea	6.09	0.21
Organophilic clay	2.29	0.83
Other non-soap (except clay)	1.52	0.20

Table 6.2 Percentage distribution of distinctive greases production on the basis of base oil [29]

Type of base oil	*World (%)*	*India (%)*
Mineral	88.97	97.53
Synthetic	5.77	1.79
Semi-synthetic	4.64	0.44
Bio-based	0.63	0.23

roughly 89% of greases produced globally were made using mineral oil. However, approximately 97% of greases in India were synthesized using mineral oil. The proportion of vegetable oil-based greases is below 1%.

6.5 SYNTHESIS PROCEDURE OF LUBRICATING GREASES

The synthesis procedure of lubricating greases is primarily determined by their composition, with some consideration given to the preparation of thickeners and the way of their dispersion. Numerous research publications have identified a significant variation in the synthesis of lubricating greases [4,5,30–32]. During the standard grease synthesis technique, a certain amount of commercially available soap is dispersed in a suitable base lube oil. This dispersion is achieved by subjecting the mixture to continuous stirring at elevated temperatures. The residual base lube oil was incorporated to regulate the viscosity of the grease. The grease was enhanced by adding appropriate additives to improve its physicochemical and tribological characteristics. The grease is subjected to intense mechanical shearing in the final phase to achieve a homogeneous structure.

6.5.1 Soap-based grease

Soap is produced through a chemical reaction between a fatty acid and a strong alkali. As a result of this reaction, water is generated as a by-product of the neutralization process. In the initial stage, the metal soap was made individually or from a commercially available source. The pre-formed soap is intended to be disseminated in a suitable base lubricating oil, while undergoing continuous stirring under specific conditions, such as maintaining a temperature 5°C–10°C higher than the melting point of the soap. The soap-based greases were synthesized by an in situ process, wherein a base lube oil was combined with a stoichiometric amount of fatty acid and a strong alkali. This mixture was continuously stirred under specific circumstances at elevated temperatures. The soap is generated in situ through a saponification reaction, and it is necessary to remove the water by subjecting it to heating at temperatures ranging from 180°C to 200°C. Moreover, the soap can crystallize at temperatures of 120°C–150°C. This process is accompanied by the addition of additional base lube oil, which serves to regulate the viscosity of the grease. The incorporation of additives into grease is a significant phenomenon. Numerous additives exhibit vulnerability to elevated temperatures, resulting in their diminished efficacy under such circumstances. By conventional practices, it is generally recommended to incorporate additives into grease while its temperature is below 80°C. If the thermal stability of additives is high or remains unaltered at elevated temperatures, these additives can be used at any stage during the synthesis of grease. During the final stage, the grease is exposed to intense mechanical shearing to achieve a homogenized structure and consistent uniformity.

6.5.2 Non-soap-based greases

Soap-based greases are sensitive to temperature variations, transitioning from a semi-solid state to a liquid form at elevated temperatures while solidifying at extremely low temperatures. Non-soap-based greases have superior lubrication performance compared to standard soap-based greases when subjected to severe temperatures during hot or cold working conditions. Generally, a non-soap-based thickener is dispersed in a suitable lubricating base oil at room temperature and subjected to continuous stirring to achieve a colloidal suspension similar to that of soap-based greases.

However, it should be noted that many non-soap thickeners, such as polyurea and polyurea complex, require similar heat and chemical reactions as soap-based thickeners. The chemical processes under consideration exhibit distinct characteristics compared to the saponification reaction.

In synthesizing polyurea grease, the grease kettle is filled with lubricant base oil and appropriate amines. The process involves slowly adding a suspension, including lubricating base oil and either isocyanate or diisocyanate, stored in separate vessels, into the solution in the grease kettle. In this context, urea synthesis occurs by chemical reactions involving amines and either isocyanate or diisocyanate compounds. The chemical reaction occurs rapidly and produces manageable amounts of thermal energy. There are no by-products resulting from this process. Polyurea greases exhibit remarkable oxidative stability due to their lack of acids and metal residues. The solution is subjected to heating at an appropriate temperature, determined by the specific chemical reactants involved. Additives are incorporated into the mixture with extra oil to regulate the consistency, after which the resulting product undergoes a process of extremely efficient milling.

While formulating organoclay-based greases, the organoclay is combined with a pregelling agent (such as methanol, ethanol, acetone, propylene carbonate, etc.) and mixed with a lubricating base oil under consistent stirring conditions. Methanol, ethanol, and acetone exhibit moderate volatility, so it is advisable to refrain from inhaling their vapors. In this context, the pregellant serves as a catalyst and can be volatilized upon completion of the procedure. The additives can be incorporated concurrently with the remaining volume of lubricating base oil. The grease undergoes rigorous milling in the final stage to achieve a homogenized structure and homogeneous consistency. Organoclay thickeners do not exhibit a distinct melting point. The grease made with organoclay does not show a dropping point, making it a widely utilized option in baking ovens.

6.6 GREASE STRUCTURE

If the saponification reaction is completed and the resulting soap is evenly dispersed in the oil, the likelihood of achieving a desirable grease structure is significantly reduced. Two fundamental factors play a critical role in this phenomenon: the soap particles must possess a significantly reduced size, and the base oil and thickening must exhibit a specific level of mutual affinity. The level of attraction between the thickening and base oil can be quantified by examining the solubility of the thickener and the solvency of the base oil. If the solubility of the thickener in the base oil is excessive, the thickening effect will decrease. Thickening can be performed using a lower solvency base oil or a less soluble thickener. There needs to be more solubility of the thickener in the base oil to prevent the thickening process.

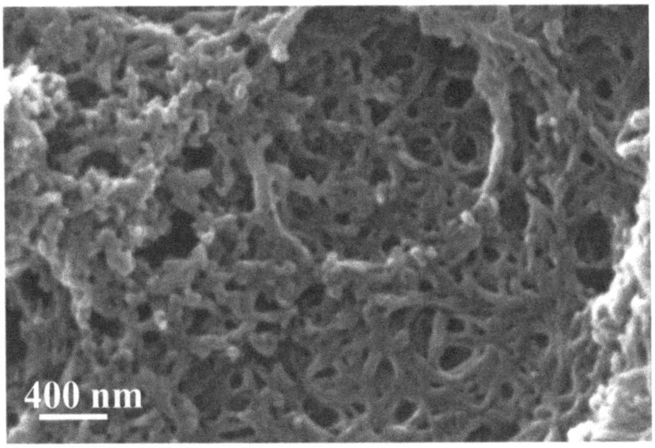

Figure 6.1 SEM images of a 12-lithium hydroxy stearate soap network after extraction of base oil. (Adapted from Ref. [39] with permission.)

The interconnection of thickening fibers is facilitated by capillary and van der Waals forces [33]. Consequently, minuscule pockets formed within the matrix, with dimensions on the order of 10^{-6}m, result in the immobilization of the base oil within these pockets [34]. Figure 6.1 illustrates the network of fibers in the soap thickener. The intertwining of fibers inside the thickening potentially contributes to developing a firmer grease texture. According to a study by researchers [35], the mean diameter and length of lithium grease soap fibers were determined to be 30 nm and 1 µm, respectively. The lubricating behavior of grease is significantly influenced by the length of fibers [36]. Non-soap-based thickeners lack a fibrous structure, necessitating more thickeners to get a comparable consistency to soap-based grease [37]. Thermal aging has a considerable impact on the fiber composition of grease. In their study, Couronne and Vergne observed that the thermal aging process had a detrimental impact on the fibrous structure of lithium and lithium complex greases. In contrast, di- and tetra-urea greases exhibited a significant positive effect under the same conditions [38].

6.7 GREASE CONSISTENCY

The consistency of grease is a crucial characteristic that signifies its relative softness and hardness in terms of its ability to flow. Greases exist in diverse states, including liquid, semi-solid, and solid blocks. The viscosity of grease plays a crucial role in selecting an acceptable lubricant for a particular application. The insufficient softness of grease can result in inadequate lubrication within the targeted region. Conversely, if a grease exhibits excessive softness, it has the potential to escape from the intended lubrication area. The pumpability of grease is also influenced by consistency to a great extent. The thin grease exhibits a high degree of pumpability.

The most critical factor of grease consistency is the concentration of thickening, whereas the viscosity of the basic stock has a relatively negligible influence on consistency. The cone penetrometer is employed to assess the rheological properties of grease by the ASTM D 217 standard [40]. The National Lubricating Institute of Grease (NLGI) has devised the NLGI grade system to categorize greases based on their penetration depth. The penetration depth exhibits a range of values, spanning from 475 (indicating a very soft material) to 85 (indicating a very hard material). Similarly, the NLGI grade ranges from 000 to 6, with each grade having an upper maximum of 30 points. The NLGI grade of grease is presented in Table 6.3. The conduction of penetration tests can be categorized into two main approaches: unworked penetration and worked penetration.

Unworked penetration test is performed by placing a cup filled with grease at a temperature of 25°C ± 0.5°C. The tip of a cone is brought into contact with the top surface of the grease. The cone assembly, which has specific dimensions and weight, is then released freely and allowed to sink into the grease for a duration of 5 ± 0.1 seconds. The evaluation of grease consistency is conducted by measuring its penetration depth in tenths of a millimeter using a dial gauge equipped with a cone penetration instrument. Consequently, a larger penetration value indicates a greasier consistency, as it suggests that the cone assembly has submerged further into the grease.

The **worked penetration test** is conducted similarly to the unworked penetration test; however, the grease sample is churned in the ASTM grease worker for 60 round-trip strokes before the penetration test. Essentially, churning grease aims to simulate high-shear or wheel-bearing situations. Compared to unworked penetration, the worked penetration results are more dependable. As a result, NLGI grade assigns grease according to work penetration.

Table 6.3 Classification of greases by the NLGI number [2]

NLGI number	*ASTM work penetration depth*	*Appearance*
000	445–475	Fluid
00	400–430	Semi-fluid
0	355–385	Very soft
1	310–340	Soft
2	265–295	Normal
3	220–250	Firm
4	175–205	Very firm
5	130–160	Hard
6	85–115	Very hard

6.8 DROPPING POINT

Grease progressively loses its viscosity and softens when heated. Grease flows more quickly at high temperatures, while oil separates and oxidizes more quickly. It also evaporates significantly. Grease loses its ability to function as a thickening lubricant and changes from semi-solid to fluid at a crucial temperature. The temperature at which grease irreversibly loses its consistency and turns into a fluid is known as the "dropping point." Put another way, the dropping point of grease is also known as the melting point. Certain non-soap greases do not show a dropping point, and their structure holds steady as the temperature rises. These greases are therefore recommended for high-temperature applications, like ovens in bakeries. If grease is heated over its drop point and allowed to recrystallize, a persistent decrease in consistency has been seen, and the grease's subsequent performance could be more satisfactory. As a result, the drop point is one of the most critical parameters that shows the temperature at which grease may be used safely.

The determination of the dropping point of lubricating grease is conducted in accordance with the ASTM D 566 [41]. In accordance with the specified experimental parameters, the grease is subjected to a steady increase in temperature within an oil bath. The dropping point is defined as the temperature at which the initial droplet of liquefied grease separates from the test cup. The temperature is the mean value obtained by averaging the grease test cup inside temperature and the oil bath temperature. The utilization of the ASTM D 566 test technique for determining the dropping point of grease at temperatures exceeding 288°C is not recommended. The test protocol ASTM D 2265 is commonly recommended for the determination of the dropping point of grease across a broad temperature range [42]. The temperature at which grease loses its consistency is also influenced by the specific thickening agent employed in its preparation. Table 6.4 illustrates the correlation between the grease thickener and its corresponding dropping point and operating temperature.

Table 6.4 The relationship between a grease thickener and dropping point [2]

Thickener	*Dropping point (°C)*	*Working limit (°C)*
Sodium soap	163–177	121
Lithium soap	177–204	135
Lithium complex soap	250–320	177
Hydrated calcium soap	70–104	60
Anhydrous calcium soap	130–140	100
Aluminum soap	100–120	79
Aluminum complex soap	200–260	150
Barium soap	120–150	100
PTFE	230–300	177

6.9 LUBRICATION MECHANISM OF LUBRICATING GREASES

The lubrication mechanism exhibited by lubricants can be classified into four categories, boundary, mixed, electrohydrodynamic, or hydrodynamic, contingent upon the specific operational circumstances. The lubrication mechanism shown by greases is intricate and distinct from that of conventional lubricants due to the unique structural composition of greases. Greases exhibit a semi-solid or gel-like structure and feature elastic and viscous properties like solids and liquids. Grease is a viscoelastic substance that shows non-Newtonian fluid behavior and thixotropic properties [43]. The interweaving of fibers within the thickening material creates a porous framework that serves as a storage medium for the lubricating base oil. It is postulated that the lubricant base oil confined within the fibrous structure was released gradually during the intermittent interactions. However, it should be noted that the leakage of lubricant base oil from the grease matrix is not considered the primary process by which lubricating greases provide lubrication. The thickening and lubricating base oil functioned synergistically, giving lubrication to contacting surfaces.

In many instances, applying a substantial quantity of grease to surfaces in operation is customary, even though only a minimal amount is necessary for lubricating purposes. The presence of an excess quantity of grease serves as a vital reservoir and facilitates lubrication between surfaces during periods of demand [37]. The thickness of the tribofilm exhibits a modest reduction as the level of frictional heat increases. The temperature in the vicinity of the contact zone causes the grease to increase, resulting in its softening and exposing a greater amount of grease on the contacting interfaces [44,45]. The experimental findings demonstrated that low-speed greases produced a thicker lubricating film than a similar lubricating base oil. Conversely, the lubricating film formed at high speeds exhibited qualities equivalent to those of the lubricating base oil alone [46].

Figure 6.2 illustrates the grease lubrication mechanism in the schematic form. Because of the grease matrix being compressed and sheared by high Hertzian contact stress, the lubricant base oil is bled with nanoadditives. The substances that are incorporated into lubricants and greases at the nanoscale are known as nanoadditives. In the past two decades, the application of nanoadditives to lubricants and greases has increased. Regarding lubrication performance, nanomaterials' dimensions, structure, and concentration are critical. The nanoadditives restrict direct asperity-to-asperity contact by penetrating readily between tribo-pairs. To comprehend the function of nanoadditives in lubrication, an extensive number of articles have documented the four most prevalent mechanisms by which they operate: (a) protective film, (b) rolling effect, (c) mending effect, and (d) polishing effect.

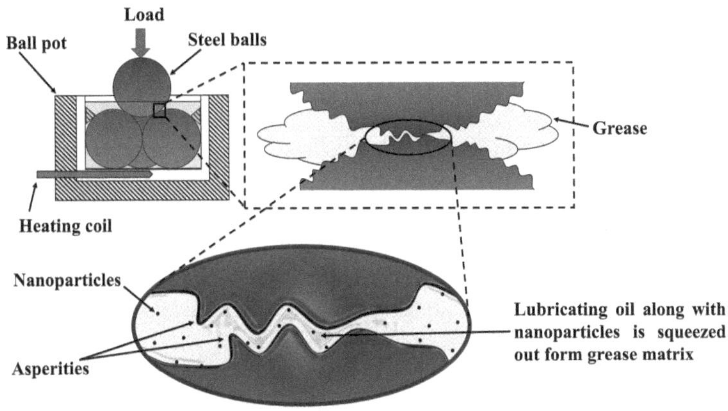

Figure 6.2 Schematic representation of a grease lubrication mechanism. (Adapted from Ref. [47] with permission.)

As a result of the interaction between surfaces carrying nanoadditives with spherical morphology, the sliding friction motion is transformed into a combination of the rolling and sliding friction motions. Consequently, a considerable decrease in friction and attrition was documented in numerous prior studies [48,49]. Self-repairing effect and self-healing effect are alternative names for the mending effect. This phenomenon occurs when nanoadditives readily deposit and settle on the scars and fissures of worn surfaces due to their nanoscale dimensions [50,51]. The application of patches to deteriorated surfaces that interact with one another leads to a decrease in both the rate of abrasion and wear. When nanoadditives meet interacting tribo-pairs, they may also produce a protective film known as tribofilm. A favorable tribofilm is formed on contacting surfaces due to physisorption and chemisorption interactions induced by elevated Hertzian contact pressure and temperature. The tribofilm effectively mitigates the direct contact between asperities, resulting in a notable enhancement in AF and AW properties. The tribofilm's stability characteristics are determined by the nanoadditives' chemical reactivity and hardness [52]. The method of surface enhancement known as the polishing effect is also referred to as the artificial smoothing or smoothing effect. As a result of this phenomenon, nanoadditives reduce surface roughness and refine the interacting surfaces like a three-body abrasive. The aforementioned refining effect significantly improves the tribological properties of greases [53].

6.10 APPLICATION OF GREASES

The choice of lubricating grease for a certain application is contingent upon various factors like load, speed, operating temperature, component type, operational lifespan, and the surrounding working conditions.

6.10.1 Ball/rolling bearing

Bearings play a crucial role in the operation of machinery since they are responsible for supporting and carrying loads. The load-carrying capacity of bearings is progressively rising because of advancements in technology. The longevity of a bearing is intricately linked to the utilization of grease lubrication, particularly under demanding operational circumstances. It is advisable to utilize NLGI grade 2 conventional lithium grease for various bearing applications, including those running at temperatures up to 120°C. NLGI grade 1 grease is the preferred lubricant for needle bearings. It is advisable to utilize complex soap or polyurea-based greases when bearings are exposed to temperatures over 120°C.

Both the speed and load factors influence the grease selection and the re-lubrication frequency. The frequency at which a standard grease needs to be reapplied is determined by the bearing's speed index ($k_f \times n \times d_m$), as specified by the F10 standard [7]. This applies to bearings operating under ambient conditions and temperatures up to 70°C, with the temperature measured at the outer ring of the bearing. Additionally, the mean bearing load (P/C) must be less than 0.1, where k_f represents a factor dependent on the type of bearing, n represents the rotational speed, d_m represents the mean diameter of the bearing, P represents the equivalent dynamic load, and C represents the dynamic load rating. According to a study [16], it has been observed that for every increase of 15 K in temperature above 75°C, the re-lubrication interval of lithium soap-based grease formed with mineral oil is reduced by half.

6.10.2 Gears

Gears are commonly subjected to severe operating conditions, necessitating the usage of greases with high AW properties capable of enduring vibrations and shock loads. Lubricating greases composed of lithium-based compounds, synthetic hydrocarbons, or polyglycols are experiencing an increasing level of recognition and acceptance. Girth gear drives are extensively utilized in numerous large-scale systems across diverse sectors. The girth gear is a type of open gear drive that necessitates using an adhesive grease composed of high-viscosity oils. Typically, these lubricants are applied through spraying onto the tooth surfaces of gears.

6.10.3 Automotive

Greases are usually used in the automotive industry. In passenger cars, car greases are used to lubricate the power transmission system [i.e., universal joints, constant velocity (CV) joints, cardan, joint, belt-pulley bearings, clutch bearing, hypoid rear axle, differential gear, fan clutch bearing], body hardware system (i.e., windshield wiper gear, door locks, seat adjustment gears, door check hinges, and window levers), the suspension system (i.e., half shaft joints, shock absorbers, and ball joints of cross-bar and anti-roll-bar), steering system (i.e., steering intermediate shafts, pivots joints of steering linkages, and rack and pinion steering gear), brake system (i.e., ABS follower and caliper pins), electrical system (alternator bearing and electrical connectors), and other components (i.e., cable actuators and water pump bearing).

The ASTM, SAE, and NLGI have collaboratively established an ASTM standard for automotive service greases. The greases used in automobile service are classified into two distinct categories: wheel-bearing greases, which are identified by the prefix "G," and chassis grease, which is denoted by the prefix "L" [54]. G-category greases are commonly employed to lubricate ball/roller bearings, while L-category greases are typically utilized for lubricating various chassis components, including steering pivots, ball joints, CV joints, universal joints, and similar components. Moreover, the G and L categories might be seen as sub-categories that have been specifically designed for service applications, which are outlined as follows:

LA: These greases are intended to appropriately lubricate chassis components, CV joints, and universal joints throughout a regular re-lubrication interval (approx. 3200 km). During its service life, this grease should be able to protect chassis components from wear and corrosion and prevent consistency degradation and oxidation under light-duty conditions and non-critical applications. NLGI grade 2 greases are commonly recommended.

LB: These greases are necessary for the adequate lubrication of chassis components, CV joints, and universal joints within a temperature range of −40°C to 120°C. This lubrication process is carried out over extended re-lubrication intervals, typically exceeding 3200 km. The grease should be able to safeguard chassis components against both wear and corrosion and defend against water contamination. Additionally, it should exhibit resistance to consistency degradation and oxidation when subjected to varying load levels, ranging from mild to severe, throughout its operational lifespan. It is normally suggested to utilize greases with NLGI grade 2 classification.

GA: These greases are employed to lubricate wheel bearings effectively in situations with moderate stress, and the applications are not critical. The temperature range for the bearing is restricted between −20°C and 70°C, necessitating regular re-lubrication intervals. There is no explicitly stated requirement for this lubricant.

GB: These lubricants effectively lubricate wheel bearings in mild to moderate loads, such as highway and off-highway service. The temperature of the bearing spans from −40°C to 120°C, necessitating less frequent intervals for re-lubrication. Typically, it is advisable to employ NLGI grade 2 greases, but in certain cases, NLGI grade 3 or 1 greases may also be suggested.

GC: Applying these greases is essential for effectively lubricating wheel bearings in various conditions, ranging from light to severe loads, including harsh braking and repeated stop-and-go service. The temperature of the bearing spans from –40°C to 160°C, and periodic re-lubrication is required. It is commonly advised to utilize NLGI grade 2 greases, although there are instances where NLGI grade 3 or 1 greases may also be suggested.

6.10.4 Locomotive

Greases are predominantly employed to lubricate the wheel flanges, traction motor gears, axle bearings, and rail curves to extend the maintenance interval. It is recommended to lube the axle bearings of locomotives with ordinary lithium EP grease. Locomotive applications often need NLGI grade 2 grease, composed of mineral oil thickened with lithium or sodium soap-based greases. The advancement of technology in current trains necessitates improving grease's high-temperature capabilities and service life. Currently, there is a growing trend toward the utilization of vegetable oil-based greases for lubricating locomotives. The utilization of soybean oil-based greases for lubricating locomotive wheel flanges and rail curves has been implemented by the United Nations [55].

6.10.5 Steel mills

In the realm of steel manufacturing, greases must possess the ability to endure high temperatures and tolerate the immense pressure exerted by large loads. The steel mill lubricant should possess water washout resistance properties to safeguard the bearings and other components against corrosion. Various types of greases, such as calcium-complex grease, calcium sulfonate complex grease, standard EP lithium grease, and lithium complex grease, are commonly employed for lubricating hot and cold rolling equipment. These greases are typically manufactured with mineral oil as their base. Currently, lithium–calcium mixed-soap-based grease is commonly favored due to its enhanced water resistance properties. Polyurea greases are experiencing increased recognition due to their elevated dropping point.

6.10.6 Mining

In the mining industry, various types of equipment are employed, including wheel bucket excavators, hydraulic excavators, dumper trucks, draglines, and conveyor systems for soil or coal transportation. These equipment's are subjected to harsh conditions involving dust, mud, water, and corrosive elements. Consequently, the longevity of the bearings utilized in these machines relies heavily on the effectiveness of their sealing mechanisms, which are further enhanced by grease. The wire ropes of draglines and excavators, which can span several kilometers in length, are subjected to significant wear, shock loads, and exposure to a humid environment. Therefore, enhancing the grease used on these ropes with a high concentration of AW and EP additives and anticorrosive agents is recommended. In the mining industry, lithium-based greases are commonly used, namely, those containing lithium complex and lithium–calcium mixed-soap formulations. These greases are enhanced with a comprehensive package of effective EP and AW additives, improving performance.

6.10.7 Food-grade applications

The primary purpose of the food-grade sector is to ensure hygiene and purity of products. Lubricating oils exhibit a pronounced propensity for leaking or splashing. The drawback associated with lubricating oils is considered undesirable in the context of the food-grade business. However, greases can effectively address this concern. The lubricants utilized in the food processing business must adhere to specific regulations about safeguarding human health, compliance with food legislation, and considerations of taste and odor. The substances utilized in the manufacture of food-grade grease, including base stocks, thickeners, and additives, are required to be disclosed in accordance with the regulations outlined in 21 CFR 178.3570 [16]. These ingredients must be approved by the Food and Drug Administration (FDA) of the United States. Approved thickeners include aluminum complex, aluminum stearate, polyurea, and organoclay. The aluminum complex is the primary thickening agent utilized in food-grade grease formulations due to its water resistance and ability to sustain elevated temperatures. In Canada, the utilization of calcium sulfonate thickeners as a thickening agent for food-grade grease has been authorized [56].

6.11 WHEN TO LUBRICATE WITH GREASE

The utilization of grease presents numerous benefits, as observed by a lubrication engineer. Greases have an intrinsic sealing capability. The proper sealing of tribo-pairs lubricated with oil-based lubricants is a complex and costly process. In the context of lubrication, using grease offers advantages in terms of reduced complexity and cost-effectiveness for sealing purposes. Another significant factor to consider is the potential for cost savings. The spatial demands and intricacy are reduced in tribo-pairs that are greased with greases. An illustrative instance is the lubrication of a bearing using an oil-based lubricant. In this scenario, it is frequently observed that the bearing is accompanied by an oil-filled sump positioned beneath it. Additionally, a pumping mechanism is installed to facilitate oil transfer from the sump to the bearings for lubrication purposes. Subsequently, the oil is returned to the sump through a drainage system. The use of greases as a lubricant eliminates the necessity for both dirt sumps and pumping devices in the lubrication of bearings. Hence, the technique of grease lubrication is characterized by reduced complexity, increased compactness, and enhanced cost-effectiveness compared to oil lubrication. Several more attributes of greases make them advantageous, as outlined below:

- To mitigate the occurrence of lubricant dripping and spattering.
- To minimize the frequency of lubrication.
- To seal out dirt and contaminations.
- To enhance the dispersion stability of solid additives.
- When machine elements are worn out badly.
- When noise minimization is an important aspect.
- Low speed combined with high load.
- When subjected to harsh operating conditions such as a sudden increase in load or high levels of pressure.

6.12 DEGRADATION OF GREASE

The proper grease storage is a critical factor, depending on whether it is stored in individual packages or larger quantities. The properties of the grease may undergo changes during storage. The grease has the potential to undergo changes in its physical properties, such as becoming softer or stiffer, altering its visual characteristics, releasing excess oil, undergoing oxidation, or accumulating impurities. The variability in grease properties is contingent upon factors such as storage length, ambient temperature, and product composition.

The consistency of the grease may vary due to lubricant base oil bleeding in a sealed stored container. Loss of consistency causes a decrease in grease's rheological and tribological performance for the intended application. The rate of bleeding increases with time and temperature. When exposed to high temperatures, lubricating greases age rapidly. Thickener fibers are temperature-sensitive. The viscosity of the lubricant base oil reduces at high temperatures, causing it to seep out of the thickening matrix. Second, at high temperatures, the thickening fiber softens or melts.

Grease oxidation stability is also a serious concern. If the greases are combined with antioxidant chemicals, the storage life of the grease is extended, and the grease is protected from oxidative breakdown even at high storage temperatures. Cosmetic changes are observed in the grease during oxidation. Contaminants are another important component that can cause grease to degrade. Moisture is the most common pollutant that can enter tightly sealed containers. Rusting of grease storage containers causes contamination and deterioration of grease. When moisture is present, soda-based grease gets softer, but calcium-complex grease and clay-based grease grow harder [2].

6.13 CURRENT AND FUTURE PROGRESSION IN GREASE LUBRICATION

Numerous tribological studies on traditional lubricating greases utilizing various additives have been conducted [57–64]. Tribological examinations of greases are becoming more specific and application-oriented. The tribological properties of α-type zirconium hydrogen phosphate (α-Zrp) as an additive of mining hoisting wire rope lubricating grease (IRIS-200BB) were evaluated, and it was discovered that 2.5% concentration of α-Zrp showed a significant improvement in AW characteristics, with the coefficient of friction (COF) reduced by 58.05% [65]. Galling is the most prevalent type of failure in metal forming operations, and it is a severe form of adhesive wear. Aluminum alloys have low formability at room temperature. Hence, increased temperatures are required for greater formability. The tribological performance of bismuth sulfide (Bi_2S_3, average size of 2 μm) as an additive of greases (commercial lithium) used for aluminum forming applications was investigated, and results showed that Bi_2S_3 improved the tribological properties of grease used in a wide range of normal loads and prevented galling up to 200°C [66]. Due to large loads, considerable shearing occurs between the workpiece and tool material during broaching operations, resulting in damaged machined surface roughness and decreased tool life. Feng et al. lubricated the textured broach surface with calcium sulfonate grease blended with carbon fiber, and the results showed that the workpiece roughness was improved by 36.7%, and the cutting force was lowered by 12% [67].

Conventional lubricants can only meet some technological advancement needs, particularly for greater temperatures and harsh operating conditions, necessitating high-performance lubricants. Synthetic oils have proven remarkable performance qualities, such as exceptional lubrication at low and high working temperatures, stability in harsh environments, and a long service life. Many studies have investigated the tribological performance of synthetic oil with various additions [68–72]. Due to its potential tribological performance, the researchers were enticed to explore synthetic oil as a basis stock for grease production with various additives [73–77]. The effect of micro and nano-sized graphite particles as additives on the tribological performance of PAO oil-based lithium grease was studied, and it was discovered that nano-sized additives significantly reduced COF by 57%, wear scar diameter (WSD) by 41% and 25%, and improved EP properties [9]. The lubricating grease was created using PAO 100 as the basis stock and lithium soap as the thickening. Chemically functionalized LaF_3 and multi-walled carbon nanotubes (MWCNTs) were blended into the grease, and their optimized concentrations reduced wear by 61% and 57%, respectively, and increased load-bearing capability from 200 to 250 kgf [78].

In 2018, the global percentage of grease businesses employing mineral oil as a base stock was 89%, while India had a share of 97% [29].

Mineral oils are biodegradable, bio-accumulative, hazardous, non-renewable, and detrimental to the environment [13]. The continuous use of mineral oil as a foundation stock in grease manufacturing will eventually reverse its environmental impact. It will contaminate groundwater and land, harming aquatic and terrestrial species and ecosystems. As a result, an alternative suitable substitute that minimizes environmental risk and has the potential to be eco-friendly must be explored and identified. Vegetable oils are a good alternative since they are renewable and highly biodegradable and have a great lubricating performance. Researchers have been investigating vegetable oil as a foundation stock in the manufacture of greases for several decades [79–83].

Rawat et al. developed castor and coconut oil-based lithium greases and compared their tribological performance to mineral oil-based lithium grease. The results revealed that coconut oil-based grease reduced mean wear volume (MWV) by 23% and COF by 9%, while castor oil-based grease lowered MWV by 60.7% and COF by 23.7% compared to mineral oil-based grease [84]. Thampi et al. created 16 grease samples using rice bran oil thickened with calcium soap and sebacic acid as a complexing agent, and they investigated the effect of thickener concentration, holding time, and saponification time on the consistency and tribological performance of the 16 grease samples. The results showed that the thickener concentration considerably affects consistency and COF, but saponification duration affects wear scar diameter (WSD) [85]. With the inclusion of rGO, GO, GO-ODA, MoS_2, and MoS_2-ODT as nanoadditives, castor oil-based lithium grease reduced COF by 56%, 54%, 35%, 34%, and 19%, respectively [86]. Saxena et al. developed soy oil-based eco-friendly greases thickened with organoclay, including biopolymers such as guar gum (GG) and gum acacia (GA) as additives. The tribological results revealed that integrating GG in grease reduced COF by 42% and enhanced AW properties by 22% [87].

Ionic liquids (ILs) have proven their potential as new lubricants in various tribological applications [88–91]. ILs are also being explored for lubricating greases as base stock or as additives. Three types of ILs, i.e., 1-octyl-3-methylimidazolium tetrafluoroborate (LB108), 1-hexyl-3-methyl imidazoliumhexafluorophosphate (L-P106), and 1-butyl-3-methylimidazolium hexafluorophosphate (L-P104), were used as additives in attapulgite clay-based grease and investigated for their tribological performance at higher temperature and attapulgite clay-based grease showed better tribological performance than bentone grease [92]. Similarly, lithium complex grease blended with non-halogenated phosphonium ILs [i.e., bis(oxalate) borate (BOB) and bis(mandelato)borate (BMB)] as additives, which reduced the wear by 50% and 60%, respectively, compared to virgin lithium complex grease. This reduction in wear is attributed to the oxygen-rich tribofilm on the worn surfaces [93]. Electrically conductive lubricating greases were synthesized using two ILs, namely, 1-ethyl-3-methylimidazolium trifluoromethane sulfonate ([EMIm][TFMS]) and 1-ethyl-3-methylimidazolium acetate ([EMIm] [Ac]) as base stock, graphite, and PTFE were used as thickeners and highly ordered pyrolytic graphite (HOPG) particles (1 vol%) as additives. Results showed that IL-based greases have smaller WSD, lower COF, and higher thermal and electrical conductivities than conventional mineral oil-based grease [94].

6.14 SUMMARY OF THIS CHAPTER

This chapter discusses the principles of greases and provides an in-depth look at their composition. The numerous additives utilized in greases and their effect on the tribological performance of greases are discussed. This chapter also discusses the domestic and worldwide market scenarios. In brief, the synthesis technique for lubricating greases is described. The structure of greases, important physicochemical features, lubrication mechanism of greases, and role of nanoadditives are all detailed in depth. This chapter then delves deeply into the applications of lubricating greases in numerous domains. Furthermore, when to use greases and how grease degrades are summarized. This chapter concludes with thorough modern tribological research on various greases to assist in understanding grease lubrication's current and future condition.

REFERENCES

1. Totten, G. E., ed., 1992, *ASM Handbook, Volume 18: Friction, Lubrication, and Wear Technology*, ASM International, Materials Park, Ohio.
2. Ehrlich, M., ed., 1984, *NLGI Lubricating Grease Guide*, NLGI, Missouri, USA.
3. Zhao, G., Zhao, Q., Li, W., Wang, X., and Liu, W., 2013, "Tribological Properties of Nano-Calcium Borate as Lithium Grease Additive," *Lubr. Sci.*, **26**(1), pp. 43–53.
4. Shen, T., Wang, D., Yun, J., Liu, Q., Liu, X., and Peng, Z., 2016, "Tribological Properties and Tribochemical Analysis of Nano-Cerium Oxide and Sulfurized Isobutene in Titanium Complex Grease," *Tribol. Int.*, **93**(January), pp. 332–346.
5. Sahoo, R. R., and Biswas, S. K., 2014, "Effect of Layered MoS_2 Nanoparticles on the Frictional Behavior and Microstructure of Lubricating Greases," *Tribol. Lett.*, **53**(1), pp. 157–171.
6. Rudnick, L. R., ed., 2006, *Synthetics, Mineral Oils, and Bio-Based Lubricants Chemistry and Technology*, CRC Press, Taylor & Francis Group, LLC, Boca Raton.
7. Lugt, P. M., 2013, *Grease Lubrication in Rolling Bearings*, John Wiley & Sons, Ltd, Chichester, United Kingdom.
8. Kumar, H., and Harsha, A. P., 2021, *Group IV Base Stock: Polyalphaolefin–A High-Performance Base Oil for Tribological Applications*, CRC Press, Taylor & Francis Group, Boca Raton.

9. Kumar, N., Saini, V., and Bijwe, J., 2020, "Tribological Investigations of Nano and Micro-Sized Graphite Particles as an Additive in Lithium-Based Grease," *Tribol. Lett.*, **68**(4), pp. 1–13.
10. Ge, X., Xia, Y., and Feng, X., 2016, "Influence of Carbon Nanotubes on Conductive Capacity and Tribological Characteristics of Poly(Ethylene Glycol-Ran-Propylene Glycol) Monobutyl Ether as Base Oil of Grease," *J. Tribol.*, **138**(1), p. 011801.
11. Shu, J., Harris, K., Munavirov, B., Westbroek, R., Leckner, J., and Glavatskih, S., 2018, "Tribology of Polypropylene and Li-Complex Greases with ZDDP and MoDTC Additives," *Tribol. Int.*, **118**(February), pp. 189–195.
12. Florea, O., Luca, M., Constantinescu, A., and Florescu, D., 2003, "The Influence of Lubricating Fluid Type on the Properties of Biodegradable Greases," *J. Synth. Lubr.*, **19**(4), pp. 303–313.
13. Fox, N. J., and Stachowiak, G. W., 2007, "Vegetable Oil-Based Lubricants: A Review of Oxidation," *Tribol. Int.*, **40**(7), pp. 1035–1046.
14. Rawat, S. S., and Harsha, A. P., 2021, "Recent Progress in Vegetable Oil-Based Lubricants for Tribological Applications," *Tribology and Sustainability*, J. K. Katiyar, M. I. U. Haq, A. Raina, S. Jayalakshmi, and R. A. Singh, eds., CRC Press, Taylor & Francis Group, Boca Raton, pp. 115–142.
15. Rawat, S. S., and Harsha, A. P., 2021, "Exploration of Bio-Greases for Tribological Applications," *Tribology, Lubrication, and Surface Engineering: Research and Applications*, C. Pruncu, and A. Aherwar, eds., CRC Press, Taylor & Francis Group, Boca Raton.
16. Mang, T., and Dresel, W., eds., 2007, *Lubricants and Lubrication*, Wiley-VCH Verlag GmbH & Co. KGaA, Weinheim.
17. Patil, P. P., Gori, Y., Kumar, A., Tyagi, M. R., and Improver, V. I., 2021, "Experimental Analysis of Tribological Properties of Polyisobutylene Thickened Oil in Lubricated Contacts," *Tribol. Int.*, **159**(July), p. 106983.
18. Chen, J., 2010, "Tribological Properties of Polytetrafluoroethylene, Nano-Titanium Dioxide, and Nano-Silicon Dioxide as Additives in Mixed Oil-Based Titanium Complex Grease," *Tribol. Lett.*, **38**(3), pp. 217–224.
19. Wu, X., Zhao, Q., Zhao, G., Liu, J., and Wang, X., 2014, "Tribological Properties of Alkylphenyl Diphosphates as High- Performance Antiwear Additive in Lithium Complex Grease and Polyurea Grease for Steel / Steel Contacts at Elevated Temperature," *Ind. Eng. Chem. Res.*, **53**(14), pp. 5660–5667.
20. Wang, Z., Xia, Y., Liu, Z., and Hu, T., 2010, "Friction and Wear Behaviour of Laser-Textured Surfaces under the Lubrication of Polyurea Grease Containing Various Additives," *Proc. Inst. Mech. Eng. Part J J. Eng. Tribol.*, **225**(3), pp. 139–150.
21. Polishuk, A. T., 1971, "Mixed Complex Aluminum Soap-Clay Grease Composition." US3620975A.
22. Adhvaryu, A., Sung, C., and Erhan, S. Z., 2005, "Fatty Acids and Antioxidant Effects on Grease Microstructures," *Ind. Crops Prod.*, **21**(3), pp. 285–291.
23. Rawat, S. S., Harsha, A. P., Khatri, O. P., and Wäsche, R., 2021, "Pristine, Reduced, and Alkylated Graphene Oxide as Additives to Paraffin Grease for Enhancement of Tribological Properties," *J. Tribol.*, **143**(2), p. 021903.
24. Rudnick, L. R., ed., 2017, *Lubricant Additives: Chemistry and Applications*, CRC Press, Taylor & Francis Group, Boca Raton.

25. Ishchuk, Y. L., 2005, *Lubricating Grease Manufacturing Technology*, New Age International (P) Limited, New Delhi, India.
26. Kumar, H., and Harsha, A. P., 2021, "Influence of Oleic Acid–Treated LaF3 Nanoparticles as an Additive on Extreme Pressure Properties of Various Grades of Polyalphaolefins," *Tribol. Trans.*, **65**(1), pp. 96–113.
27. Kumar, H., and Harsha, A., 2021, "Taguchi Optimization of Various Parameters for Tribological Performance of Polyalphaolefins Based Nanolubricants," *Proc. Inst. Mech. Eng. Part J J. Eng. Tribol.*, **235**(6), pp. 1262–1280.
28. Rawat, S. S., Harsha, A. P., Das, S., and Deepak, A. P., 2020, "Effect of CuO and ZnO Nano-Additives on the Tribological Performance of Paraffin Oil–Based Lithium Grease," *Tribol. Trans.*, **63**(1), pp. 90–100.
29. Singh, T., 2020, "Grease Production Survey Report for Calendar Year 2018," *22nd Lubricating Grease Conference, NLGI India-Chapter (2020)*, NLGI, Missouri.
30. Rawat, S. S., Harsha, A. P., and Deepak, A. P., 2018, "Tribological Performance of Paraffin Grease with Silica Nanoparticles as an Additive," *Appl. Nanosci.*, **9**(3), pp. 305–315.
31. Ge, X., Xia, Y., and Cao, Z., 2015, "Tribological Properties and Insulation Effect of Nanometer TiO_2 and Nanometer SiO_2 as Additives in Grease," *Tribol. Int.*, **92**(December), pp. 454–461.
32. Ji, X., Chen, Y., Zhao, G., Wang, X., and Liu, W., 2011, "Tribological Properties of $CaCO_3$ Nanoparticles as an Additive in Lithium Grease," *Tribol. Lett.*, **41**(1), pp. 113–119.
33. Lugt, P. M., 2009, "A Review on Grease Lubrication in Rolling Bearings," *Tribol. Trans.*, **52**(4), pp. 470–480.
34. Stachowiak, G. W., and Batchelor, A. W., 2013, *Engineering Tribology*, Butterworth-Heinemann, Oxford.
35. Salomonsson, L., Stang, G., and Zhmud, B., 2008, "Oil/Thickener Interactions and Rheology of Lubricating Greases," *Tribol. Trans.*, **50**(3), pp. 302–309.
36. Yamamoto, Y., and Gondo, S., 2002, "Frictional Performance of Lithium 12-Hydroxystearate Greases with Different Soap Fibre Structures in Sliding Contacts," *Lubr. Sci.*, **14**(3), pp. 349–362.
37. Lansdown, A. R., 2004, *Lubrication and Lubricant Selection: A Practical Guide*, Professional Engineering Publishing Limited, London.
38. Couronne, I., and Vergne, P., 2000, "Rheological Behavior of Greases: Part II — Effect of Thermal Aging, Correlation with Physico-Chemical Changes," *Tribol. Trans.*, **43**(4), pp. 788–794.
39. Rawat, S. S., Harsha, A. P., Chouhan, A., and Khatri, O. P., 2020, "Effect of Graphene-Based Nanoadditives on the Tribological and Rheological Performance of Paraffin Grease," *J. Mater. Eng. Perform.*, **29**(4), pp. 2235–2247.
40. ASTM-D217, 2003, "Standard Test Methods for Cone Penetration of Lubricating Grease," *ASTM Int.*
41. ASTM-D566, 2016, "Standard Test Method for Dropping Point of Lubricating Grease."
42. ASTM-D2265, 2015, "Standard Test Method for Dropping Point of Lubricating Grease Over Wide Temperature Range."

43. Singh, J., Kumar, D., and Tandon, N., 2019, "Rheological and Film Forming Behavior of the Developed Nanocomposite Greases under Elastohydrodynamics Lubrication Regime," *J. Tribol.*, **141**(2), pp. 1–10.
44. Gonçalves, D., Vieira, A., Carneiro, A., Campos, A. V., and Seabra, J. H. O., 2017, "Film Thickness and Friction Relationship in Grease Lubricated Rough Contacts," *Lubricants*, **5**(3), p. 34.
45. Fukunaga, K., 1988, "Allowable Surface Durability in Grease Lubricated Gears," *Tribol. Trans.*, **31**(4), pp. 454–460.
46. Kanazawa, Y., Sayles, R. S., and Kadiric, A., 2017, "Film Formation and Friction in Grease Lubricated Rolling-Sliding Non-Conformal Contacts," *Tribol. Int.*, **109**(May), pp. 505–518.
47. Rawat, S. S., and Harsha, A. P., 2019, "Current and Future Trends in Grease Lubrication," *Automotive Tribology*, J. K. Katiyar, S. Bhattacharya, V. K. Patel, and V. Kumar, eds., Springer Nature, Singapore, pp. 147–182.
48. Kumar, R., and, Gautam, R. K., 2022, "Tribological Investigation of Sunflower and Soybean Oil with Metal Oxide Nanoadditives," *Biomass Convers. Biorefinery*, 14(2), pp. 2389–2401.
49. Gupta, R. N., and Harsha, A. P., 2018, "Tribological Evaluation of Calcium - Copper-titanate/Cerium Oxide-based Nanolubricants in Sliding Contact," *Lubr. Sci.*, **30**(4), pp. 175–187.
50. Gupta, R. N., and Harsha, A. P., 2018, "Antiwear and Extreme Pressure Performance of Castor Oil with Nano-Additives," *Proc. Inst. Mech. Eng. Part J J. Eng. Tribol.*, **232**(9), pp. 1055–1067.
51. Talib, N., Nasir, R. M., and Rahim, E. A., 2017, "Tribological Behaviour of Modified Jatropha Oil by Mixing Hexagonal Boron Nitride Nanoparticles as a Bio-Based Lubricant for Machining Processes," *J. Clean. Prod.*, **147**(March), pp. 360–378.
52. Shafi, W. K., and Charoo, M. S., 2021, "An Overall Review on the Tribological, Thermal and Rheological Properties of Nanolubricants," *Tribol. – Mater. Surfaces Interfaces*, **15**(1), pp. 20–54.
53. Sajeeb, A., and Rajendrakumar, P. K., 2020, "Tribological Assessment of Vegetable Oil Based CeO_2/CuO Hybrid Nano-Lubricant," *Proc. Inst. Mech. Eng. Part J J. Eng. Tribol.*, **234**(12), pp. 1940–1956.
54. ASTM-D4950, 2019, "Standard Classification and Specification for Automotive Service Greases," pp. 1–6.
55. Honary, L. A. T., and Richter, E., 2011, *Biobased Lubricants and Greases: Technology and Products*, John Wiley & Sons, Ltd., Chichester.
56. Mackwood, W., Muir, R., and Dunn, W., 2003, "Calcium Sulfonate Complex Grease the next Generation Food Machinery Grease," *NLGI Spokesm.*, **67**(2), pp. 17–22.
57. Chang, H., Lan, C. W., Chen, C. H., Kao, M. J., and Guo, J. B., 2014, "Anti-Wear and Friction Properties of Nanoparticles as Additives in the Lithium Grease," *Int. J. Precis. Eng. Manuf.*, **15**(10), pp. 2059–2063.
58. Peña-Parás, L., Taha-Tijerina, J., García, A., Maldonado, D., Nájera, A., Cantú, P., and Ortiz, D., 2015, "Thermal Transport and Tribological Properties of Nanogreases for Metal-Mechanic Applications," *Wear*, **332–333**(May), pp. 1322–1326.
59. Wang, L., Wang, B., Wang, X., and Liu, W., 2007, "Tribological Investigation of CaF_2 Nanocrystals as Grease Additives," *Tribol. Int.*, **40**(7), pp. 1179–1185.

60. Akhtar, K., Khalid, H., Haq, I. U., and Malik, A., 2016, "Improvement in Tribological Properties of Lubricating Grease with Quartz-Enriched Rice Husk Ash," *Tribiology Int.*, **93**(January), pp. 58–62.
61. Kamel, B. M., Mohamed, A., El Sherbiny, M., and Abed, K. A., 2016, "Tribological Behaviour of Calcium Grease Containing Carbon Nanotubes Additives," *Ind. Lubr. Tribol.*, **68**(6), pp. 723–728.
62. Qiang, H., Anling, L., Yangming, Z., Liu, S., and Yachen, G., 2017, "Experimental Study of Tribological Properties of Lithium-Based Grease with Cu Nanoparticle Additive," *Tribol. – Mater. Surfaces Interfaces*, **11**(2), pp. 75–82.
63. Rawat, S. S., Harsha, A. P., Agarwal, D. P., Kumari, S., and Khatri, O. P., 2019, "Pristine and Alkylated MoS_2 Nanosheets for Enhancement of Tribological Performance of Paraffin Grease under Boundary Lubrication Regime," *J. Tribol.*, **141**(7), p. 072102.
64. Singh, J., Kumar, D., and Tandon, N., 2017, "Development of Nanocomposite Grease: Microstructure, Flow, and Tribological Studies," *J. Tribol.*, **139**(5), pp. 052001–9.
65. Zhang, Q., Peng, Y., Chang, X., Zhou, X., Zhang, Y., Chen, G., and Lu, H., 2022, "Tribological Performances of α-Zrp as the Additive of the Grease Used for Mine Hoisting Wire Rope," *Wear*, **504–505**(June), p. 204428.
66. Prieto, G., Kafexhiu, F., Tuckart, W. R., and Podgornik, B., 2022, "Tribological Properties of Bismuth Sulfide (Bi_2S_3) Particles as Grease Additive for Aluminum Forming Operations," *Wear*, 506–507, p. 204442.
67. Feng, K., Ni, J., Zhang, H., Lu, D., Chen, Z., and He, Q., 2022, "Assessment of Textured Broach Lubricated with Carbon Fiber and Calcium Sulfonate Composite Grease," *J. Manuf. Process.*, **83**(May), pp. 536–546.
68. Battez, A. H., Rico, J. E. F., Arias, A. N., Rodriguez, J. L. V., Rodriguez, R. C., and Fernandez, J. M. D., 2006, "The Tribological Behaviour of ZnO Nanoparticles as an Additive to PAO6," *Wear*, **261**(3–4), pp. 256–263.
69. Kumar, H., and Harsha, A. P., 2021, "Enhanced Lubrication Ability of Polyalphaolefin and Polypropylene Glycol by COOH-Functionalized Multiwalled Carbon Nanotubes as an Additive," *J. Mater. Eng. Perform.*, **30**(2), pp. 1075–1089.
70. Kumar, H., and Harsha, A. P., 2020, "Investigation on Friction, Anti-Wear, and Extreme Pressure Properties of Different Grades of Polyalphaolefins with Functionalized Multi-Walled Carbon Nanotubes as an Additive," *J. Tribol.*, **142**(8), p. 081702.
71. Zhang, M., Wang, X., and Liu, W., 2013, "Tribological Behavior of LaF3 Nanoparticles as Additives in Poly-Alpha-Olefin," *Ind. Lubr. Tribol.*, **65**(4), pp. 226–235.
72. Hou, K., Gong, P., Wang, J., Yang, Z., Wang, Z., and Yang, S., 2014, "Structural and Tribological Characterization of Fluorinated Graphene with Various Fluorine Contents Prepared by Liquid-Phase Exfoliation," *RSC Adv.*, **4**(100), pp. 56543–56551.
73. Fan, X., Xia, Y., Wang, L., and Li, W., 2014, "Multilayer Graphene as a Lubricating Additive in Bentone Grease," *Tribol. Lett.*, **55**(3), pp. 455–464.
74. Dai, Y., Niu, W., Zhang, X., Xu, H., and Dong, J., 2017, "Tribological Investigation of Layered Zirconium Phosphate in Anhydrous Calcium Grease," *Lubricants*, **5**(4), pp. 22–31.

75. Chen, T., Xia, Y., Liu, Z., and Wang, Z., 2014, "Preparation and Tribological Properties of Attapulgite – Bentonite Clay Base Grease," *Ind. Lubr. Tribol.*, **66**(4), pp. 538–544.
76. Kumar, N., Saini, V., and Bijwe, J., 2021, "Exploration of Talc Nanoparticles to Enhance the Performance of Lithium Grease," *Tribiology Int.*, p. 107107.
77. Schultheiss, H., Tobie, T., and Stahl, K., 2016, "The Effect of Selected Grease Components on the Wear Behavior of Grease-Lubricated Gears," *J. Tribol.*, **138**(1), p. 011602.
78. Kumar, H., Harsha, A. P., and Rawat, S. S., 2023, "Tribological Evaluation of PAO 100 Oil-based Lithium Greases with Chemically Functionalized Nanoadditives," *Lubr. Sci.*, **35**(7), pp. 528–548.
79. Sharma, B. K., Adhvaryu, A., Perez, J. M., and Erhan, S. Z., 2006, "Biobased Grease with Improved Oxidation Performance for Industrial Application," *J. Agric. Food Chem.*, **54**(20), pp. 7594–7599.
80. Rawat, S. S., Harsha, A. P., and Khatri, O. P., 2021, "Synergistic Effect of Binary Systems of Nanostructured MoS_2/SiO_2 and GO/SiO_2 as Additives to Coconut Oil-Derived Grease : Enhancement of Physicochemical and Lubrication Properties," *Lubr. Sci.*, **33**(5), pp. 290–307.
81. Padgurskas, J., Rukuiža, R., Kupcinskas, A., and Kreivaitis, R., 2015, "Lubrication Properties of Modified Lard and Rapeseed Oil Greases with Sodium and Lithium Thickeners," *Ind. Lubr. Tribol.*, **67**(6), pp. 557–563.
82. Nagendramma, P., and Kumar, P., 2015, "Eco-Friendly Multipurpose Lubricating Greases from Vegetable Residual Oils," *Lubricants*, **3**(4), pp. 628–636.
83. Wang, W., Qian, S., Gong, L., Ni, Z., and Ren, H., 2021, "Effects of Carbon Nano Onions on the Tribological Performance of Food-Grade Calcium Sulfonate Complex Grease," *Lubr. Sci.*, **33**(8), pp. 460–470.
84. Rawat, S. S., and Harsha, A. P., 2022, "The Lubrication Effect of Different Vegetable Oil - Based Greases on Steel - Steel Tribo - Pair," *Biomass Convers. Biorefinery*, pp. 1–13.
85. Thampi, A. D., Biju, G., Alokkan, D., Sasidharan, B., and Rani, S., 2021, "Formulation and Tribological Property Evaluation of Rice Bran Oil Based Lubricating Grease with Sebacic Acid as a Complexing Agent," *Biomass Convers. Biorefinery*, **13**(8), pp. 6715–6726.
86. Rawat, S. S., Harsha, A. P., and Khatri, O. P., 2022, "Tribological Investigations of Two-Dimensional Nanostructured Lamellar Materials as Additives to Castor-Oil-Derived Lithium Grease," *J. Tribol.*, **144**(9), p. 091902.
87. Saxena, A., Kumar, D., and Tandon, N., 2022, "Development of Lubricious Environmentally Friendly Greases Using Synergistic Natural Resources: A Potential Alternative to Mineral Oil-Based Greases," *J. Clean. Prod.*, **380**(P1), p. 135047.
88. Gan, C., Liang, T., Li, W., Fan, X., and Zhu, M., 2019, "Amine-Terminated Ionic Liquid Modi Fi Ed Graphene Oxide/Copper Nanocomposite toward Efficient Lubrication," *Appl. Surf. Sci.*, **491**(March), pp. 105–115.
89. Maurya, U., Vasu, V., and Saini, V., 2024, "Nano-Innovations in Lubrication: Unveiling the Synergistic Boundary Lubrication Potential of Zinc Titanate with Phosphonium Ionic Liquid," *Tribol Int.*, **199**, pp. 109987. https://doi.org/10.1016/J.TRIBOINT.2024.109987.

90. Upendra, M., and Vasu, V., 2020, "Synergistic Effect between Phosphonium-Based Ionic Liquid and Three Oxide Nanoparticles as Hybrid Lubricant Additives," *J. Tribol.*, **142**(5), pp. 052101–12.
91. Maurya, U., Vasu, V., and Kashinath, D., 2022, "Ionic Liquid-Nanoparticle-Based Hybrid-Nanolubricant Additives for Potential Enhancement of Tribological Properties of Lubricants and Their Comparative Study with ZDDP," *Tribol Lett.*, **70**, p. 11. https://doi.org/10.1007/s11249-021-01551-6.
92. Wang, Z., Xia, Y., and Liu, Z., 2012, "Comparative Study of the Tribological Properties of Ionic Liquids as Additives of the Attapulgite and Bentone Greases," *Lubr. Sci.*, **24**(4), pp. 174–187.
93. Ploss, M., Tian, Y., Yoshikawa, S., Westbroek, R., Leckner, J., and Glavatskih, S., 2020, "Tribological Performance of Non-halogenated Phosphonium Ionic Liquids as Additives to Polypropylene and Lithium-Complex Greases," *Tribol. Lett.*, **68**(3), pp. 1–13.
94. Millar, W., Aman, Z. M., Atkin, R., and Li, H., 2022, "Graphite Infused Ionic Liquid Greases," *Colloids Surfaces A Physicochem. Eng. Asp.*, **653**(August), p. 130017.

Chapter 7

Prospects of bio-lubricants in the area of tribology

Sachin G. Ghalme, Mahesh Harne, and Sandeep Thorat

7.1 INTRODUCTION

7.1.1 Global energy scenario and bio-lubricants

With the increasing awareness of environment and sustainability, the national and local government bodies are promoting the use of bio-based lubricants. In line with this, the United Nations (UN) established the Sustainable Development Goal (SDG), to be met by the end of 2030; various guidelines are defined by different regional and national governing bodies for promoting circular economy and sustainable development [1]. Bio-based products are replacing petroleum products and fostering sustainable growth in the developing countries. In the category of bioproducts, this review mainly concentrates on bio-lubricants. Edible and non-edible oil-based bio-lubricants have various advantages and are replacing mineral oils. Bio-lubricants are biodegradable with a higher flash point compared to petroleum-based lubricants. Bio-lubricants are environmentally friendly and safer in terms of storage and transport. Edible and non-edible oil-based bio-lubricants and their production by biorefineries promote sustainable industries.

A major part of engine energy is consumed by the automotive industry to overcome the friction in engines. Quality lubricants contribute to cost savings by minimizing the frictional losses and improving service life. Use of conventional lubricants is challenging nowadays because they are non-renewable and lead to environmental pollution. Re-use of mineral oil is beneficial due to the increasing price of energy, but re-use of lubricants is not possible in the case of commercial vehicles.

An improper way of dumping mineral oil causes [2]:

- Increased cost of sewage treatment,
- Damage to environmental and marine life,
- Deteriorating air quality with increased air pollution.

Depletion of fossil fuels and environmental awareness have promoted the search for bio-based oils since 1973. Bio-lubricants have the ability to

 DOI: 10.1201/9781003642848-7

biodegrade effectively and quickly reduce the risk of long-term environmental damage. In contrast, traditional lubricants, which often contain complex synthetic chemical additives, can persist in the environment for extended periods of time, leading to pollution. Bio-lubricants are significantly advantageous because they lead to less environmental issues. They are biodegradable and dispose naturally in the environment and do not persist as harmful pollutants. This feature minimizes the risk of air, water, and soil pollution, especially in the case of spills or leakages, making them more eco-friendly. Bio-lubricants are prepared from various natural resources like vegetable oils, animal fats, and other plant-based material extract. Unlike petroleum-based lubricants, which rely on limited finite fossil fuel reserves, the use of natural resources ensures a continuous and sustainable supply of lubricants. The production of bio-lubricants typically generates lower greenhouse gas emissions compared to petroleum-based lubricants. The cultivation of plants for feedstock absorbs carbon dioxide from the atmosphere, offsetting some of the emissions associated with their production. Bio-lubricants are typically non-toxic compared to their petroleum-based counterparts. This property makes them safer for human health and reduces the impact on ecosystems in case of accidental release.

The growing demand for bio-lubricants encourages research and development in green technologies. This drive for innovation leads to the creation of new and advanced lubricant formulations that can further improve their performance and environmental benefits. Businesses that embrace green lubricants and sustainable practices often gain a positive reputation and increased brand value among environment-conscious consumers and stakeholders. Green lubricants can enhance energy efficiency in industrial processes. Their high lubricity and lower friction characteristics can reduce energy consumption and improve overall equipment performance. As the world becomes more environment-conscious, businesses are increasingly focusing on sustainable practices. Adopting bio-lubricants can contribute to a company's sustainability efforts, demonstrating environmental responsibility to customers, investors, and regulators.

Adoption of stricter environment regulations by many countries and regions minimizes the use of traditional lubricants. By switching to bio-lubricants, industries can ensure compliance with these evolving regulations and help avoid potential threats.

North America holds the largest bio-lubricant market share and is anticipated to keep that position throughout the forecast period. Along with increased regulatory intervention by the U.S. government regarding the use of conventional lubricants, a rejuvenated automobile industry in the U.S. and Canada is anticipated to drive the global market. The U.S. Air Force is also promoting biodegradable products prepared with natural resources as a fundamental approach toward the national security and the economy, and this is another market-guiding factor. Government spending on bio-lubricants in the maritime and automotive sector is expected to increase, opening up attractive market potential. Implementing the Vessel General Permit (VGP) in 2013 by the U.S. Environmental Protection Agency compelled vessels flying in and around the U.S. to use environment-friendly lubricants in oil-to-sea interfaces, which are non-toxic and non-bioaccumulative. This increased the use of bio-lubricants and created awareness, prompting the International Maritime Organization (IMO) to use environmentally friendly lubricants in the Polar region from January 2017.

The global lubricant market is a vast and growing industry, with a wide range of products and applications. Lubricants play a crucial role in minimizing friction between moving parts, which can contribute to extending the life of machinery and improve efficiency. The overall efficiency of the global industrial sector depends upon the performance of lubricants. Out of the total global lubricant demand, more than half of that is attributed to the automotive industry. The promising popularity of renewable energy is also expected to drive growth in the global lubricants market with the use of lubricants in their parts to minimize friction. This is expected to lead to increased demand for lubricants in the coming years. In 2020, as depicted in Figure 7.1, there was a remarkable surge in the total global demand for lubricants, reaching an unprecedented peak of 37 million metric tonnes. This figure is anticipated to remain high, hovering around 37.4 million metric tonnes by 2023. The global market is poised to achieve a valuation of $182.6 billion USD by 2025. In 2019, China claimed the top position in the global lubricants market in terms of volume, recording 7.3 million tonnes. Following closely, the United States and India secured the second and third positions with volumes of 6.05 million tonnes and 1.7 million tonnes, respectively [3].

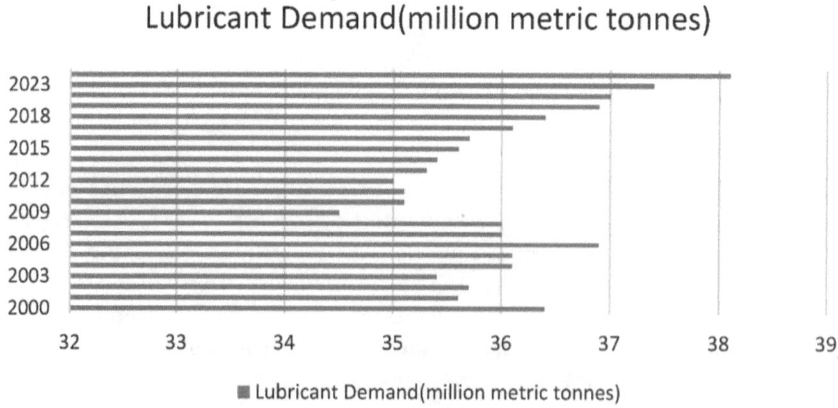

Figure 7.1 Lubricant demand from 2000 to 2020, with projection for 2023 [4].

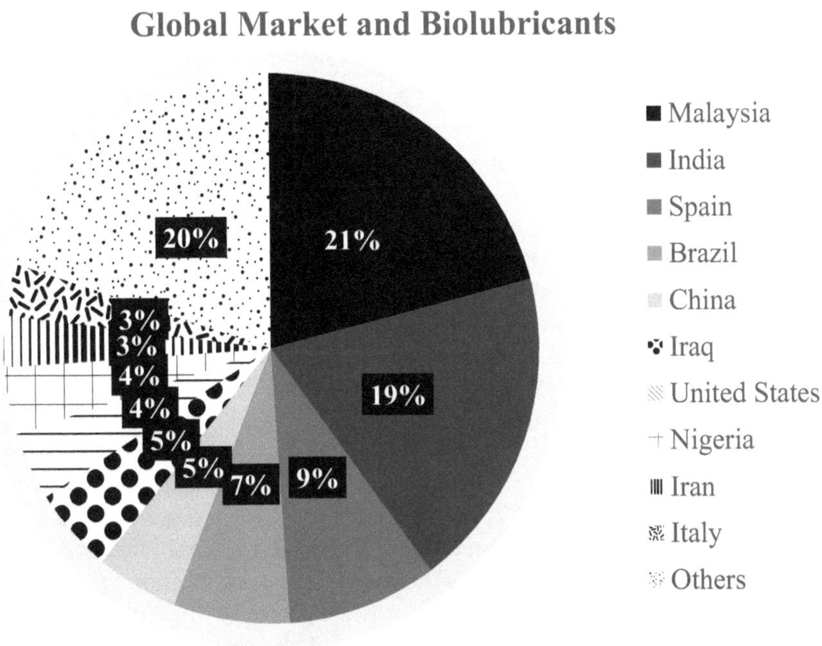

Figure 7.2 Current market scenarios of bio-lubricants.

The demand for bio-lubricants in the global market is continuously increasing; it was about 2.92 billion USD in 2021, and it is predicted to grow up to 4.26 billion USD by 2029, with a compound annual growth rate (CAGR) of 4.7%. The tight environmental laws, growing concerns about the environment and sustainability, and increasing acceptance among end users are some of the driving forces for growth of the bio-lubricants market. Industrial sectors including agriculture, marine, and construction industries will be the leading lubricants market in coming years. Bio-lubricants, also known as ecologically acceptable lubricants (EAL), are mandated for use in marine boats in the United States. In contrast, Europe requires a VGP and eco-label for such applications. Throughout the forecast period, Europe is anticipated to dominate the bio-lubricant market, holding the majority share. It is also expected to lead the global bio-lubricants market from 2020 to 2025 [3–5]. Figure 7.2 presents the distribution of global market for bio-lubricants based on geographical area, and it is clear that the Asia/Pacific region is the major consumer for bio-lubricants.

7.2 LIFECYCLE ASSESSMENT OF LUBRICANTS

Although bio-lubricants are biodegradable and environmentally friendly, considering food security, the economic cost for bio-lubricants is higher. Lifecycle assessment (LCA) of mineral-based and bio-based lubricants is important for understanding their role in environmental sustainability and the economy. LCA can be conducted through scope determination, inventory lifecycle, and assessment of lifecycle impact and result interpretation. For mineral-based oil, it starts with crude oil exploration and ends with disposal into the environment after use, while intermediately it also includes oil refining and production of lubricants. The LCA for bio-lubricants starts with farming and with environment after use, while intermediately, it passes through milling, refining, and production of lubricants [6,7]. The production cost of conventional lubricants is lower than that of bio-lubricants. Comparative performance may be evaluated by adopting certain metrics, which can be beneficial to all stakeholders and eventually for the environment. The environmental footprints can be analyzed with these metrics [6]. According to the Living Planet Report 2010 global footprint network, there is a requirement of approximately 1.5 planets to fulfill humanity's annual needs. Figure 7.3a and b presents the lifecycle for both conventional and bio-lubricants. In the case of bio-lubricants, their LCA is a kind of circular economy with end disposal at landfill sites and CO_2 emissions used in the synthesis of agricultural raw materials. With conventional lubricants, their LCA is a linear economic kind, as their waste disposal cannot be reutilized. A linear economy targets profitability irrespective of lifecycle, whereas a circular economy emphasizes sustainability of the natural sources.

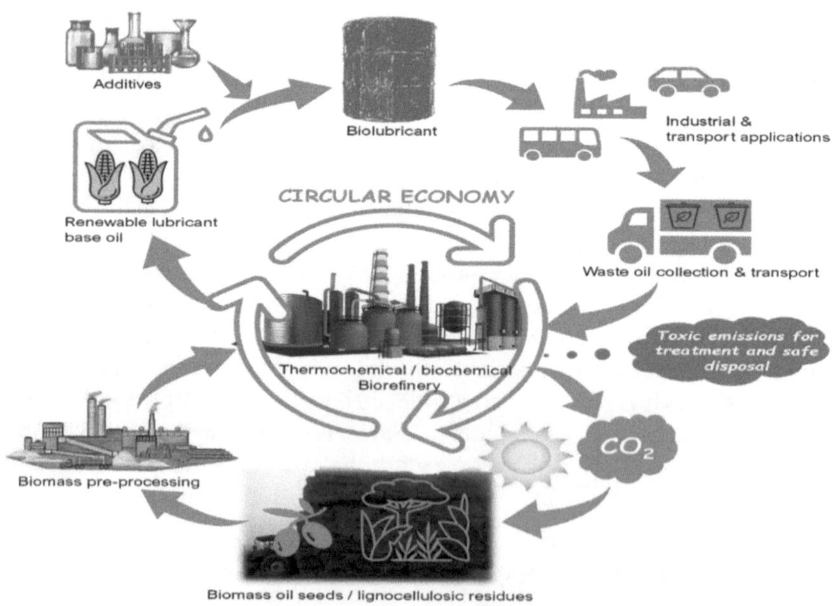

Figure 7.3 (a) Lifecycle of conventional mineral oils [7]. (b) Lifecycle of bio-lubricants [7].

(*Continued*)

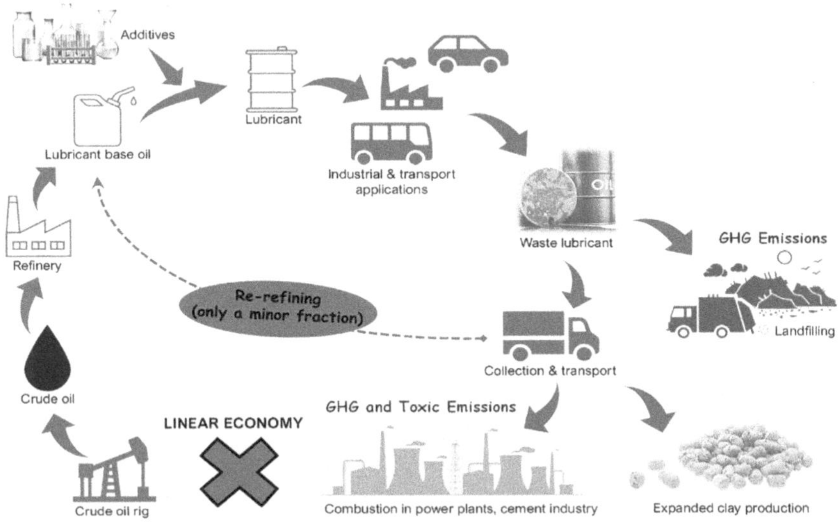

Figure 7.3 (Continued) (a) Lifecycle of conventional mineral oils [7]. (b) Lifecycle of bio-lubricants [7].

7.3 PROPERTIES OF BIO-LUBRICANTS

Bio-lubricants, also recognized as biodegradable or green lubricants, are an eco-friendly replacement to conventional mineral oil-based lubricants. The lubricating performance of oils is affected by physio-chemical properties. Bio-lubricants possess several unique properties that make them highly advantageous for various applications, including viscosity, flash point, biodegradability, and thermo-oxidative stability.

7.3.1 Viscosity

The viscosity of any lubricant is its resistance to flow and shear. It is essential for ensuring proper lubrication in different operating conditions. Bio-lubricants can be formed to have a wide range of viscosities, making them suitable for a wide range of applications. This flexibility allows them to maintain consistent performance across varying temperature conditions in many applications. High viscosity index bio-lubricants are particularly desirable, as they exhibit minimal changes in viscosity with temperature fluctuations, ensuring optimal lubrication efficiency throughout their service life. The viscosity index is modified with the addition of polyols and the presence of hydroxyl groups in the lubricant formulation [8].

7.3.2 Flash point

The lowest temperature at which a lubricant's vapors can ignite when exposed to an open flame is known as its flash point. A higher flash point indicates greater safety during handling, storage, and application. Bio-lubricants typically have higher flash points when compared to the flash point of conventional mineral oil-based lubricants. This attribute reduces the risk of fire hazards, making them safer for use in industries. The higher flash points also contribute to their improved thermal stability, ensuring better performance under high-temperature conditions [9].

7.3.3 Pour point

The pour point temperature is the temperature at which the lubricants fail to maintain their liquid state and flowability. Trimethylolpropane (TMP), a ternary alcohol used in bio-lubricants helps to keep a lower pour point with less resistance to thermo-oxidation, but larger branched alcohols such as neopentyl polyols help to improve oxidation stability [10]. It is an important characteristic for applications in low-temperature conditions.

7.3.4 Thermo-oxidative stability

The ability of a lubricant to resist degradation due to heat and oxidative reactions is known as thermo-oxidative stability. Bio-lubricants are formulated to withstand elevated temperatures and exposure to oxygen, making them suitable for high-temperature applications. Their superior thermo-oxidative stability ensures longer service life, reduces the need for frequent lubricant replacements, and minimizes maintenance costs. This property is particularly valuable in demanding industrial processes and machinery that operate under extreme conditions. Thermo-oxidative stability of bio-lubricant can be tested using the rotating pressure vessel oxidation test (RPVOT) test (ASTM D2272).

7.3.5 Anti-wear properties

Lubricating properties are well observed at low speed and low-pressure conditions, under which lubricants perform well. In some situations, the viscosity of an oil is insufficient to separate the contacting surface and leads to boundary lubrication conditions. Anti-wear additives help to generate a protective film coating on the contact surface. Petroleum-based lubricants are inferior to bio-lubricants when considered for anti-wear performance [11].

7.3.6 Synergy of properties

The combination of these properties makes bio-lubricants highly suitable for a number of applications, encompassing automotive engines and industrial equipment to hydraulic systems and marine applications. In automotive engines, for example, bio-lubricants with their high viscosity index provide better protection and reduced friction, improving fuel efficiency and lowering emissions. In hydraulic systems, their superior biodegradability minimizes the risk of environmental contamination in case of leaks. Moreover, their excellent thermo-oxidative stability ensures reliable performance in heavy-duty machinery, where lubricant breakdown due to high temperatures can lead to equipment failure. The unique properties of bio-lubricants, including viscosity, flash point, biodegradability, and thermo-oxidative stability, collectively contribute to their economic and environmental significance. As industries strive to adopt more sustainable practices and comply with stringent regulations, the use of bio-lubricants continues to gain momentum. The continued research and development of bio-lubricants will likely lead to even more advanced formulations and broader applications in various industrial sectors. Some of the physio-chemical properties of edible and non-edible oils are presented in Table 7.1.

Table 7.1 Physio-chemical properties of edible and non-edible oils [12–14]

Oils	*Density (kg/m³)*	*Kinematic viscosity at 40°C (mm²/s)*	*Oxidation stability 110°C, h*	*Cloud point °C*	*Flash point °C*
Edible oils					
Palm	875	5.72	4.0	13.0	165
Sunflower	878	4.45	0.9	3.42	185
Coconut	805	2.75	35.4	0	112
Soybean	885	4.05	2.1	1.0	176
Linseed	890	3.74	0.2	−3.8	178
Olive	892	4.52	3.4	–	179
Peanut	882	4.92	2.1	5.0	177
Rape seed	880	4.45	7.5	−3.3	62
Rice bran	886	4.95	0.5	0.3	–
Non-edible oils					
Jatropha	878	4.82	2.3	2.75	136
Karanja	917	4.80	6.0	9.0	150
Mahua	850	3.40	–	–	210
Neem	885	5.20	7.2	14.5	44
Castor	898	15.25	1.2	−13.5	260
Tobacco	887	4.25	0.8	–	166

7.4 ECOTOXICITY/BIODEGRADABILITY

The ecotoxicity of bio-lubricants refers to their potential impact on the environment and organisms when released or disposed of improperly. Bio-lubricants are deemed to be degradable because as per CECL-33-A-93 standards, bio-lubricants dissolve almost 80% in 28 days and about 60% after 28 days as per OECD 301B. Lethal Dose (LD50) is considered to evaluate the ecotoxicity lubricant. LD50 is the required amount to kill 50% of the population. A lubricant is considered to be non-ecotoxic if its LD50 exceeds 1000 ppm [8]. Bio-lubricants are often promoted as non-toxic alternatives to conventional petroleum-based lubricants due to their biodegradability. However, it is essential to assess their ecotoxicity to ensure their overall environmental safety.

Biodegradability is a critical factor in determining ecotoxicity. Bio-lubricants that are readily biodegradable break down into harmless substances by natural microbial activity, reducing their potential to persist in the environment and cause long-term harm. High biodegradability means that these lubricants are less likely to accumulate in ecosystems, minimizing adverse effects on aquatic life and terrestrial organisms. Bio-lubricants should be non-toxic or have lower toxicity levels compared to petroleum-based lubricants. Toxicity tests are conducted to evaluate the effects of these lubricants on aquatic organisms, such as fish, algae, and crustaceans, as well as on terrestrial organisms like earthworms and plants. These tests help assess the potential harm they may cause to ecosystems and organisms in case of accidental spills or disposal. When bio-lubricants meet stringent ecotoxicity criteria and exhibit biodegradability, they are considered more environmentally friendly and pose lower risks to ecosystems. It is essential for manufacturers, regulators, and end users to prioritize the selection and use of bio-lubricants with favorable ecotoxicity profiles to minimize their environmental impact and contribute to sustainable practices in various industries.

7.5 TYPES OF BASE STOCKS FOR BIO-LUBRICANTS

Base stocks are a fundamental component of lubricant oils, and they are important for evaluating the overall performance and characteristics of the lubricant. In bio-lubricants, base stocks are derived from renewable resources, making them non-toxic environmentally friendly sustainable substitutes to conventional mineral oil-based lubricants. Let us explore the different types of base stocks used in bio-lubricant oils in detail.

7.5.1 Vegetable oils

Vegetable oils are among the most commonly used base stocks for bio-lubricants. They are obtained from various oil-rich plants, such as soybeans, rapeseed, sunflower seeds, palm fruits, and coconut. Vegetable oils are attractive as base stocks due to their renewable nature, high biodegradability, and low toxicity. They are also readily available, making them economically viable options for bio-lubricant formulations.

However, pure vegetable oils have some limitations that affect their suitability as lubricant base stocks. They often exhibit poor low-temperature properties, including high pour points and viscosities, which can result in difficulties during cold start-up. Additionally, they are susceptible to oxidative degradation at elevated temperatures, leading to reduced lubricant life and performance. To address these issues, vegetable oils can undergo various chemical modifications, such as esterification and hydrogenation, to improve their properties. Esterification involves reacting the vegetable oil with alcohols to form esters, known as biodiesel, which exhibits better low-temperature performance and oxidative stability. Hydrogenation involves the addition of hydrogen to the vegetable oil to reduce the number of double bonds, resulting in reduced reactivity and improved thermal stability.

7.5.2 Animal fats

Fish oils, lard, and beef tallow are some of the animal fats used as base stock for bio-lubricants. Like vegetable oils, animal fats are renewable resources and exhibit high biodegradability. Animal fats have more content of saturated fatty acids, which improves its oxidative stability. However, like vegetable oils, animal fats also suffer from poor low-temperature properties and may require similar chemical modifications to improve their performance.

7.5.3 Synthetic esters

Bio-lubricants also use synthetic esters as a base stock due to their excellent performance characteristics and are generated through the organic acids and alcohols reactions. Synthetic esters can be further modified to achieve specific properties so that esters can be used for a variety of applications:

a. Polyol esters: Polyol esters are prepared with polyols and fatty acids. They offer superior thermal stability, oxidative resistance, and lubricity. Polyol esters can withstand high temperatures and are commonly used in high-performance applications, such as aviation and racing engines.
b. Diesters: Diesters are formed by the reaction of dicarboxylic acids with alcohols. They exhibit good oxidative stability, low volatility, and high biodegradability. Diesters are commonly used in compressor oils, hydraulic fluids, and environmentally friendly refrigeration lubricants.
c. Complex esters: Complex esters are a blend of various ester molecules, each contributing specific properties to the base stock. They offer a balance of performance characteristics, including oxidative stability, lubricity, and low-temperature behavior. Complex esters are versatile and find applications in a wide range of industries.

7.5.4 Polyalphaolefins (PAOs)

Although not derived directly from renewable resources, PAOs can be synthesized from alpha-olefins, which can be derived from renewable feed stocks like vegetable oils. PAOs are synthetic hydrocarbons and offer excellent thermal stability, oxidative resistance, and low-temperature performance than synthetic oils. They are compatible with existing lubricant formulations and equipment, making them a feasible option for bio-based lubricants in certain applications.

7.5.5 Polyalkylene glycols (PAGs)

PAGs are another class of synthetic base stocks used in bio-lubricants. They offer outstanding thermal stability, biodegradability, and lubricity. PAGs are commonly used in applications requiring superior performance under extreme temperatures or high loads, such as in industrial gears and compressors.

7.5.6 Alkylated naphthalenes

Alkylated naphthalenes are synthesized by alkylating naphthalene, a hydrocarbon compound found in coal tar or derived from plant-based sources. Alkylated naphthalenes offer high thermal stability, low volatility, and good viscosity-temperature characteristics. Alkylated naphthalenes are particularly suitable for applications where high-temperature stability and oxidation resistance are critical, such as in high-temperature greases and industrial gear oils.

7.5.7 Ionic liquids

Ionic liquids are a unique class of base stocks that are still in the early stages of research and development for use in bio-lubricants. They consist of ions and offer promising potential to improve lubricant performance, especially under extreme conditions. While ionic liquids are not derived from renewable resources, ongoing research aims to develop greener and more sustainable sources for these specialized base stocks.

Base stocks for bio-lubricant oils encompass a diverse range of options, each with its specific advantages and applications. Vegetable oils and animal fats provide a renewable and biodegradable foundation for bio-lubricants but may require chemical modifications to improve their performance. Synthetic esters, such as polyol esters and diesters, offer excellent thermal stability and oxidative resistance, while PAOs and PAGs provide synthetic alternatives with exceptional properties. Alkylated naphthalenes and ionic liquids represent emerging options that offer unique benefits in specific applications.

The continued research and development in bio-lubricants aim to optimize the performance of base stocks, enhance their environmental sustainability, and expand their applicability across diverse industries. As demand for environmentally friendly and sustainable lubricants grows, base stocks derived from renewable resources will play a crucial role in shaping the future of bio-lubricants and promoting greener and more responsible industrial practices.

7.6 LIMITATIONS OF BIO-LUBRICANTS

Apart from the various advantages of bio-lubricants, they have certain limitations for use of them in their natural form.

1. Oxidation and stability: Bio-lubricants based on natural oils contain fatty acids and are liable to oxidation and degradation over time. When exposed to oxygen and heat, the double bonds in the fatty acid and chain can break, leading to the formation of peroxides and other harmful compounds [15–17]. The presence of methylene interrupted polyunsaturation and caused bio-lubricants to be more prevalent for oxidation [18,19]. This oxidation leads to the thickening of the lubricants, formation deposits, and sludge. These changes over time decreases shelf life and lubricating performance and causes damage to the machinery. These limitations can be mitigated through the addition of antioxidants, anti-wear agents, and corrosion inhibitors to enhance their service life and performance.
2. Viscosity variability: Viscosity is an important property, which directly affects the lubrication performance of the lubricants. Bio-lubricants are derived from natural sources, which may exhibit inherent variability in their composition. Viscosity is temperature-dependent, and the viscosity of bio-lubricants varies more compared to synthetic lubricants with change in temperature. Variations in viscosity can affect their lubrication properties and make it challenging to maintain consistent performance [20].
3. Compatibility: Natural bio-lubricants have polar molecules and affinity for certain materials. They are not compatible with certain materials commonly used in machinery, such as seals and gaskets made from elastomer and synthetic rubber. Continuous exposure to the bio-lubricants for these materials may lead to swelling and deterioration, leading to leakage and other operational problems.
4. Cost: Bio-based lubricants are costlier to produce than mineral oil-based lubricants.
5. Limited applicability: Bio-lubricants are not useful in all conditions, especially where extreme pressure and anti-wear characteristics are expected.
6. Working temperature limitation: Bio-lubricants cannot work at a wide range of temperatures as compared to petroleum-based lubricants [21,22]. The low-temperature fluidity, as well as lower volatility, at higher temperature of bio-lubricants is managed by presence of the ester group [23]. They are not suitable for extreme temperature conditions.

7. Pour point: The temperature at which the lubricant loses its flowability is known as its pour point temperature. The presence of ternary alcohols like TMP in bio-lubricants diminishes its pour point but simultaneously also reduces the thermo-oxidative stability [24].

7.7 BIO-LUBRICANTS AND CHEMICAL MODIFICATIONS

Although bio-lubricants are a good alternative to conventional lubricants in the process of sustainable development, but they need certain chemical modifications to overcome their limitations. Researchers in the field of bio-lubricants are actively trying to evaluate the relationship between chemical compositions and molecular structure in view of their chemical modifications. Figure 7.2 shows the overview for the chemical modifications required for bio-lubricants.

7.7.1 Esterification/transesterification reactions

Due to acyl ester moieties of triacylglycerols, the plant oils are very much prone to oxidation. In the transesterification process, the glycerol moieties of the triacylglycerols (commonly found in plant oils) are replaced with a branched-chain or long-chain alcohols (Figure 7.4).

While in the esterification process, the fatty acid of natural oils reacts with branched-chain and long-chain alcohols and from their respective esters,

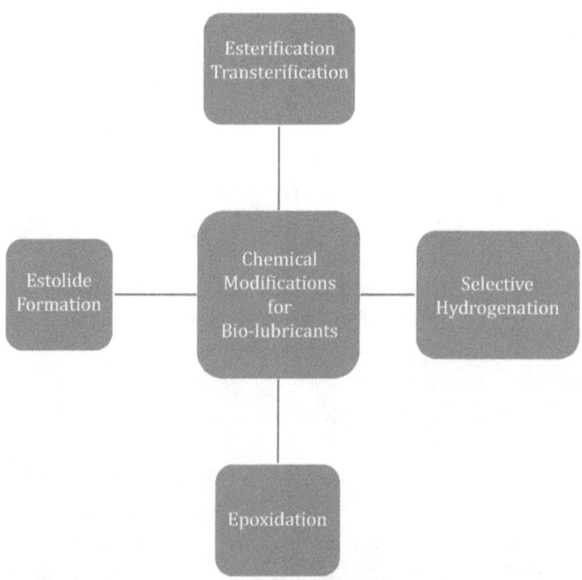

Figure 7.4 Chemical modifications for bio-lubricants [25].

Table 7.2 Transesterification process for non-edible oils

Oil	*Alcohol*	*Catalyst*	*Yield (%)*	*Oil/alcohol*	*Ref.*
Bitter almond	Methanol	Potassium acetate supported over calcium oxide [PA/CaO]	88	1:4	[33]
Rubber seed	Trimethylolpropane	p-Toluene sulfonic acid	71	1:3	[34]
Palm kernel	Isoamyl alcohol	Lipase *B. cepacia* immobilized on SiO_{2}- HEC	NA	1:6	[35]
Castor	2-Ethyl 1-hexanol	Titanium isopropoxide	93	1:3	[36]
Rapeseed	2-Ethyl 1-hexanol	Titanium isopropoxide	93	1:3	[36]
Palm stearin	Methanol	*Candida antarctica* lipase B	95.26	1:4	[37]
Cardon	2–2-Dimethyl-1, 3-propanediol	Ti(IV)isopropoxide	NA	1:1	[38]
Microalgal	Fusel oil	*Burkholderia cepacia* (Lipase LBC)	70.8–89	1:8	[39]
Dessert date	Trimethylolpropane	Sodium methoxide	NA	1:3	[40]
Jatropha	Trimethylolpropane	Sodium methoxide	NA	1:0.025	[41]
Karanja	Trimethylolpropane	Sulfuric acid	90.97	1:4	[42]
Neem	Methanol	Calcium oxide	96.7	1:6	[43]

which can be used as bio-lubricants [26]. These reactions can be catalyzed using homogeneous or heterogeneous (acids or alkalis) [27–29] or biochemical (lipases) catalysts [30]. The esterification and transesterification process uses mineral acids (HCL, H_2SO_4, or p-toluenesulfonic acid). These processes have limitations due to strong acid corrosivity with product isolation and significant mineral oil output. This process requires basic catalysts such as calcium oxide, CaO, for full transesterification conversion. During reaction, high temperature is also required to activate and renew the catalyst [31]. The different alcohols and acids used for transesterification and esterification directly affect the lubricating characteristics of the synthetic ester bio-lubricants. As a result of this, the synthetic esters perform well in terms of lubricating characteristics of the corresponding plant oils. The use of alkalis, alkali earth oxides, alkoxides, carbonates, and hydroxides has been reported to result in a higher yield of up to 90% in many cases with a less number of catalysts and less reaction time than solid acid catalysts [32]. Table 7.2 provides the details about some transesterification processes.

7.7.2 Selective hydrogenation

The bio-based oils are mixtures of triacylglycerol with 12–22 carbon atoms and 0–3 double bonds in each hydrocarbon, thereby making them more susceptible to oxidation attack [44]. The hydrogenation of plant oil helps in modifying physical characteristics and improving stability in terms of oxidation and decomposition [45]. Selective hydrogenation is a kind of chemical process used to change the unsaturated fatty acids like linoleic and linolenic acids present in natural fats and oils into single unsaturated fatty acids without increasing the saturated part of the substance so that the aging stability can be improved [46]. Although selective hydrogenation improves the aging behavior of the oils, it also leads to the isomerization and forms cis- and trans-acids from monoenoic acids, affecting the physical properties of the oil. During selective hydrogenation, care should be taken for the generation of 80% cis-oleic acid to meet the lubricating requirements of the industries [47].

7.7.3 Epoxidation

Oxidation stability is important for lubricating oils, and the presence of unsaturation degree/number significantly affects the oxidation stability of plant oils. Unsaturation functionality is more liable for hydrolytic breakdown and thermal degradation during the oxidation process [48]. Epoxidation is an important chemical process for modifying unsaturation present in bio-lubricants at –C=C– bonds for improving their oxidation stability, pour point, metal surface adsorption, lubricity, and viscosity index [49,50]. This process helps double bonds result in production of oxirane rings [51,52]. The epoxidation of unsaturated bonds takes place in the presences of hydrogen peroxides and peroxy acids, resulting in peracids that epoxidates double bonds [53]. The epoxidation advancements for various non-edible oils are presented in Table 7.3.

Although epoxidized oils show improved oxidation stability, they have poor properties at low temperatures. So these oils further need esterification or re-esterification or by alkylation, acylation, and acetoxylation for improving their physio-chemical and lubrication properties [63]. The unsaturated functional group or their fatty acid, such as oleic and linoleic acids, is important to initiate the modification process for the production of bio-lubricants.

7.7.4 Estolide formation/oligomerization

Estolides are oligomeric esters generated during one fatty acid links to the site of unsaturation of another fatty acid. Estolides are better in oxidation stability and cold flow than vegetable oils and are easily biodegradable in the environment [64]. Estolides are a compound class produced when altering the plant oils and their derivatives using strong acids to change the ester functionality of the alkene group of fatty acids. Methanesulfonic acid (CH_3SO_3H), sulfuric acid (H_2SO_4), and perchloric acid ($HCIO_4$) were used as a catalyst for the activation of the oleic acids alkene group to generate estolide [65]. Although estolides are produced, a small amount of alkene group remains in estolides because perchloric acids cannot covert completely alkene groups into other functional groups. These alkene groups are vulnerable to oxidative destruction. So, it is always preferred for preparing estolides from oleic acid using a less dangerous reagent. Estolides produced from naturally occurring hydroxy

Table 7.3 Epoxidation of various non-edible oils

Oil	*Acid*	H_2O_2(% wt)	*Yield*	*Ref.*
Neem	Formic acid	30	NA	[54]
Palm stearin	Candida antarctica lipase and glacial acetic acid	30	90.8%	[55]
Michelia champaca	Peracetic acid+sulfuric acid	30	NA	[56]
Pongamia	Acetic acid and sulfuric acid	30	71.5%	[57]
Tung	Formic acid	30	93.6%	[58]
Jatropha	Formic acid	30	NA	[59]
Castor	Acetic acid	30	95.84%	[60]
Crude palm	Acetic acid and sulfuric acid	60	87.59%	[61]
Euphorbia lathyris	Acetic acid and sulfuric acid	30	NA	[62]

acids like ricinoleic acid and lesquerolic acid show the presence of the alkene group, which is vulnerable for oxidative degradation. So, they require additional chemical processing for removing alkene groups. 100% conversion of free fatty acid (FFA) into estolides using a three-step process was proposed by Juan Carlos et al. [66]. It was proposed that $HClO_4$, a cheap, strong acid, can be used for the production of estolides, which can be further converted into estolide esters and the unreacted FFA into biodiesel using methanol in the presence of cheap ionic liquid through the esterification process. These products are in line with international standards of SAE J306 and EN 14214 for lubricants and biodiesel. The conversion of oleic acid into polyhydroxy estolide through its reaction with hydrogen peroxide was proposed by Hoong et al. [67–69]. The generated estolide esters were further treated with alcohols and amines, producing byproducts which have phytochemical characteristics comparable to commercial lubricants.

7.8 APPLICATIONS

With growing concerns for the environment and sustainability, edible and non-edible oil-based bio-lubricants have created their own space in the field of mechanical and industrial applications [70]. Bio-lubricants are of great importance in the areas where machines directly lose oils and also in the areas that are environmentally sensitive, like food processing industries and water treatment plants [71]. Bio-lubricants are renewable and non-toxic in nature, due to which they can be used in a variety of applications like lubrication oils, engine oils, hydraulic oils, metal working fluids, compressor oils, gear oil, chainsaw lubricants, and grease. Table 7.4 presents the various bio-lubricants, their properties, and their major applications.

Table 7.4 Bio-lubricants, properties, and major applications

Sr. No.	*Oil*	*Properties*	*Major applications*	*Ref.*
1	Castor	High VI, low volatility, high antioxidants, low deposit formation	Greases, gear lubricants	[72,73]
2	Canola	High VI, high flash point than some conventional oils, high lubricity, low evaporative loss, low coefficient of friction, better lubricity, non-toxicity	Metal working fluids (MWFs), transmission fluids, food-grade lubes, hydraulic fluids, penetrating oils, transmission fluids	[72,73]
3	Soybean		Hydraulic oils, biodiesel fuel, engine oils, transmission fluids, printing inks, paints, detergents, coatings, pesticides, shampoos	[72,73]
4	Safflower		Resins, diesel fuels, enamels	[72,73]
5	Linseed		Stains, coatings, vanishes, paints	[72,73]
6	Olive		Engine oils	[72,73]
7	Sunflower		Diesel fuels, greases	[72,73]
8	Crambe	n.a.	Greases, surfactants, cosmetics, chemicals	[73]
9	Coconut	High anti-wear, better lubricity, low coefficient of friction	Engine oils	[72,73]
10	Palm	Less corrosive, low coefficient of friction, high viscosity	Greases, MWFs	[72,73]
11	Pongamia	Low frictional losses, low emissions, minimum break-specific fuel consumption and high break thermal efficiency at medium loads	Power transformer applications, anticorrosive coating	[72,74,75]
12	Tallow		Soaps, cosmetics, plastics, hydraulic oils	[73]
13	Cuphea		Motor oils, cosmetics	[73]
14	Jojoba		Greases, cosmetics, lubricants	[73]
15	Jatropha	High VI, low wear loss, low cumulative weight loss, low coefficient of friction	Biodiesel	[72,76,77]

7.9 CONCLUSION

Petroleum-based mineral oils are the driving force for the economic growth of any country, but they are getting depleted rapidly with increasing demand. About 30–40 million tons of lubricants are utilized for various industrial applications, and from this, about 50%–75% of lubricant products go into the environment. These petroleum-based lubricants consist of paraffinic, olefinic, naphthenic, and aromatic hydrocarbons, which are non-renewable and toxic in nature, posing a threat to human beings and the environment. To comply with the industrial requirements and environmental regulations, bio-lubricants have emerged as potential substitutes for traditional mineral oils. Bio-lubricants have higher viscosity index, flash point, lubricity, high biodegradability, non-toxicity, and good resistance to shear, while they exhibit higher pour point and lower oxidative stability and moisture content. Furthermore, there is a need for research to develop a low cost process for production of bio-lubricants from base stock; chemical modifications are also essential for improving the bio-lubricant properties and applicability of bio-lubricants in the industrial sector.

REFERENCES

1. United Nation General Assembly. 2015. *Transforming Our World: The 2030 Agenda for Sustainable Development. Resolution 70/1*. New York, NY, USA: United Nation General Assembly. https://www.un.org/en/development/desa/population/migration/generalassembly/docs/globalcompact/A_RES_70_1_E.pdf.
2. Chowdary, K., Kotia, A., Lakshmanan, V., Elsheikh, A.H., Ali, M.K.A. 2021. A review of the tribological and thermophysical mechanisms of bio-lubricants based nanomaterials in automotive applications. *J. Mol. Liq.* 339, 116717. https://doi.org/10.1016/j.molliq.2021.116717.
3. Sarno, M., Iuliano, M., Cirillo, C. 2019. Optimized procedure for the preparation of an enzymatic nanocatalyst to produce a bio-lubricant from waste cooking oil. *Chem. Eng. J.* 377, 120273. https://doi.org/10.1016/j.cej.2018.10.210.
4. Lubricants Demand Worldwide. 2020. *Statista*. https://www.statista.com/statistics/411616/Lubricants-demand-worldwide/ (Accessed 25 March 2022).
5. Abdullah, B.M., Zubairi, S.I., Huri, H.Z., Hairunisa, N., Yousif, E., Basu, R.C. 2016. Polyesters based on linoleic acid for biolubricant basestocks: low-temperature, tribological and rheological properties. *PLoS One.* 11(3), e0151603. https://doi.org/10.1371/journal.pone.0151603.
6. Damjanović, N., Davidović, Z., Spasojević Šantić, T. 2016. Ecological and economical aspects of use of bio-lubricants and conventional lubricants. Fuels and lubricants: journal for tribology, lubrication, application of liquid and gaseous fuels and combustion engineering. *Goriva Maziva.* 55(3), 188–199.
7. Narayana Sarma, R., Vinu, R. 2022. Current status and future prospects of biolubricants: properties and applications. *Lubricants.* 10, 70. https://doi.org/10.3390/lubricants10040070.

8. Cecilia, J.A., Ballesteros Plata, D., Alves Saboya, R.M., Tavares de Luna, F.M., Cavalcante Jr, C.L., Rodríguez-Castellón, E. 2020. An overview of the biolubricant production process: challenges and future perspectives. *Processes.* 8(3), 257. https://doi.org/10.3390/pr8030257.
9. Rizvi, S.Q.A. (Ed.). 2009. *A Comprehensive Review of Lubricant Chemistry, Technology, Selection, and Design.* West Conshohocken, PA: ASTM International. https://doi.org/10.1520/MNL59-EB.
10. Suhane, A., Rehman, A., Khaira, H.K. 2012. Potential of non edible vegetable oils as an alternative lubricants in automotive applications. *Int. J. Eng. Res. Afr.* 2(5), 1330–1335.
11. Kotturu, C.M.V., Srinivas, V., Vandana, V., Chebattina, K.R.R., Seetha Rama Rao, Y. 2020. Investigation of tribological properties and engine performance of polyol ester–based bio-lubricant: commercial motorbike engine oil blends. *P. I. Mech. Eng. D.-J. Aut.* 234(5), 1304–1317. https://doi.org/10.1177/0954407019878359.
12. Gui, M.M., Lee, K.T., Bhatia, S. 2008. Feasibility of edible oil vs. non-edible oil vs. waste edible oil as biodiesel feed stock. *Energy Convers. Manag.* 33, 1646–53.
13. Leung, D, Wu, X, Leung, M. 2010. A review on biodiesel production using catalyzed transesterification. *Appl. Energy.* 87, 1083–95.
14. Karmakar, A., Karmakar, S., Mukherjee, S. 2010. Properties of various plants and animals feed stocks for biodiesel production. *Bioresour. Technol.* 101, 7201–10.
15. Mobarak, H.M., Mohamad, E.N., Masjuki, H.H., Kalam, M.A., Al Mahmud, K.A.H., Habibullah, M., Ashraful, A.M. 2014. The prospects of biolubricants as alternatives in automotive applications. *Renew. Sustain. Energy Rev.* 33, 34–43.
16. Nagendramma, P., Kaul, S. 2012. Development of ecofriendly/biodegradable lubricants: an overview. *Renew. Sustain. Energy Rev.* 16(1), 764–774. https://doi.org/10.1016/j.rser.2011.09.002.
17. Erhan, S.Z., Sharma, B.K., Perez, J.M. 2006. Oxidation and low temperature stability of vegetable oil-based lubricants. *Ind. Crop. Prod.* 24(3), 292–299. https://doi.org/10.1016/j.indcrop.2006.06.008.
18. Dabai, M.U., Owuna, F.J., Sokoto, M.A., Abubakar, A.L. 2018. Assessment of quality parameters of ecofriendly biolubricant from waste cooking palm oil. *Asian J. Appl. Chem. Res.* 1, 1–11.
19. Bilal, S., Mohammed-Dabo, I.A., Nuhu, M., Kasim, S.A., Almustapha, I.H., Yamusa, Y.A. 2013. Production of biolubricant from jatropha curcas seed oil. *J. Chem. Eng. Mater. Sci.* 4, 72–79.
20. Quinchia, L.A., Delgado, M.A., Valencia, C., Franco, J.M., Gallegos, C. 2010. Viscosity modification of different vegetable oils with EVA copolymer for lubricant applications. *Ind. Crop. Prod.* 32(3), 607–612. https://doi.org/10.1016/j.indcrop.2010.07.011.
21. Li, W., Wang, X. 2015. Bio-lubricants derived from waste cooking oil with improved oxidation stability and low-temperature properties. *J. Oleo Sci.* 64, 367–374.
22. Liu, Z., Sharma, B.K., Erhan, S.Z., Biswas, A., Wang, R., Schuman, T.P. 2015. Oxidation and low temperature stability of polymerized soybean oil-based lubricants. *Thermochim. Acta.* 601, 9–16. https://doi.org/10.1016/j.tca.2014.12.010.

23. Jumat, S., Bashar, M.A., Rahimi, M.Y., Nadia, S. 2014. Synthesis, reactivity and application studies for different biolubricants. *Chem. Cent. J.* 8, 8–16.
24. Saboya, R.M.A., Cecilia, J.A., García-Sancho, C., Sales, A.V., de Luna, F.M.T., Rodríguez-Castellón, E., Cavalcante Jr, C.L. 2017. Assessment of commercial resins in the biolubricants production from free fatty acids of castor oil. *Catal. Today.* 279, 274–285.
25. Salih, N., Salimon, J. 2021. A review on new trends, challenges and prospects of ecofriendly friendly green food-grade biolubricants. *Biointerface Res. Appl. Chem.* 12(1), 1185–1207. https://doi.org/10.33263/BRIAC121.11851207.
26. Salih, N., Salimon, J., Yousif, E. 2011. The physicochemical and tribological properties of oleic acid based triester biolubricants. *Ind. Crops. Prod.* 34, 1089–1096. https://doi.org/10.1016/j.indcrop.2011.03.025.
27. Lam, M.K., Lee, K.T., Mohamed, A.R. 2010. Homogeneous, heterogeneous and enzymatic catalysis fortransesterification of high free fatty acid oil (waste cooking oil) to biodiesel: a review. *Biotechnol. Adv.* 28, 500–518.
28. Fukuda, H., Kondo, A., Noda, H. 2001. Biodiesel fuel production by transesterification of oils. *J. Biosci. Bioeng.* 92, 405–416.
29. Macario, A., Giordano, G., Onida, B., Cocina, D., Tagarelli, A., Giuffrè, A.M. 2010. Biodiesel production process by homogeneous/heterogeneous catalytic system using an acid–base catalyst. *Appl. Catal. A Gen.* 378, 160–168.
30. Panchal, T.M., Patel, A., Chauhan, D.D., Thomas, M., Patel, J.V. 2017. A methodological review on bio-lubricants from vegetable oil based resources. *Renew. Sustain. Energy Rev.* 70, 65–70. https://doi.org/10.1016/j.rser.2016.11.105.
31. Nor, M.N., Salih, N., Salimon, J. 2020. Optimization of the ring opening of epoxidized palm oil using D-optimal design. *Asian J. Chem.* 33(1), 67–75. https://doi.org/10.14233/ajchem.2021.22938.
32. García-Sancho, C., Saboya, R.M.A., Cecilia, J.A., Sales, A.V., Luna, F.M.T., Rodríguez-Castellón, E., Cavalcante Jr, C.L. 2017. Influence of pore size and loading for Nb_2O_5/SBA-15 catalysts on synthetic ester production from free fatty acids of castor oil. *Mol. Catal.* 436, 267–275. https://doi.org/10.1016/j.mcat.2017.04.036.
33. Fadhil, A.B., Al-Tikrity, E.T.B., Khalaf, A.M. 2018. Transesterification of non-edible oils over potassium acetate impregnated CaO solid base catalyst. *Fuel.* 234, 81–93. https://doi.org/10.1016/j.fuel.2018.06.121.
34. Kamarudin, N.S., Veny, H., Sidek, N.F., Abnisa, F., Sazali, R.A., Aziz, N. 2020. Investigation on synthesis of trimethylolpropane (TMP) ester from non-edible oil. *Bull. Chem. React. Eng. Catal.* 15(3), 808–817. https://doi.org/10.9767/bcrec.15.3.8862.808-817.
35. Ceron, A.A., Vilas Boas, R.N., Biaggio, F.C., de Castro, H.F. 2018. Synthesis of biolubricant by transesterification of palm kernel oil with simulated fusel oil: batch and continuous processes. *Biomass Bioenergy.* 119, 166–172. https://doi.org/10.1016/j.biombioe.2018.09.013.
36. Encinar, J., Nogales-Delgado, S., Sanchez, N., Gonzalez, J. 2020. Biolubricants from rapeseed and castor oil transesterification by using titanium isopropoxide as a catalyst: production and characterization. *Catalysts.* 10(4), 366. https://doi.org/10.3390/catal10040366.

37. Afifah, A.N., Syahrullail, S., Azlee, N.I.W., Sidik, N.A.C., Yahya, W.J., Abd Rahim, E. 2019. Biolubricant production from palm stearin through enzymatic transesterification method. *Biochem. Eng. J.* 148, 178–184. https://doi.org/10.1016/j.bej.2019.05.009.
38. Nogales-Delgado, S., Encinar Martín, J.M. 2021. Cardoon biolubricant through double transesterification: assessment of its oxidative, thermal and storage stability. *Mater. Lett.* 302, 130454. https://doi.org/10.1016/j.matlet.2021.130454.
39. Da Silva, A.P.T., Bredda, E.H., de Castro, H.F., Da Ros, P.C.M. 2020. Enzymatic catalysis: an environmentally friendly method to enhance the transesterification of microalgal oil with fusel oil for production of fatty acid esters with potential application as biolubricants. *Fuel.* 273, 117786. https://doi.org/10.1016/j.fuel.2020.117786.
40. Singh, Y., Sharma, A., Singh, N.K., Chen, W.H. 2020. Development of bio-based lubricant from modified desert date oil (balanites aegyptiaca) with copper nanoparticles addition and their tribological analysis. *Fuel.* 259, 116259. https://doi.org/10.1016/j.fuel.2019.116259.
41. Ramesh, G.C., Nagaraju, T., Chowgi, S. 2018. Synthesis and study of bio-lubricant from Jatropha vegetable oil. *SSRN Electron. J.* https://doi.org/10.2139/ssrn.3328410.
42. Sharma, U.C., Sachan, S., Trivedi, R.K. 2018. Viscous flow behaviour of karanja oil based bio-lubricant base oil. *J. Oleo Sci.* 67(1), 105–111. https://doi.org/10.5650/jos.ess17140.
43. Olufemi, B.A., Essien, R.A. 2020. Production of biolubricants from neem seed oil catalyzed by calcium oxide from snail shell. *Acta Tech. Corviniensis Bull. Eng.* 13, 107–113.
44. Gunstone, F.D. 2009. *The Chemistry of Oils and Fats: Sources, Composition, Properties and Uses.* Oxford: Wiley-Blackwell Publishing Ltd.
45. Wagner, H., Luther, R., Mang, T. 2002. Lubricant base fluids based on renewable raw materials-their catalytic manufacture and modification. *Appl. Catal. A. Gen.* 221, 429–442. https://doi.org/10.1016/s0926-860x(01)00891-2.
46. Nesrain, F., Salih, N., Salimon, J. 2021. Physiochemical properties of Saudi Nigella sativa L. ('Black cumin') seed oil. *Oilseeds Fats Crops Lip.* 28. https://doi.org/10.1051/ocl/2020075.
47. Ahmed, W.A., Yarmo, M.A., Salih, N., Derawi, M.D., Yusop, M.R. 2015. Synthesis and lubricity properties analysis of branched dicarboxylate esters based lubricant. *Malaysian J. Anal. Sci.* 19, 106–117.
48. Salimon, J., Salih, N., Yousif, E. 2012. Improvement of pour point and oxidative stability of synthetic ester basestocks for biolubricant applications. *Arab. J. Chem.* 5(2), 193–200. https://doi.org/10.1016/j.arabjc.2010.09.001.
49. Abdullah, B.M., Zubairi, S.I., Huri, H.Z., Hairunisa, N., Yousif, E., Basu, R.C. 2016. Polyesters based on linoleic acid for biolubricant basestocks: low-temperature, tribological and rheological properties. *PLoS One.* 11(3), e0151603. https://doi.org/10.1371/journal.pone.0151603.
50. McNutt, J., Sophia, Q., He. 2016. Development of biolubricants from vegetable oils via chemical modification. *J. Ind. Eng. Chem.* 36, 1–12. https://doi.org/10.1016/j.jiec.2016.02.008.

51. do Valle, C.P., Rodrigues, J.S., Fechine, L.M.U.D., Cunha, A.P., Malveira, J.Q., Luna, F.M.T., Ricardo, N.M.P.S. 2018. Chemical modification of Tilapia oil for biolubricant applications. *J. Clean. Prod.* 191, 158–166. https://doi.org/10.1016/j.jclepro.2018.04.062.
52. Hernández-Cruz, M.C., Meza-Gordillo, R., Domínguez, Z., Rosales-Quintero, A., Abud-Archila, M., Ayora-Talavera, T., Villalobos-Maldonado, J.J. 2021. Optimization and characterization of in situ epoxidation of chicken fat with peracetic acid. *Fuel.* 285, 119127. https://doi.org/10.1016/j.fuel.2020.119127.
53. Bashiri, S., Ghobadian, B., Dehghani Soufi, M., Gorjian, S. 2021. Chemical modification of sunflower waste cooking oil for biolubricant production through epoxidation reaction. *Mater. Sci. Technol.* 4, 119–127. https://doi.org/10.1016/j.mset.2021.03.001.
54. Kareemullah, M., Afzal, A., Rehman, K.F., KC, V., Khan, H., Soudagar, M.E.M., Kaladgi, A.R. 2021. Preparation and physicochemical properties evaluation of epoxidized neem oil-based bio-lubricant. *Aust. J. Mech. Eng.* 1–10. https://doi.org/10.1080/14484846.2021.1920141.
55. Afifah, A.N., Syahrullail, S., Wan Azlee, N.I., Rohah, A.M. 2021. Synthesis and tribological studies of epoxidized palm stearin methyl ester as a green lubricant. *J. Clean. Prod.* 280, 124320. https://doi.org/10.1016/j.jclepro.2020.124320.
56. Singh, Y., Abd Rahim, E. 2020. Michelia Champaca: sustainable novel non-edible oil as nano based bio-lubricant with tribological investigation. *Fuel.* 282, 118830. https://doi.org/10.1016/j.fuel.2020.118830.
57. Sahana, V.N., Chandra Kumar, R. 2019. Manufacturing of pongamia oil based bio-lubricant for machining application. *IRJET.* 6(7), 1639–1645.
58. Ju, Y.H., Sari, N.N.F., Go, A.W., Wang, M.J., Agapay, R.C., Ayucitra, A. 2020. Preparation of epoxidized fatty acid ethyl ester from tung oil as a bio-lubricant base-stock. *Waste Biomass Valor.* 1(8), 4145–4155. https://doi.org/10.1007/s12649-019-00749-z.
59. Taung Mai, L.L., Aung, M.M., Muhamad Saidi, S.A., H'ng, P.S., Rayung, M., Jaafar, A.M. 2021. Non edible oil-based epoxy resins from jatropha oil and their shape memory behaviors. *Polymers.* 13(13), 2177. https://doi.org/10.3390/polym13132177.
60. Borugadda, V.B., Goud, V.V. 2015. In-situ epoxidation of castor oil using heterogeneous acidic ion-exchange resin catalyst (IR-120) for bio-lubricant application. *Tribol. Online.* 10(5), 354–359. https://doi.org/10.2474/trol.10.354.
61. Hartawan, R.A., Said, M., Faizal, M., Prianto, J., Juwita, M., Aprianti, N. 2022. Synthesis of epoxide as intermediate compounds for biolubricant production from crude palm oil. *Indones. J. Fundam. Appl. Chem.* 7(2), 42–50. https://doi.org/10.24845/ijfac.v7.i2.42.
62. Singh, Y., Singh, N.K., Sharma, A., Singla, A., Singh, D., Abd Rahim, E. 2021. Effect of ZnO nanoparticles concentration as additives to the epoxidized Euphorbia Lathyris oil and their tribological characterization. *Fuel.* 285, 119148. https://doi.org/10.1016/j.fuel.2020.119148.
63. Karmakar, G., Ghosh, P., Sharma, B. 2017. Chemically modifying vegetable oils to prepare green lubricants. *Lubricants.* 5, 1–17. https://doi.org/10.3390/lubricants5040044.

64. Salimon, J., Salih, N. 2009. Improved low temperature properties of 2-ethylhexyl 9(10)-hydroxy-10(9)-acyloxystearate derivatives. *Eur. J. Sci. Res.* 31, 583–591.
65. Sharma, B.K., Biresaw, G. (Eds.). 2016. *Environmentally Friendly and Biobased Lubricants.* CRC Press. https://doi.org/10.1201/9781315373256.
66. de Haro, J.C., del Garrido, P.M., Perez, A., Carmona, M., Rodriguez, J.F. 2018. Full conversion of oleic acid to estolides esters, biodiesel and choline carboxylates in three easy steps. *J. Clean. Prod.* 184, 579–585. https://doi.org/10.1016/j.jclepro.2018.02.190.
67. Arslan, A., Rancke-Madsen, A., Brask, J. 2020. Enzymatic synthesis of estolides from castor oil. *Catalysts.* 10(8), 835. https://doi.org/10.3390/catal10080835.
68. Hoong, S.S., Arniza, M.Z., Mariam, N.M.D.N.S., Armylisas, A.H.N., Ishak, S.A., Ismail, T.N.M.T., Yeong, S.K. 2020. Synthesis of estolide ester and amide from acetylated polyhydroxy estolide for lubricant base oil. *Eur. J. Lipid Sci. Technol.* 122(10), 2000098. https://doi.org/10.1002/ejlt.202000098.
69. Hoong, S.S., Arniza, M.Z., Mariam, N.M.D.N.S., Armylisas, A.H.N., Yeong, S.K. 2019. Synthesis and physicochemical properties of novel lauric acid capped estolide esters and amides made from oleic acid and their evaluations for biolubricant basestock. *Ind. Crop. Prod.* 140, 111653. https://doi.org/10.1016/j.indcrop.2019.111653.
70. Sammaiah, A., Padmaja, K.V., Prasad, R.B. 2016. Synthesis and physical properties of novel estolides from dicarboxylic acids and methyl ricinoleate. *Eur. J. Lipid Sci. Technol.* 118(3), 486–494. https://doi.org/10.1002/ejlt.201500109.
71. Abdullah, B.M., Zubairi, S.I., Huri, H.Z., Hairunisa, N., Yousif, E., Basu, R.C. 2016. Polyesters based on linoleic acid for biolubricant basestocks: low-temperature, tribological and rheological properties. *PLoS One.* 11, 1–15.
72. Gobinda, K., Pranab, G., Bragendra, K.S. 2017. Chemically modified vegetable oils to prepare green lubricants. *Lubricants.* 44, 1–17.
73. Mobarak, H.M., Mohamad, E.N., Masjuki, H.H., Kalam, M.A., Al Mahmud, K.A.H., Habibullah, M., Ashraful, A.M. 2014. The prospects of biolubricants as alternatives in automotive applications. *Renew. Sustain. Energy Rev.* 33, 34–43.
74. Thanigaiselvan, R., Raja, T.S.R., Karthik, R. 2015. Investigations on eco friendly insulating fluids from rapeseed and pongamia pinnata oils for power transformer applications. *J. Electr. Eng. Technol.* 10, 2348–2355.
75. Ahmad, S., Ashraf, S.M., Naqvi, F., Yadav, S., Hasnat, A. 2003. A polyesteramide from pongamia glabra oil for biologically safe anticorrosive coating. *Prog. Org. Coat.* 47, 95–102.
76. Sahoo, P.K., Das, L.M. 2009. Combustion analysis of Jatropha, Karanja and Polanga based biodiesel as fuel in a diesel engine. *Fuel.* 88, 994–999.
77. Habibullah, M., Masjuki, H.H., Kalam, M.A., Ashraful, A.M., Habib, M.A., Mobarak, H.M. 2014. Effect of bio-lubricant on tribological characteristics of steel. *Procedia Eng.* 90, 740–745.

Chapter 8

Next-generation nanolubricants

From colloidal stability to tribological performance

Vinay Saini

8.1 INTRODUCTION

Resource conservation and lowering emissions have gained prominence in the automotive sector due to elevated energy demands, fuel consumption, and environmental issues [1]. New rigorous restrictions for emissions in the transportation and industrial sectors have limited the use of sulfur (S) and phosphorus (P) additives in engine oils to 0.50% and 0.08%, respectively. The presence of sulfur and phosphorus in additives impairs the efficacy of emission after-treatment systems by diminishing the activity of catalytic surfaces. NOx catalysts are poisoned, and ash may obstruct after-treatment particle filters. This has compelled oil researchers and manufacturers to create an entirely novel group of environmentally acceptable additives with minimal or negligible SAPS (sulfated ash, phosphorus, and sulfur) [2]. Modern engines, designed to operate under extreme conditions such as high temperatures, heavy loads, and lower speeds, impose additional demands on lubricants. Downsized engines, aimed at improving fuel efficiency, often subject mechanical components to higher contact pressures and wear, necessitating lubricants that provide superior antiwear (AW) properties and extended oil drain intervals. Given that up to one-third of fuel energy in vehicles is lost to friction and heat, the formulation of energy-efficient lubricants becomes paramount. The potential impact of improved lubrication is immense, with studies suggesting an 18% reduction in frictional energy losses in the coming decade. Effective lubricant formulations can extend the service life of mechanical components, reduce greenhouse gas (GHG) emissions, and contribute significantly to global climate change mitigation efforts Consequently, there is an imperative necessity to formulate energy-efficient lubricants that possess superior antiwear properties and extended engine oil drain intervals compared to traditional lubricants, since nearly 1/3 of the fuel energy is wasted to overcome friction heat in these vehicles [3,4]. Moreover, lubrication has been pushed to the extreme limits of loads, speeds, and torque transfer in several industrial segments such as wind turbines (>10 MW) and heavy-duty engines. Increasing power transmission efficiency with lower surface distress requires a combination

DOI: 10.1201/9781003642848-8

of advanced lubricants and surface engineering (coatings and texturing). In light of this, the incorporation of nanoparticles (NPs) as additives has risen over the past decade, as shown in Figure 8.1, showing the lubricant additive timeline. In subsequent sections, this chapter will explore the integration of NPs into lubricants, focusing on their tribological performance, dispersion stability, and the challenges associated with formulating next-generation nanolubricants.

It is estimated that effective lubrication advancements could nearly lead to decrement of parasitic frictional energy losses in engines by 18% reduction in the next decade. Overall, the potential savings of GHG emission (in tons) from passenger cars can be achieved through the use of energy-efficient oil additives capable of extending the life of vehicle components [6,7]. These savings can significantly boost global efforts to combat climate change by overcoming tribological issues, especially related to friction and wear, that impact adverse effects on energy efficiency and operational longevity. Typically, the lubricants (oils/grease) have a multitude of functions in addition to friction and wear mitigation, such as cooling and corrosion control [8]. In the subsequent section, nanolubricants are covered in detail.

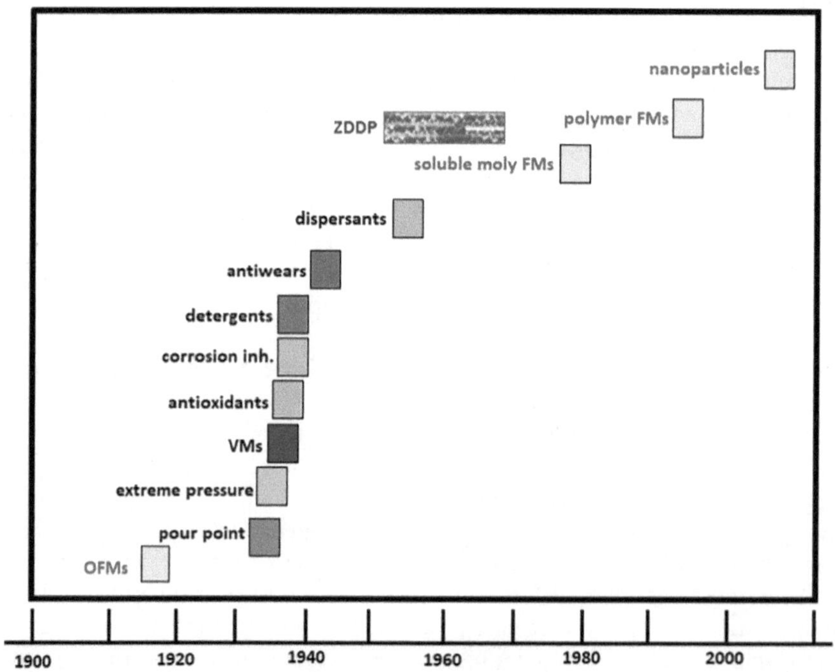

Figure 8.1 Evolution timeline of various additives. (Adapted from Ref. [5].)

8.1.1 Nanolubricants

The term "Nanofluids," first introduced by Choi and Eastman [9], refers to fluids containing NPs, which are particles with at least one dimension smaller than 100 nm. Since their inception, NPs have garnered immense interest due to their small size and high surface area, which significantly enhances their surface interaction potential and strength. High aspect ratios and surface-to-volume ratios contribute to strong particle–particle and particle–surface interactions in various applications.

NPs suspended in base fluids are often referred to as nano-oils or nanolubricants [10]. In the context of lubricants, these particles, irrespective of whether they are metallic, metal oxides, or carbon-based, can enhance the tribological properties of oils and greases. A decade ago, Gulzar et al. [11] classified these NPs into seven broad categories:

- Metals [12]: Cu [13,14], Fe [15,16], Al, Pb, Ti [17], etc.
- Metal oxides [12,18]: CuO [19], ZnO [20], Al_20_3, TiO_2 [21], etc.
- Chalcogenides [22]: MoS_2 [23], WS_2, (Fullerenes [24]), etc.
- Carbon-based NPs [25,26]: Graphite [27,28], graphene [29,30], graphene oxide [31], carbon onions [32], multiwalled carbon nanotube (CNT)-based lubricants, etc.
- Ceramics: Al_2O_3 [18], hBN [33,34], SiO_2 [35], talc [36–38], etc.
- Nanocomposites/core shell [39]: Al_2O_3/SiO_2, ZrO_2/SiO_2, zinc titanate ($ZnTiO_3$) [40], etc.
- Polymers: PTFE (polytetrafluoroethylene) [41,42], polyaniline [43], etc.

These NPs have demonstrated their ability to significantly enhance the AW, anti-friction (AF), and extreme pressure (EP) properties of oils when added in optimal concentrations [44]. However, the size and concentration of the NPs are critical in determining the quality of the lubricating film at the boundary interface. This film provides lower friction and prevents wear by reducing direct metal-to-metal contact.

It is noteworthy that numerous nano-oils and nano-greases can be synthesized through different combinations of NPs [45], emerging ionic liquids, and functionalized compounds and NPs [46]. Fatty acids, such as oleic acid [47], have been used to functionalize NPs, thereby improving their dispersion in oils. Hybrid oils [23,48], comprising two or more types of NPs, have also been explored, demonstrating synergistic properties that surpass the performance of individual NPs.

However, the formulation of nanolubricants remains a critical challenge. The suspension and dispersion of NPs in base oils require meticulous strategies to prevent aggregation and sedimentation. Greases, which are highly structured colloidal semi-solid suspensions, demand even more careful formulation when eco-friendly NPs like nanoclay, talc, graphite, or cellulose are incorporated. These additives not only influence the microstructure of the grease but also enhance its tribological properties.

Ensuring long-term colloidal stability is crucial for maintaining the performance of nanolubricants. A stable suspension requires minimizing particle size, ideally below 100 nm, to prevent sedimentation and aggregation. Furthermore, achieving stable dispersion can be enhanced by employing steric or electrostatic repulsion, surfactant addition, or selecting base oils with higher polarity. Ultrasonic-based methods, such as probe sonication, have proven to be the most effective for dispersing NPs uniformly into base oil.

Nevertheless the formulation process of these nanolubricants is very crucial and involves multiple suspensions and dispersion of NPs through different strategies in order to fully disperse NPs in aggregated states. Greases that are highly structural and colloidal with semi-solid suspension need more careful articulation when being developed through the addition of eco-friendly NPs such as nanoclay, talc, graphite, and cellulose, whether as thickener additives that influence its microstructure distribution or simply as triboadditives.

A colloidal solution is often a liquid or gel that contains scattered/dispersed solid particles in oil phase for nanolubricants. To achieve a homogeneous and stable system, the NP size should be minimized, ideally below 100 nm. A suspension consists of a continuous liquid phase (oil) and a scattered solid phase (NPs). Various phenomena contribute to the instability of dispersion, including the following:

- Aggregation (or flocculation) of the particles.
- Sedimentation of the particles and aggregates of particles.

The majority of NPs are challenging to be dispersed in oil. For instance, graphene platelets, talc, and inorganic fullerenes (IF) exhibit spherical or cylindrical shapes with dimensions ranging from 50 to 150 nm. They possess little affinity toward oil, rendering their dispersion challenging without surfactants or dispersants. The stability of a these colloidal suspension of nano-oils and nanogrease can be enhanced by the following:

- By using steric repulsion between particles: through functionalization by polymers.
- By using electrostatic repulsion: adsorption of ionic surfactants and dispersant.
- By using appropriate base oil having a higher polarity.

The ultrasonic-based method (probe sonication) is the most popular and effective means for dispersion of NPs. When high-frequency ultrasonic waves are passed through liquids, the high pressure of the order of 500 atm is generated for a microsecond at the localized points due to the cavitation phenomenon. The ultrasonic energy has also been used for dispersing particles into aqueous or hydrocarbon solvents. The dispersant is a vital integral part of finished oil, as it keeps impurities and dirt in the suspended form. It also helps in the stability of the suspension of NPs in nanolubricants. The hydrophilic heads will orient themselves toward NPs in suspension. A lipophilic tail component has a strong affinity for oil.

Figure 8.2 shows detailed parameters involved that influence the nanolubricant formulation and provides a complete understanding of the nanolubricant performance.

- Dynamic light scattering (DLS) was carried out for measuring particle size distribution and stability in solutions or suspensions. Transmission electron microscopy (TEM) was carried out for high-resolution images to determine NP size and morphology. Secondary electron microscopy (SEM_EDX) was performed to determine the wear surface topography and its composition analysis for decoding surface

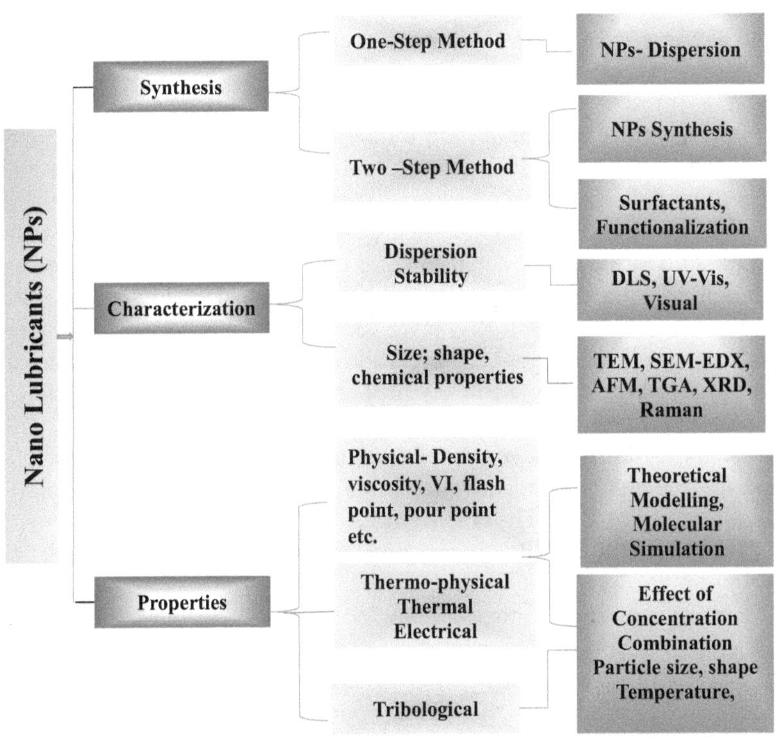

Figure 8.2 Approach for effective nanolubricants synthesis and performance optimization nanolubricants. (Adapted from Ref. [21].)

films over the metallic. 3D surface profilometry and atomic force microscopy (AFM) were carried out to determine surface topography roughness and frictional analysis at the nanoscale. Ultraviolet-visible spectroscopy (UV-Vis) was done to measure the gradual absorbance of oil colloidal suspension. X-ray diffraction (XRD) was done to identify the crystalline structure and phase composition of the developed NPs, especially the core shell and composite structures.

Additive chemistry plays a crucial role in forming robust boundary lubrication films and achieving effective lubrication under extreme conditions [5,49]. Boundary lubrication is a critical aspect of lubrication science, particularly in situations where traditional lubrication mechanisms prove to be inadequate. Through the collaborative efforts of researchers and tribology experts, a deeper understanding of the mechanisms and factors influencing boundary lubrication has been achieved. Given below are a couple of detailed works previously reported.

8.1.2 Role of NPs as additives

NPs offer many advantages over micro-particles (MPs) due to their smaller particle size, resulting in more particles for the same wt% compared to MPs. They present several significant advantages and are often efficient at ambient temperature also, and no induction period is necessary to obtain the desired action. If solid lubricants (SLs) are present in oils, efficient lubrication of such surfaces becomes feasible. SLs are materials that generally possess an inter-lamellar layer lattice structure and thus can slide easily in the sliding direction due to their inherent very low shear strength. However, it offers high compressive strength in the uniaxial direction of the applied load. Some SLs (graphite, h BN, talc, MoS_2, PTFE, etc.) are widely used in a plethora of lubrication operations where liquid lubrication is not possible and not desirable. Some of the key properties of these NPs responsible for excellent lubricity are mentioned in Figure 8.3.

These nanoadditives have been documented to enhance tribological performance through mechanisms such as mending, polishing, film formation, and rolling (Figure 8.4). The effectiveness of these mechanisms is influenced by the specific type of NPs, their geometric shapes, and their size distribution (Figure 8.4) [11,37,50].

The major tribo-mechanisms are as follows.

a. Protective film: NPs of PTFE or those with a layer lattice structure are smeared between the sliding contacts. A very thin, coherent, and smooth film is immediately transferred, covering the asperities almost entirely. The film has very low shear resistance, thereby reducing friction and wear significantly.

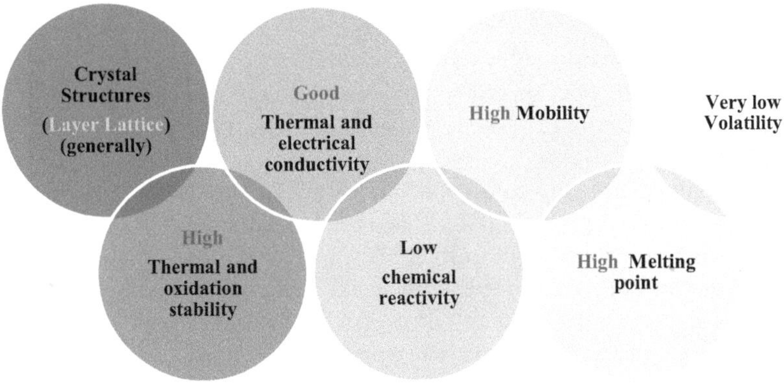

Figure 8.3 Approach for effective nanolubricants synthesis and performance optimization nanolubricants.

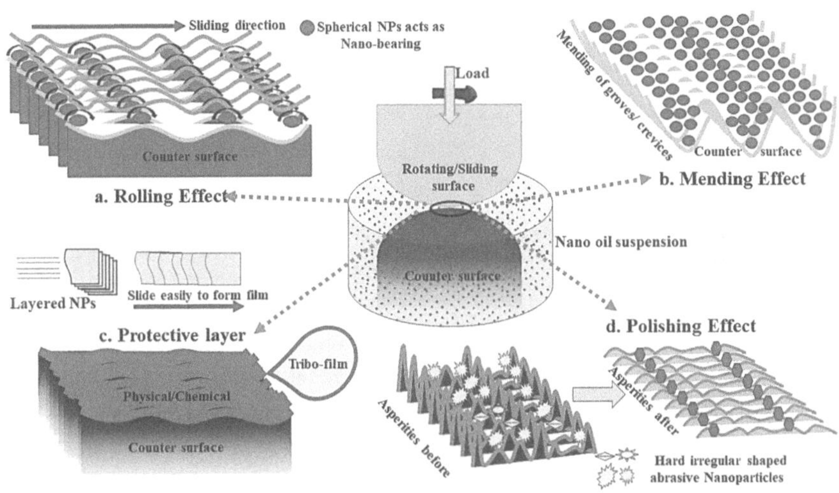

Figure 8.4 Typical lubrication mechanisms with NPs. (Reproduced with permission [51].)

b. Rolling effect: Tao et al. [52] exhibited the nano-bearing or ball-bearing effect by diamond NPs trapped between the rubbing surfaces. They proposed that spherical-shaped hard NPs may act like nano-bearings (Spacers) in between two friction surfaces. This applies to nanotubes also. Since the area contact is converted into point contacts or line contacts, friction and wear gets reduced drastically.
c. Polishing effect: When hard NPs have an irregular shape and sharp edges, they act like nano-abrasives and polish the surfaces. The smoothened surfaces further lead to a large reduction in friction and wear.

d. Mending effect: Liu et al. [15] reported the mending of grooves and valleys by copper particles. When the size of NPs is smaller than that of grooves and valleys on the contacting surfaces, the NPs get deposited on the friction tribo-surfaces or fill the grooves and valleys, thereby smoothening the surfaces and reducing friction and wear.

These NPs may operate by solo or combo-mechanisms [8–10]. Similarly, Chinas-Castillo and Spikes [53] had reported mechanisms for dispersed NPs in oils in electrohydrodynamic (EHD) contacts that NPs penetrated inside the contact zone due to mechanical entrapment. The spherical NPs are likely to act as nano-bearings during tribo-action firstly; then, with sliding time, NPs also get trapped in-between the surfaces and smeared flat under higher temperature-load conditions, finally form a film that reduces μ and wear [44].

8.2 POPULAR NPS AS ADDITIVES FOR NANOLUBRICANTS

A diverse range of NPs and sub-micron particles (SMPs) have been investigated over the past decade for their potential applications in lubricants. These materials encompass a wide spectrum of chemical compositions, including metals (Fe, Cu, Co, etc.) [13,16]; metal oxides (Fe_3O_4, CuO, TiO_2, etc.) [14,19,54–56]; metal chalcogenides (sulfides of MoS_2, WS_2, etc.) [22,57–59]; carbon (Bucky ball, graphite, graphene, and CNTs) [60–64]; ceramics (talc, hBN, SiO_2, and Al_2O_3) [36,65,66]; hybrid nanocomposite (SiO_2/Al_2O_3) [35,57]; polymers (PTFE and UHMPHE) [41,42,67]; etc. The performance of these NPs/SMPs in tribological systems is significantly influenced by their size, shape, and concentration. Smaller particles often exhibit superior tribological properties due to their increased surface area-to-volume ratio, which facilitates enhanced interaction with the lubricant and contacting surfaces [11,27,50]. The shape of NPs/SMPs can also play a crucial role in their effectiveness. For instance, spherical particles may exhibit different tribological behaviors compared to plate-like 2D particles such as graphene and talc [38]. In addition to their individual properties, the synergistic or antagonistic effects of combining different types of NPs/SMPs have also been investigated [43,68]. By carefully selecting and combining appropriate materials, it may be possible to achieve enhanced tribological performance through complementary or synergistic interactions. The development of effective techniques for incorporating NPs/SMPs into lubricants remains a challenging task. These materials often have a tendency to agglomerate or settle, which can reduce their effectiveness. Various strategies, such as the use of dispersants or surface modification techniques, have been explored to improve the dispersion and stability of NPs/SMPs in lubricants. Overall, the effectiveness of the nanolubricant lies in their extent of dispersion and stability, which is linked to compatibility with base oil polarity, sizes, and shapes and added dosage of NPs. Nanomaterials tend to agglomerate quickly; thus, stability is a crucial aspect for oils and nanogrease, since these nanoadditives govern the flow and viscoelastic properties of oils in these colloidal structures. Gupta et al. [21] reported a detailed study to explore the multi-functionality of dispersants and their interactions with other additives present in oils and greases, which influence and impair the functionality of other additives in the system. Some of the most widely reported nanomaterials that have been reported to significantly influence the nanolubricant performance are discussed in the upcoming sections.

8.2.1 Polytetrafluoroethylene (PTFE) particles in oils

PTFE is a crystalline fluoropolymer discovered in 1938 by Dr. Plunkett of Dupont, commercially sold under trademark "Teflon®." PTFE is a linear polymer made by free-radical polymerization of monomer tetrafluoroethylene [$CF_{2=}CF_2$] in aqueous media. It consists of long parallel macromolecules, each made of a carbon chain surrounded by fluorine atoms forming a linear repeating structure -CF_2-CF_2-CF_2- CF_2-, as shown in Figure 8.5 The fluorine atoms cover and sheath the carbon chain (C-C) and protect it from chemical activity. This uniform fluorine sheath also provides electrical inertness to its molecule. They also repel other PTFE molecules and different substances, resulting in high chemical resistance and exceptional non-sticking properties. It is estimated that fluorine content in PTFE is around (74%–76%). Due to its cost-effectiveness, PTFE is widely used in almost all industrial sectors such as oils, chemical processing, electricals, etc. Likewise, it is the most preferred material for low load bushings and bearings for dry lubrication.

PTFE has exceptional characteristics owing to its highly stable C-F covalent bond and covalent bond and distinctive intramolecular and intermolecular interactions among fluorine atoms and the major chains. Stable un-flocculated dispersions can be developed in oils/water owing to its low surface energy (20 mN/m). It is extensively utilized in numerous applications,

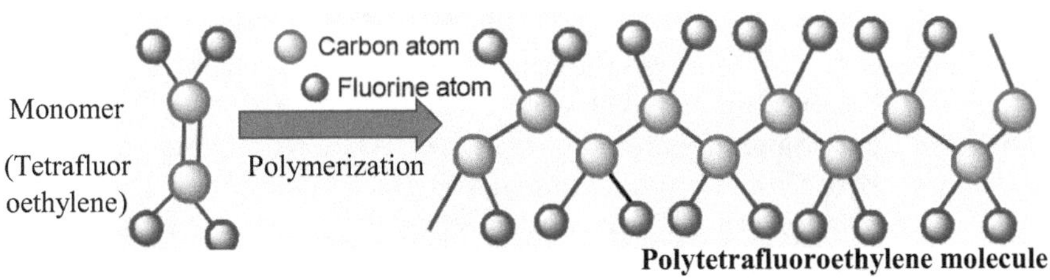

Figure 8.5 Schematic representing the molecular structure of PTFE.

including bearings, AF coatings, gaskets, piston rings, chemical vessel linings, and polymer-based composite components. Additives such as thickeners in grease enhance the viscosity of lubes.

The PTFE-based nanolubricant oils and grease have been reported to perform better as extreme pressure additives (EPA) additives. With the highest EP weld load reported as per IP-2309 standard for ball bearing, Dubey et al. [67] explored the size effect of NPs utilizing PTFE particles of four different sizes (50, 150, and 400 nm and 12 µm) in API group II base oil. The authors reported the nanosized particles performed best among particles of all other sizes. In a subsequent study, the concentration effect was studied by varying PTFE NPs from 1 to 6 wt% [41]; PTFE NPs proved better as EPA with nearly 454% increment with 3 wt% dosage. Also, the EP performance was nearly 2 times better than that of ZDDP-added oils. However, they showed minimal performance as AW additives, with only (13%–16%) improvement at lower loads (392 N). Interestingly, the formulated PTFE nano-oils were stable only for 28 days. The addition of polyisobutylene succinimide (PIBSI) dispersant improved the shelf life of PTFE oils by ~4 times [69]. However, PIBSI interfered in the film-forming process of PTFE and affected the tribo-performance in comparison to bare PTFE as an additive. The activation temperature for EP additives containing chlorine is around 200°C. Nonetheless, PTFE undergoes a transformation chain scission to form a chemically reactive film [51]. The size and concentration optimization strategy followed by authors is adapted and reported in Figure 8.6, which reveals that NPs at an optimal concentration perform better at EP.

8.2.2 Talc NPs in oils

Talc ($Mg_3Si_4O_{10}(OH)_2$) is a 2:1 layered magnesium hydro-silicate consisting of an octahedral magnesium hydroxide layer interposed between two tetrahedral silica layers. The schematic in Figure 8.7 reveals the layer lattice structure of talc. The layers are loosely linked, endowing talc with its exceptional softness during shear deformation. Talc is the softest mineral, possessing a hardness of 1 on the Mohs scale [15]. The composition of talc consists of Si_2O_5 sheets with Mg atoms intercalated between layers in octahedral positions. Talc exhibits excellent cleavage along planes between its loosely linked sheets. The sheets are interconnected solely by van der Waals forces, enabling them to readily slide over one other.

Rudenko et al. [36] reported the tribo properties of talc MPs (1 µm) as an additive in 5W30 oil in different amounts (0.05–1.5 wt%). Friction reduction was higher at a temperature above 100°C. At 0.15 wt%, µ decreased by 30%. Similarly, the MPs (9 µm) of talc in MO and 5W30 led to deterioration in AW performance owing to inability of talc MPs to enter the contact zone as the R_a of the surface was lesser than the size of talc MPs In contrast, the Talc NPs, have been reported to enter contacts zone during sliding and found to form metalo- oxide film resulting in superb Anti- wear performance [36].

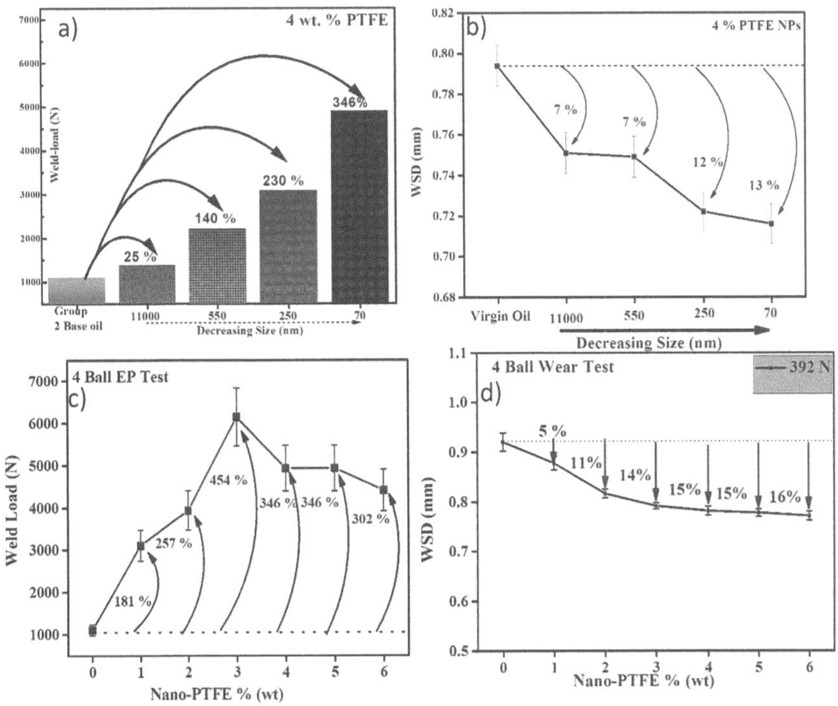

Figure 8.6 PTFE particles as additives: (a) effect of size addition over the weld load, (b) impact on AW properties, and (c) concentration effect of NPs. (Adapted from Refs. [42,43].)

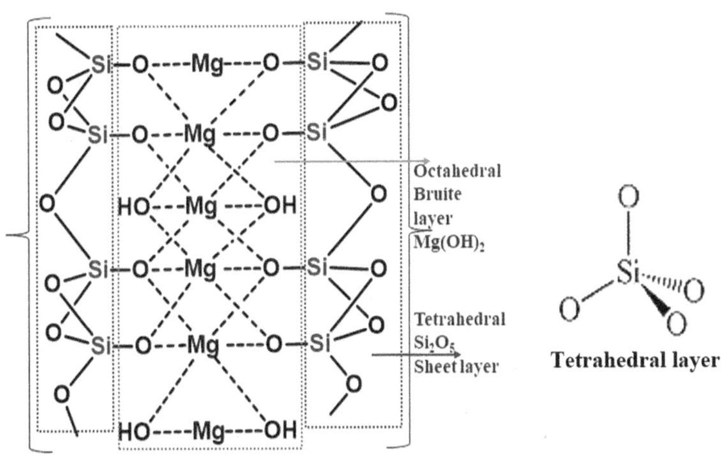

Figure 8.7 Schematic showing the chemical structure of talc [15].

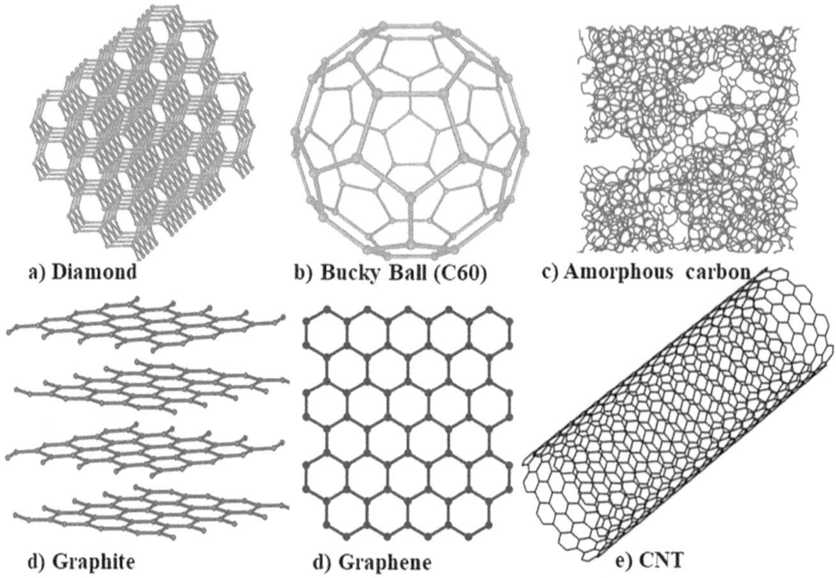

Figure 8.8 Different forms of carbon allotropes [12].

8.2.3 Carbonaceous NPs in oils

Most of the research conducted has focused on various forms of carbonaceous particles, which are graphite, diamond, carbon nano onions, buckyballs (C60), CNTs (single-walled, SWCNTs, and multiwalled, MWCNTs), and graphene. The most frequently used additives in oils are graphite, MWCNTs, graphene flakes, and their functionalized chemical derivatives such as reduced graphene oxide (rGO). Commonly explored graphene allotropes in oils for lubrication are shown in Figure 8.8.

8.2.3.1 Graphene NPs

Graphene is the strongest and thinnest material known to exist. It has an atomic-scale honeycomb lattice made of carbon atoms through sp^2 hybridization. It has a lamellar crystalline structure. The bonding between molecules within each layer is strong covalent bonding, while the bonding between layers is almost entirely through weak van der Waals forces. A one-atom thick layer of C atoms with 2-dimensional crystalline allotrope of carbon C-C bond length is 0.142 nm at a spacing of 0.335 nm.

Owing to excellent properties and honey comp structure, as shown in Figure 8.7d, graphene particles are widely being explored by researchers for various applications. Graphene flakes are being studied worldwide as an additive for coolants and lubricants due to their excellent thermo-physical and tribo properties. However, obtaining stable suspensions of graphene-based mineral oils is difficult due to the amphiphilic nature of graphene platelets and is still a challenge.

8.2.3.2 Graphite NPs

Graphite is one of the widely used carbon allotropes. It has a layered lattice hexagonal structure, which involves rings of six carbon atoms arranged in a layer and stacked on top of one another to form a 3-dimensional structure. In each layer, one C atom is linked to three other C atoms through covalent bonds, i.e., sp^3-hybridized. The fourth valence C atom is connected to the C atom of the surrounding layers by a weak van der Waals force between two layers. The C-C bond length is 0.142 nm within the layer, and the C–C distance between two adjacent layers is 0.34 nm, as shown in Figure 8.8. Owing to a layered structure and weak van der Waals bond between layers, it offers low shear strength along the sliding direction. Thus, it is one of the widely used dry SLs in various industrial applications.

8.2.3.3 Carbon nanotubes

CNTs are also one of the carbon allotropes. They comprise a rolled-up sheet of graphene (C–C) layer in a cylindrical tube structure with a diameter in nm. They can be further classified based upon a number of rolled layers: SWCNTs, with one rolled layer (avg. dia. ~1 nanometer), and MWCNTs, comprising of several concentrically interlinked SWCNTs, weakly bound together by van der Waals interactions in a tree ring-like structure. The length of SWCNT or MWCNT can vary from several μm to mm depending upon the development technique utilized. These CNTs offer excellent tensile strength, electrical conductivity, and thermal conductivity due to their nanostructure and strength of the C–C bonds. CNTs can be chemically modified and made suitable for different applications requiring high durability, strength-weight ratio, electrical conductivity, and thermal conductivity.

Lin et al. [68] reported the tribo-behavior of graphene platelets modified by stearic and oleic acid in mineral base oil on a four-ball tester for carbonaceous particles. Eswaraiah et al. [29] reported 80%, 33%, and 40% improvement in μ, AW, and EP performance, respectively, with the addition of exfoliated ultrathin graphene. Joly-pottuz et al. [32] investigated the carbon onions at the 0.1 wt% amount in PAO base oil and reported improvement in wear properties; however, dispersions stability was limited to 12–24 hours. Huang et al. [70] added graphite nano-sheets in paraffin oil and reported that 1% was the optimal concentration for the highest friction and wear reduction. Zhang et al. [71] reported a comparative study of nanosized graphite particles (10–50 nm) with diesel soot (20–70 nm) particles in PAO4. Better dispersion of diesel soot was observed in the presence of a dispersant (Span80). It was found that 0.01 wt% diesel soot had better AW properties than that containing nano-graphite. Vidal and Avila [72] achieved a reduction of 24% in wear and 17% in friction with nano-graphite platelets in mineral oil. Cursaru et al. [26] compared different CNTs used as additives in sunflower vegetable oil, viz., SWCNTs, functionalized SWCNTs (SWCNTf), and MWCNTs. Addition of 0.2 wt% SWCNTs improved AF by 13% and the wear rate of the ball by 47%. Similarly, Vakili-Nezhaad and Dorany [73] reported the effect of MWNTs (d~40–60 nm, l 5–15 μm) in SAE40 oil in different concentrations (0.01–0.2) wt% on the viscosity index of oils. Increment in VI by 14% (highest) for 0.2 wt% MWNTs was observed.

8.2.4 Metallic NP nanolubricants

Numerous metallic NPs such as (Fe, Cu, Co, Ag, Au, Ni, Bi, Mo, W, and Sn) have been widely studied in recent years [18]. Padgurskas et al. [74] explored three metallic NPs, namely, Fe, Cu, and Co as an additive in SAE 10. The Cu NPs proved to be most effective as anti-wear additives (AWA) with nearly 47% improvement. The rolling effect of NPs and the formation of the thin low shear film were responsible for the enhanced tribo-performance. However, the shelf life reported for the Cu nano-oils varied between 5 and 24 days since the density of Cu is ~8.96 g/cm^3.

A similar comparative study involving Cu, Fe, and Zn NPs in different vegetable oils (canola oil, olive oil, rapeseed oil, and soya oil), mineral oil, and synthetic oil is reported [75]. The Fe NPs showed better dispersion stability in comparison to Cu and Zn NPs. It was reported that Fe NPs improved the AW capability of rapeseed oil, whereas Zn NPs AW improved the AW capability of mineral oil [75]. Similarly, the addition of Ni NPs in PAO6 resulted in nearly 5%–45% and 7%–30% enhancement in friction and wear reduction [76], respectively. Similarly, Tarasov et al. [77] conducted the tribo-studies on Cu NPs in mineral base oil at different loads and sliding speeds and reported that mending of Cu NPs was responsible for the improvement in behavior. However, Cu NPs are known to readily react with other additives with active S compounds to form copper sulfide Cu_2S 18, which exhibits corrosive behavior and might corrode the tribo-pairs during its working life. Zhang et al. [78] examined the Sn and Fe metallic NPs and reported that Sn is better anti-friction additives (AFA), while Fe is better AWA. They reported the formation of low shear strength tribo-film on the wear track, mainly responsible for reduced friction and wear.

Similarly, a protective film composed of Pd of low shear strength is reported by Abad and Sanchez-Lopez [79]. There is tribo-potential of titanium NPs in oils or greases, especially when titanium metal has a close-packed hexagonal crystal structure.

Overall, the combination of NPs has soared over the past few years, with different NPs contributing to film formation in synergism if used in an optimized concentration [43]. However, it is important to tailor the nanolubricants.

8.3 CONCLUSIONS/FUTURE OUTLOOK

The future of nanolubricants and nanogrease hinges on the optimization of NP properties and the development of sustainable formulations. To maximize tribological performance while minimizing costs, researchers should focus on the following:

- Tailoring NP properties: Determining optimal size, concentration, and morphology for different applications.
- Leveraging simulation tools: Utilizing simulation software to optimize NP concentrations and reduce costs.
- Enhancing dispersion stability: Employing surface modification and ultrasonication to ensure long-term dispersion stability, especially when using polar oils.
- Prioritizing sustainability: Developing eco-friendly nanolubricants using biodegradable additives and NPs like nanoclay and cellulose.

By addressing these key areas, researchers can pave the way for the widespread adoption of nanolubricants and nanogrease in various industries while minimizing environmental impact.

REFERENCES

1. U. Maurya, V. Vasu, Ionic liquids as a potential sustainable green lubricant for machining in the era of industry 4.0. In K.K. Gajrani, A. Prasad, and A. Kumar (Eds.), *Advances in Sustainable Machining and Manufacturing Processes*. CRC Press, Boca Raton (2022): pp. 135–155. https://doi.org/10.1201/9781003284574-9.
2. H. Spikes, Low-and zero-sulphated ash, phosphorus and sulphur anti-wear additives for engine oils, *Lubr Sci* 20 (2008) 103–136.
3. K. Holmberg, A. Erdemir, Influence of tribology on global energy consumption, costs and emissions, *Friction* 5 (2017) 263–284. https://doi.org/10.1007/s40544-017-0183-5.
4. K. Holmberg, P. Andersson, A. Erdemir, Global energy consumption due to friction in passenger cars, *Tribol Int* 47 (2012) 221–234. https://doi.org/10.1016/j.triboint.2011.11.022.
5. H. Spikes, Friction modifier additives, *Tribol Lett* 60 (2015). https://doi.org/10.1007/s11249-015-0589-z.

6. T. Nakamura, Improvement of fuel efficiency of passenger cars by taking advantage of tribology, *Tribol Online* 12 (2017) 76–81. https://doi.org/10.2474/trol.12.76.
7. A. Arora, S. Jha, V. Saini, Aspects of green-sustainable tribology and its impacts on future product development: A review, *Ecol Environ Conserv* 25 (2019) S146–S157.
8. I. Minami, Molecular science of lubricant additives, *Appl Sci* 7 (2017) 445.
9. S.U.S. Choi, J.A. Eastman, *Enhancing Thermal Conductivity of Fluids with Nanoparticles*. No. ANL/MSD/CP-84938; CONF-951135-29. Argonne National Lab (ANL), Argonne, IL (1995).
10. J.M. Martin, N. Ohmae, *Nanolubricants*. John Wiley & Sons (2008).
11. M. Gulzar, H.H. Masjuki, M.A. Kalam, M. Varman, N.W.M. Zulkifli, R.A. Mufti, R. Zahid, Tribological performance of nanoparticles as lubricating oil additives, *J Nanopart Res* 18 (2016) 223.
12. F. Du, C. Li, D. Li, X. Sa, Y. Yu, C. Li, Y. Yang, J. Wang, Research progress regarding the use of metal and metal oxide nanoparticles as lubricant additives, *Lubricants* 10 (2022). https://doi.org/10.3390/lubricants10080196.
13. Y. Choi, C. Lee, Y. Hwang, M. Park, J. Lee, C. Choi, M. Jung, Tribological behavior of copper nanoparticles as additives in oil, *Curr Appl Phys* 9 (2009) e124–e127. https://doi.org/10.1016/j.cap.2008.12.050.
14. M. Asnida, S. Hisham, N.W. Awang, A.K. Amirruddin, M.M. Noor, K. Kadirgama, D. Ramasamy, G. Najafi, F. Tarlochan, Copper (II) oxide nanoparticles as additve in engine oil to increase the durability of piston-liner contact, *Fuel* 212 (2018) 656–667. https://doi.org/10.1016/j.fuel.2017.10.002.
15. G. Liu, X. Li, B. Qin, D. Xing, Y. Guo, R. Fan, Investigation of the mending effect and mechanism of copper nano-particles on a tribologically stressed surface, *Tribol Lett* 17 (2004) 961–966. https://doi.org/10.1007/s11249-004-8109-6.
16. J. Padgurskas, R. Rukuiza, I. Prosyčevas, R. Kreivaitis, Tribological properties of lubricant additives of Fe, Cu and Co nanoparticles, *Tribol Int* 60 (2013) 224–232. https://doi.org/10.1016/j.triboint.2012.10.024.
17. V. Saini, J. Bijwe, S. Seth, S.S.V. Ramakumar, Unexplored solid lubricity of titanium nanoparticles in oil to modify the metallic interfaces, *Appl Surf Sci* 580 (2022) 152127. https://doi.org/10.1016/J.APSUSC.2021.152127.
18. I.E. Uflyand, V.A. Zhinzhilo, V.E. Burlakova, Metal-containing nanomaterials as lubricant additives: State-of-the-art and future development, *Friction* 7 (2019) 93–116. https://doi.org/10.1007/s40544-019-0261-y.
19. S.M. Alves, V.S. Mello, E.A. Faria, A.P.P. Camargo, Nanolubricants developed from tiny CuO nanoparticles, *Tribol Int* 100 (2016) 263–271. https://doi.org/10.1016/j.triboint.2016.01.050.
20. M. Upendra, K. Ajay, S. Vinay, T. Gananath, H. Gulam Asharaf, Unravelling the effect of chain and branch content on viscosity of polyisobutylene-mineral oil blends by modelling and its tribological properties. *Tribol Ind* 2 (2024) 45–56. https://doi.org/10.61552/JME.2024.03.005.
21. M.K.A. Ali, H. Xianjun, A. Elagouz, F.A. Essa, M.A.A. Abdelkareem, Minimizing of the boundary friction coefficient in automotive engines using Al_2O_3 and TiO_2 nanoparticles, *J Nanopart Res* 18 (2016) 377.
22. M.F. Sgroi, M. Asti, F. Gili, F.A. Deorsola, S. Bensaid, D. Fino, G. Kraft, I. Garcia, F. Dassenoy, Engine bench and road testing of an engine oil containing MoS_2 particles as nano-additive for friction reduction, *Tribol Int* 105 (2017) 317–325. https://doi.org/10.1016/j.triboint.2016.10.013.

23. T. Nagarajan, N. Sridewi, W.P. Wong, R. Walvekar, V. Khanna, M. Khalid, Synergistic performance evaluation of MoS_2–hBN hybrid nanoparticles as a tribological additive in diesel-based engine oil, *Sci Rep* 13 (2023) 12559. https://doi.org/10.1038/s41598-023-39216-0.
24. L. Rapoport, N. Fleischer, R. Tenne, Fullerene-like WS2 nanoparticles: Superior lubricants for harsh conditions, *Adv Mater* 15 (2003) 651–655.
25. V. Saini, J. Bijwe, S. Seth, S.S.V. Ramakumar, J. Bijwe, Carbon nanoparticles of varying shapes as additives in mineral oil assessment of comparative performance potential, *ACS Appl Mater Interfaces* (2021).
26. D.L. Cursaru, I. Ramadan, C. Tanasescu, R. Ripeanu, Study of tribology behavior of different carbonaceous nanomaterials such as additives for an environmentally friendly lubricant, *Digest J Nanomater Biostruct* 8 (2013) 805–815.
27. M.K. Gupta, J. Bijwe, Exploration of potential of graphite particles with varying sizes as EPA and AWA in oils, *Tribol Int* 127 (2018) 264–275. https://doi.org/10.1016/j.triboint.2018.06.009.
28. N. Kumar, V. Saini, J. Bijwe, Tribological investigations of nano and micro-sized graphite particles as an additive in lithium-based grease, *Tribol Lett* 68 (2020) 124. https://doi.org/10.1007/s11249-020-01362-1.
29. V. Eswaraiah, V. Sankaranarayanan, S. Ramaprabhu, Graphene-based engine oil nanofluids for tribological applications, *ACS Appl Mater Interfaces* 3 (2011) 4221.
30. V. Khare, M.-Q. Pham, N. Kumari, H.-S. Yoon, C.-S. Kim, J.-I. Park, S.-H. Ahn, Graphene–ionic liquid based hybrid nanomaterials as novel lubricant for low friction and wear, *ACS Appl Mater Interfaces* 5 (2013) 4063.
31. B. Gupta, N. Kumar, K. Panda, S. Dash, A.K. Tyagi, Energy efficient reduced graphene oxide additives: Mechanism of effective lubrication and antiwear properties, *Sci Rep* 6 (2016) 18372.
32. L. Joly-Pottuz, N. Matsumoto, H. Kinoshita, B. Vacher, M. Belin, G. Montagnac, J.M. Martin, N. Ohmae, Diamond-derived carbon onions as lubricant additives, *Tribol Int* 41 (2008) 69–78. https://doi.org/10.1016/j.triboint.2007.05.001.
33. M.K. Gupta, J. Bijwe, M. Padhan, Role of size of hexagonal boron nitride particles on tribo-performance of nano and micro oils, *Lubr Sci* (2018) 1–16. https://doi.org/10.1002/ls.1431.
34. N. Kumar, V. Saini, J. Bijwe, Dependency of lithium complex grease on the size of hBN particles for enhanced performance, *Tribol Lett* 71 (2023) 20. https://doi.org/10.1007/s11249-022-01691-3.
35. H. Xie, B. Jiang, X. Hu, C. Peng, H. Guo, F. Pan, Synergistic effect of MoS_2 and SiO_2 nanoparticles as lubricant additives for magnesium alloy–steel contacts, *Nanomaterials* 7 (2017) 154. https://doi.org/10.3390/nano7070154.
36. P. Rudenko, A. Bandyopadhyay, Talc as friction reducing additive to lubricating oil, *Appl Surf Sci* 276 (2013) 383–389. https://doi.org/10.1016/j.apsusc.2013.03.102.
37. N. Kumar, V. Saini, J. Bijwe, Exploration of talc nanoparticles to enhance the performance of lithium grease, *Tribol Int* 162 (2021) 107107.
38. V. Saini, J. Bijwe, S. Seth, S.S. V Ramakumar, Potential exploration of nano-talc particles for enhancing the anti-wear and extreme pressure performance of oil, *Tribol Int* 151 (2020) 106452.

39. H. Chen, L. Zhang, M. Li, G. Xie, Synthesis of core–shell micro/nanoparticles and their tribological application: A review, *Materials* 13 (2020). https://doi.org/10.3390/ma13204590.
40. M. Upendra, V. Vasu, Synergistic effect between phosphonium-based ionic liquid and three oxide nanoparticles as hybrid lubricant additives, *J Tribol* 142 (2020). https://doi.org/10.1115/1.4045769.
41. M.K. Dubey, J. Bijwe, S.S.V. Ramakumar, Nano-PTFE: New entrant as a very promising EP additive, *Tribol Int* 87 (2015) 121–131. https://doi.org/10.1016/j.triboint.2015.01.026.
42. R. Solanki, V. Saini, J. Bijwe, Exploration of PTFE sub-micron particles for enhancing the performance of commercial oils, *Surf Topogr Metrol Prop* 9 (2021) 025005. https://doi.org/10.1088/2051-672X/abf327.
43. V. Saini, J. Bijwe, Polyaniline nanoparticles: A novel additive for augmenting thermal conductivity and tribo-properties of mineral oil and commercial engine oil, *Lubricants* 10 (2022) 300. https://doi.org/10.3390/lubricants10110300.
44. L. Kong, J. Sun, Y. Bao, Preparation, characterization and tribological mechanism of nanofluids, *RSC Adv* 7 (2017) 12599–12609. https://doi.org/10.1039/C6RA28243A.
45. Z. Jiang, Y. Sun, B. Liu, L. Yu, Y. Tong, M. Yan, Z. Yang, Y. Hao, L. Shangguan, S. Zhang, W. Li, Research progresses of nanomaterials as lubricant additives, *Friction* 12 (2024) 1347–1391. https://doi.org/10.1007/s40544-023-0808-9.
46. F. Mariño, J.M. Liñeira del Río, E.R. López, J. Fernández, Chemically modified nanomaterials as lubricant additive: Time stability, friction, and wear, *J Mol Liq* 382 (2023) 121913. https://doi.org/10.1016/j.molliq.2023.121913.
47. B. Wang, K. Gao, K. Wang, Y. Wang, Q. Chang, H. Yang, Induced decomposition and slip interface transformation of oleic acid enables ultralow wear in boundary lubrication, *Langmuir* 40 (2024) 1941–1949. https://doi.org/10.1021/acs.langmuir.3c03767.
48. V. Saini, S. Seth, S.S. V Ramakumar, J. Bijwe, Carbon nanoparticles of varying shapes as additives in mineral oil assessment of comparative performance potential, *ACS Appl Mater Interfaces* 13 (2021) 38844–38856. https://doi.org/10.1021/acsami.1c09478.
49. V. Saini, J. Bijwe, S. Seth, S. Ramakumar, Augmenting the lubrication performance of nano-oils through synergistic co-functioning of nanoparticles, *Tribol Int* 182 (2023) 108332. https://doi.org/10.1016/J.TRIBOINT.2023.108332.
50. W. Dai, B. Kheireddin, H. Gao, H. Liang, Roles of nanoparticles in oil lubrication, *Tribol Int* 102 (2016) 88–98. https://doi.org/10.1016/j.triboint.2016.05.020.
51. V. Saini, J. Bijwe, S. Seth, S.S.V. Ramakumar, Interfacial interaction of PTFE sub-micron particles in oil with steel surfaces as excellent extreme-pressure additive, *J Mol Liq* 325 (2021). https://doi.org/10.1016/j.molliq.2020.115238.
52. X. Tao, Z. Jiazheng, X. Kang, The ball-bearing effect of diamond nanoparticles as an oil additive, *J Phy D Appl Phys* 29 (1996) 2932–2937. https://doi.org/10.1088/0022-3727/29/11/029.
53. F. Chiñas-Castillo, H.A. Spikes, Mechanism of action of colloidal solid dispersions, *J Tribol* 125 (2003) 552–557. https://doi.org/10.1115/1.1537752.

54. J.M.L. Del Río, E.R. López, M.G. Gómez, S.Y. Vilar, Y. Piñeiro, J. Rivas, D.E.P. Gonçalves, J.H.O. Seabra, J. Fernández, Tribological behavior of nanolubricants based on coated magnetic nanoparticles and trimethylolpropane trioleate base oil, *Nanomaterials* 10 (2020). https://doi.org/10.3390/nano10040683.
55. A. Hernández Battez, R. González, D. Felgueroso, J.E. Fernández, M. del Rocío Fernández, M.A. García, I. Peñuelas, Wear prevention behaviour of nanoparticle suspension under extreme pressure conditions, *Wear* 263 (2007) 1568–1574. https://doi.org/10.1016/j.wear.2007.01.093.
56. U. Maurya, V. Vasu, V. Saini, Nano-innovations in lubrication: Unveiling the synergistic boundary lubrication potential of zinc titanate with phosphonium ionic liquid. *Tribol Int* 199 (2024) 109987. https://doi.org/10.1016/J.TRIBOINT.2024.109987.
57. K. Gong, X. Wu, G. Zhao, X. Wang, Nanosized MoS_2 deposited on graphene as lubricant additive in polyalkylene glycol for steel/steel contact at elevated temperature, *Tribol Int* 110 (2017) 1–7. https://doi.org/10.1016/j.triboint.2017.01.024.
58. U. Maurya, V. Vasu, Boehmite nanoparticles for potential enhancement of tribological performance of lubricants. *Wear* 498–499 (2022) 204311. https://doi.org/10.1016/J.WEAR.2022.204311.
59. Z. Jiang, Y. Zhang, G. Yang, C. Gao, L. Yu, S. Zhang, P. Zhang, Synthesis of oil-soluble WS2 nanosheets under mild condition and study of their effect on tribological properties of poly-alpha olefin under evaluated temperatures, *Tribol Int* (2019). https://doi.org/10.1016/j.triboint.2019.05.036.
60. U. Maurya, V. Vasu, D. Kashinath, Ionic liquid-nanoparticle-based hybrid-nanolubricant additives for potential enhancement of tribological properties of lubricants and their comparative study with ZDDP. *Tribol Lett* 70 (2022) 11. https://doi.org/10.1007/s11249-021-01551-6.
61. D. Berman, S.A. Deshmukh, S.K.R.S. Sankaranarayanan, A. Erdemir, A. V. Sumant, Macroscale superlubricity enabled by graphene nanoscroll formation, *Science* 348 (2015) 1118–1122. https://doi.org/10.1126/science.1262024.
62. U. Maurya, V. Vasu, D. Kashinath, Three-way compatibility study among Nanoparticles, Ionic Liquid, and Dispersant for potential in lubricant formulation. *Mater Today Proc* (2022). https://doi.org/10.1016/J.MATPR.2022.03.329.
63. J. Ota, S.K. Hait, M.I.S. Sastry, S.S. V Ramakumar, Graphene dispersion in hydrocarbon medium and its application in lubricant technology, *RSC Adv* 5 (2015) 53326.
64. W. Zhai, N. Srikanth, L.B. Kong, K. Zhou, Carbon nanomaterials in tribology, *Carbon* (2017). https://doi.org/10.1016/j.carbon.2017.04.027.
65. A. Kotia, S. Borkakoti, S.K. Ghosh, Wear and performance analysis of a 4-stroke diesel engine employing nanolubricants, *Particuology* 37 (2018) 54–63. https://doi.org/10.1016/j.partic.2017.05.016.
66. S. Li, H. Qin, R. Zuo, Z. Bai, Tribological performance of Mg/Al/Ce layered double hydroxides nanoparticles and intercalated products as lubricant additives, *Appl Surf Sci* 353 (2015) 643–650. https://doi.org/10.1016/j.apsusc.2015.05.091.
67. M. Kumar Dubey, J. Bijwe, S.S.V. Ramakumar, PTFE based nano-lubricants, *Wear* 306 (2013) 80–88. https://doi.org/10.1016/j.wear.2013.06.020.

68. L.Y. Lin, D.E. Kim, W.K. Kim, S.C. Jun, Friction and wear characteristics of multi-layer graphene films investigated by atomic force microscopy. *Surf Coat Technol* 205(20) (2011) 4864–4869. https://doi.org/10.1016/j.surfcoat.2011.04.092.
69. M.K. Dubey, J. Bijwe, S.S.V. Ramakumar, Effect of dispersant on nano-PTFE based lubricants on tribo-performance in fretting wear mode, *RSC Adv* 6 (2016) 22604–22614. https://doi.org/10.1039/C5RA16795G.
70. H.D. Huang, J.P. Tu, L.P. Gan, C.Z. Li, An investigation on tribological properties of graphite nanosheets as oil additive, *Wear* 261 (2006) 140–144. https://doi.org/10.1016/j.wear.2005.09.010.
71. Z.C. Zhang, Z.B. Cai, J.F. Peng, M.H. Zhu, Comparison of the tribology performance of nano-diesel soot and graphite particles as lubricant additives, *J Phys D Appl Phys* 49 (2016) 045304, 1–9. https://doi.org/10.1088/0022-3727/49/4/045304.
72. F.A.C. Vidal, A.F. Avila, Tribological investigation of nanographite platelets as additive in anti-wear lubricant- a top-down approach, *J Tribol* 136 (2014) 031603, 1–9.
73. G.R. Vakili-Nezhaad, A. Dorany, Investigation of the effect of multiwalled carbon nanotubes on the viscosity index of lube oil cuts, *Chem Eng Commun* 196 (2009) 997–1007. https://doi.org/10.1080/00986440902797865.
74. J. Padgurskas, R. Rukuiza, I. Prosyčevas, R. Kreivaitis, Tribological properties of lubricant additives of Fe, Cu and Co nanoparticles. *Tribol Int* 60 (2013) 224–232. https://doi.org/10.1016/j.triboint.2012.10.024.
75. S.J. Asadauskas, R. Kreivaitis, G. Bikulčius, A. Grigucevičienė, J. Padgurskas, Tribological effects of Cu, Fe and Zn nano-particles, suspended in mineral and bio-based oils. *Lubri Sci* 28(3) (2016) 157–176. https://doi.org/10.1002/ls.1307.
76 R. Chou, A.H. Battez, J.J. Cabello, J.L. Viesca, A. Osorio, A. Sagastume, Tribological behavior of polyalphaolefin with the addition of nickel nanoparticles. *Tribol Int* 43(12) (2010) 2327–2332. https://doi.org/10.1016/j.triboint.2010.08.006.
77. S. Tarasov, A. Kolubaev, S. Belyaev, M. Lerner, F. Tepper, Study of friction reduction by nanocopper additives to motor oil. *Wear* 252(1–2) (2002) 63–69. https://doi.org/10.1016/S0043-1648(01)00860-2.
78. S. Zhang, L. Hu, D. Feng, H. Wang. Anti-wear and friction-reduction mechanism of Sn and Fe nanoparticles as additives of multialkylated cyclopentanes under vacuum condition. *Vacuum* 87 (2013) 75–80. https://doi.org/10.1016/j.vacuum.2012.07.009.
79. M.D. Abad, J.C. Sánchez-López, Tribological properties of surface-modified Pd nanoparticles for electrical contacts. *Wear* 297(1–2) (2013) 943–951. https://doi.org/10.1016/j.wear.2012.11.009.

Chapter 9

Ionic liquid as lubricant additives

Upendra Maurya and Gourav Khajuria

9.1 INTRODUCTION

With the rapid industrialization of the 20th century, the demand for high-performance lubricants surged [1]. The aviation and automotive industries, in particular, drove the relentless search for more efficient lubricants to boost productivity. Various chemical blends were tested using a "trial and error" approach, revealing that adding chlorine, sulfur, phosphorus, or boron compounds enhanced performance under high pressure [2–5]. However, many of these common compounds once used in lubricants are now banned due to their harmful effects on health and the environment [6]. For example, additive groups like amines, chlorinated compounds, and boric acid are now prohibited in lubricants because of their carcinogenic properties and other health risks [7]. Studies on animals have shown that nitrosamines and certain polycyclic aromatic hydrocarbons also possess carcinogenic properties, leading to their exclusion from lubricant formulations [6,8]. Currently, several regulatory bodies, including the REACH (Registration, Evaluation, Authorization, and Restriction) of Chemicals and EHS (Environment Health and Safety), have implemented strict regulations, significantly limiting the application of harm causing biocides and volatile organic compounds [6,7,9,10].

Automotive industry is aspiring for lower viscosity oil for reducing viscous losses in engine and improving fuel economy by up to 5.5% [11,12]. However, lowering viscosity may increase asperity contact and shift lubrication regime toward hydrodynamic to boundary lubrication regime [12,13]. Ionic liquids (ILs) are one of the additive groups that may facilitate development of effective low viscosity lubricants.

Most current lubricants are unsustainable due to their performance limitations and adverse environmental effects, such as soil, water, and air pollution resulting from leaks, disposal, and exhaust emissions. They can also pose significant health risks, including cancer, skin irritation, and chronic respiratory issues. To address these concerns, new, effective additives must be formulated for future-ready metalworking fluids (MWFs).

DOI: 10.1201/9781003642848-9

Room temperature ILs (RTILs) are emerging as a potential alternative, offering effective lubrication at the tool–workpiece interface and serving as a safer substitute for harmful additives.

9.2 A BRIEF EVOLUTION OF ILs

ILs are composed entirely of ions, both anions and cations. Logically, molten table salt could be classified as an IL, given its melting point of 801°C, but under standard temperature and pressure, it remains in a solid crystalline form. In contemporary scientific literature and common usage, the term "ionic liquids" generally refers to ionic compounds that are in the liquid phase below 100°C. In contrast, ionic compounds with melting points above 100°C are typically referred to as "molten salts" [14–16]. Paul Walden in the year 1914 synthesized the first-ever RTIL (ethyl ammonium nitride) by metathesis reaction between hydrochloride salt and lipophilic counterions [17]. The compound was water-like with a melting temperature of 12°C. In the following decades, several other first-generation ILs were synthesized using derivatives of imidazolium and pyridinium cations with metal halide anions [18]. Chloroaluminate and other metal halide anions of first-generation ILs exhibited toxicity, non-biodegradability, water reactivity, and air sensitivity, making them difficult to handle outside of an inert atmosphere [19–21].

The limitations of first generational ILs prompted exploration of water- and air-stable, second-generation ILs. This was achieved by replacing water- and air-sensitive metal halide anions with halides (Cl^-, Br^-, I^-) and other sets of anions such as tetrafluoroborate (BF_4^-), hexafluorophosphate (PF_6^-), etc. [19,22]. Two more cation groups, i.e., ammonium and phosphonium derivatives were also introduced along with earlier imidazolium and pyridinium-based cations. These second-generation ILs can be used in novel applications in various disciplines of scientific and engineering research due to their excellent physicochemical properties. These ILs were still economically restrictive due to higher initial material cost (particularly fluorinated anions) and costly final product purification methods [20]. One of the major limitations of these ILs was their toxicity and corrosive nature, which could be attributed to halides. To eliminate toxicity and for achieving excellent biodegradability, third-generation ILs were synthesized from biodegradable cations and anions, replacing harmful halide anions. Environment-benign and stable anions such as fatty acids, amino acids, sulfates, phosphates, etc. are being explored in the third-generation ILs, providing several advantages such as excellent biodegradability, lower cost, lower toxicity, no purification requirement, etc. Generally, ILs are categorized by their cation group, and alkyl-imidazolium, alkyl-ammonium, alkyl-pyridinium, and alkyl-phosphonium are the most widely studied ILs (Figure 9.1). In particular, functionalized alkyl-imidazolium has been widely reported in

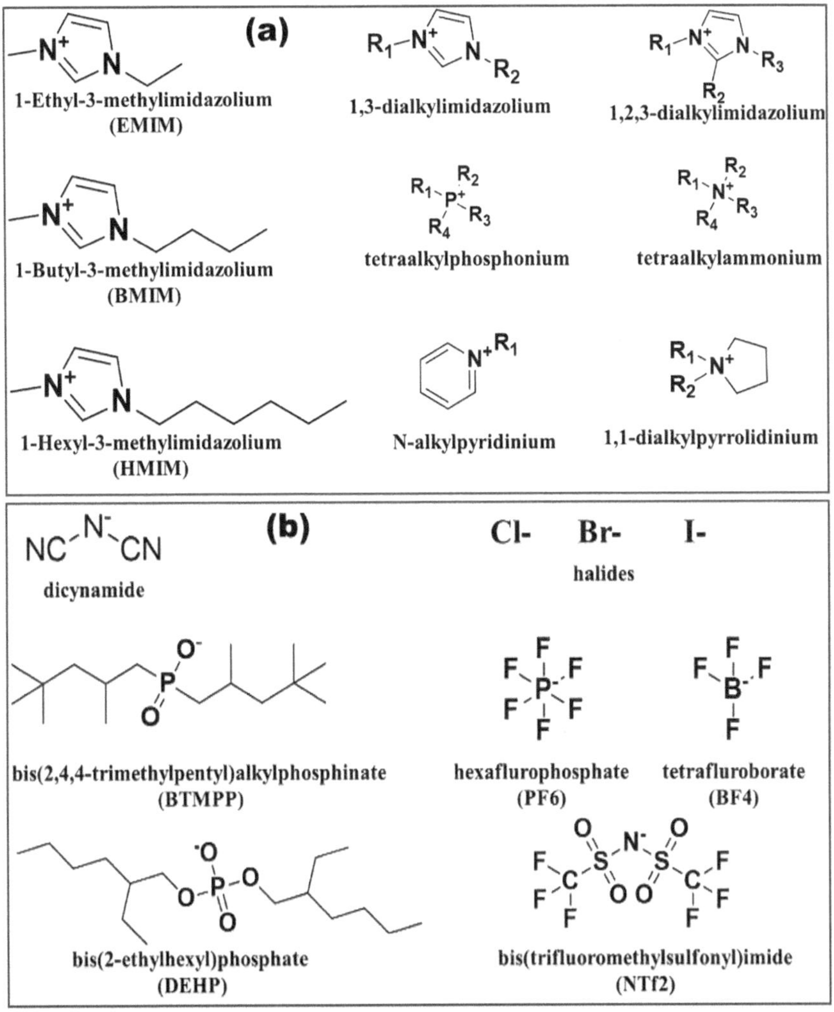

Figure 9.1 Most common (a) cation and (b) anion groups in the IL research.

the literature due to its higher thermal stability and lower melting points. RTILs are synonymy used for these ILs, generally composed of organic or inorganic anions and organic cations.

9.3 PHYSICOCHEMICAL PROPERTIES OF ILs

Physical properties of ILs such as melting point, miscibility, viscosity, density, hydrophobicity, and thermal stability have shown a strong correlation with cation and anion structures. That means depending on the objective,

IL properties can be tuned by altering the molecular structure of the cation and anion [23,24]. By attaching functionalized groups on both the ions of ILs, the specific properties can be induced in the ILs. For instance, tribo-active elements and groups such as phosphates, sulfate, etc. can be added in ILs by replacing imidazolium and pyridinium-containing cations with phosphonium and ammonium groups to enhance the protective film formation capability of the lubricants [23,25–30].

Researchers have reported that the melting point of ILs highly depends on the composition, symmetry, and chain length of both the ions of ILs [31–34]. It was observed that with the same anion group, the melting temperature of ILs was decreased by replacing shorter cations with larger cation groups [35]. In general, it was reported in the published literature that bulky unsymmetric cations reduce the melting point of the ILs [32,34]. Zheng et al. varied the symmetry and chain length of imidazolium ILs and reported that symmetric ILs with a shorter chain length (N=4, 6) normally crystallized, while asymmetric ILs with longer chain length gave a comparatively lower melting temperature [34]. Holbrey et al. classified ILs in three categories based on their thermal behavior and also observed that a similar increase in chain length decreased the melting point due to decreased lattice energy, disordered crystal packing, and prolongation of the molecular structure [31]. Similarly, anion's molecular structure also influences the melting point of the ILs. Fredlake et al. explored 13 imidazolium-based ILs having different anions and observed increased melting points with anion order $Br>Cl>BF_4>PF_6>Tf_2N$ [36]. In addition, they observed a decrease in density with an increase of cation chain length and an increase in density with an increase in molecular weight of anions.

A direct correlation between the IL structure and viscosity can be established. It was reported that an increase in chain length, symmetry, and branching of cation or anion increased the viscosity of ILs [37–39]. Xue and coworkers studied the effect of alkyl chain branching on physicochemical properties of ILs and reported that for the same carbon number, the branched structure registered a higher viscosity than the linear structure [37]. They attributed this to a relatively higher stable packing between both the ions (cation and anion) due to symmetry. An increase in alkyl chain length increased the viscosity due to stronger van der Waals interaction between ions, as reported by Zeng et al. [38]. Viscosity also influences the fluid film formation during machining, and generally, a higher viscosity may provide a better surface finish. The tunable properties of ILs may be used to obtain the desirable viscosity by altering the molecular structure of cations and using higher molecular weight for anions.

The inherent properties, such as excellent thermal stability and lower vapor pressure, are some of the biggest advantages of using ILs in machining. ILs may have broader fluidity, ranging from –100°C to 450°C, suitable for application in many disciplines, particularly in lubricants [40]. As per the published literature for cations with the symmetric structure, ILs' thermal stability was relatively enhanced, while on the other hand, increased chain length decreased the thermal degradation temperature [32,41]. Tao et al. explored imidazolium cations by altering them with varying functional groups such as n-alkyl, cyclo-alkyl, aromatic, and aliphatic [42]. Further, they reported that ILs composed of the n-alkyl group are most stable among the functionalized groups with a degradation temperature of 436°C. Shirota et al. reported that di-cationic imidazolium-based ILs showed higher decomposition temperature than mono-cationic ILs with anions such as NTf_2, BF_4, NPf_2, and NO_3 [43]. Several studies have also reported the effect of the anion structure on the stability of ILs having the same cation [36,40,44–47].

9.4 ILs IN LUBRICANTS

The exploration into the application of ILs as lubricants began in 2001 when Ye et al. first investigated their potential as neat lubricants [48]. Their group assessed the tribological properties of 1-methyl-3-hexylimidazolium tetrafluoroborate and 1-ethyl-3-hexylimidazolium tetrafluoroborate using an optical SRV ball-on-disc configuration. Despite the promising tribological performances, the use of ILs as base oils faced economic constraints due to their relatively high cost. Hence, efforts shifted toward exploring ILs as lubricant additives. Iglesias et al. experimented by blending a 1 wt% crystal [$N_{12,H,H,H}$] [Cl] in a neutral base oil, exceeding the solubility limit [49]. Nevertheless, they reported improved tribological properties for aluminum-steel tribopairs. Up until 2011, several researchers attempted to blend ILs as additives in base oils but encountered limited success, mainly because earlier ILs were immiscible in base oils due to shorter anion chain lengths (PF_6, sulfate, BF_4, Cl) [50].

A breakthrough occurred in 2012 when Qu's group used oil-miscible phosphonium-based ILs in base oil. They introduced two ILs, namely, trihexyltetradecylphosphonium bis(2-ethylhexyl) phosphate ([$P_{6,6,6,14}$][DEHP]) and trihexyltetradecylphosphonium bis(2,4,4-trimethylpentyl) alkylphosphinate ([$P_{6,6,6,14}$][BTMPP]), which exhibited solubility with common nonpolar base oils [51,52]. Subsequently, various oil-miscible ILs, including those based on phosphonium, ammonium, pyrrolidium, and imidazolium, have been explored since 2012 (Table 9.1). Notably, phosphonium and ammonium-based ILs have garnered attention for their enhanced affinity toward metallic surfaces when employed as lubricant additives [53–55]. ILs can be designed to be miscible with a variety of base oils, including mineral, synthetic, and biodegradable oils. This compatibility is crucial for their application as additives in the existing lubrication systems. Table 9.1 gives selected ILs blended in various base oil and respective friction and wear performances.

Table 9.1 Selected oil-miscible ILs as lubricant additives

Cation	*Anion*	*BO*	*Conc%*	*Tribopair*	*Fric.*	*Wear*	*Surf. char.*	*References*
[P66614]	[DEHP]	PAO and 5W30	5 and 50	CI vs steel	≈ 50%	≈ 70%	SEM, TEM, EDS	[51]
[P4444]	[DEHP]	PAO-4	800 ppm of phosphorus	CI vs steel		92	SEM, TEM, EDS, XPS	[27]
[P66614]	[DEHP]					96		
	[i-C7H15COO]					≈ 90		
	[n-C17H35COO]					≈ 91		
	[RSO3]					82		
	[DEHP]					76		
[P88814]	[DEHP]	MAC	1	Steel vs steel	31%	71%	SEM, EDS, XPS	[56]
([P8888]	[DEHP]	GTL-4	800 ppm of phosphorus	CI vs steel	≈ 10%	≈ 91		[57]
[P66614]						≈ 85		
[P1444]	DPP	Vegetable oil	0.25	Steel vs steel	Lower than BO	≈ 50%	SEM, AFM, EDS, XPS	[58]
P66614						≈ 50%		
P1888	Dibutyl phosphate (DBP)	MO		Steel vs steel	10%	42	XPS	[59]
	BTMPP				0	6		
N1888	DBP				0	–21		
	BTMPP				10%	35		
N1,12,12,12	DBP				0	45		
	BTMPP				0	21		
P66614	BTMPP				0	8.82		

(*Continued*)

Table 9.1 (Continued) Selected oil-miscible ILs as lubricant additives

Cation	*Anion*	*BO*	*Conc%*	*Tribopair*	*Fric.*	*Wear*	*Surf. char.*	*References*
choline	[DEHP]	MO	1000 ppm	Steel vs steel	≈ 19%	≈ 64%	SEM, SPM, XANES, nano-indentation	[28]
	[DBDTP]				≈ 16%	≈ 46%		
[P66614]	[BTMPP]				≈ 9%	≈ 43%		
	[TFSI]				≈ 14%	≈ 65%		
[P1444]	[DMP]				≈ 11%	≈ 30%		
[P4444]	[DEDTP]				≈ 4%	≈ 30%		
P66614	[DEHP]	PAO 32	1%	Steel vs steel	≈ 13%	≈ 65%		[60]
P66614	BTMPP							
P2444	DEP							
N11112	DOSS	PAO 10	1%	Steel vs steel and steel vs copper	≈47	≈84	M-XAM 3D-SEM	[61]
N111818								
N1888								
N1121212								

SEM, scanning electron microscopy; TEM, transmission electron microscopy; EDS, energy-dispersive X-ray spectroscopy; XPS, X-ray photoelectron spectroscopy; AES, Auger electron spectroscopy; FIB, focused ion beam; XANES, X-ray absorption near-edge structure; AFM, atomic force microscopy.

9.5 MECHANISM OF ILs AS A LUBRICANT ADDITIVES

ILs are known for their ability to form stable boundary films on metal surfaces, which is one of the primary mechanisms by which they reduce friction and wear [62,63]. The formation of these films is driven by the strong electrostatic interactions between the ions and the surface. ILs can adsorb onto the surface of materials through both physical and chemical adsorption [23]. Physical adsorption is typically governed by van der Waals forces, while chemical adsorption involves the formation of bonds between the ions and the surface atoms. This dual adsorption capability allows ILs to create a durable protective layer that mitigates direct contact between surfaces. The strong electrostatic forces between the cation and anion of the IL contribute to the formation of an ordered structure at the interface. This ordered structure can manifest as an electric double layer, which acts as a cushion during sliding, thereby reducing friction [23]. The formation of this double layer is particularly important in minimizing friction in boundary lubrication regimes, where the thickness of the lubricant film is comparable to the surface roughness. Under operating conditions, such as high pressure and temperature, ILs can undergo tribochemical reactions with the surface material, leading to the formation of a protective tribofilm. For example, ILs containing phosphorus or fluorine can react with metal surfaces to form metal phosphates or metal fluorides, which are known for their wear-resistant properties [23].

The high viscosity of many ILs provides a cushioning effect between surfaces, reducing friction by allowing the viscous layer to shear easily under load. Some ILs also exhibit shear-thinning behavior, where their viscosity decreases with the increasing shear rate [23]. This property is particularly advantageous in high-speed applications, as it allows for smoother operation and lower energy consumption. In addition to their lubricating properties, certain ILs have been found to offer excellent corrosion inhibition [23]. This is particularly useful in environments where both lubrication and corrosion resistance are needed. ILs can react with surface contaminants or oxidation products to form a passivating layer, protecting the underlying material from further degradation. One particular study reported that the IL achieved over 90% corrosion inhibition efficiency in 1M HCl, demonstrating its effectiveness in protecting carbon steel from corrosion [64]. The IL adsorbed onto the carbon steel surface, forming a protective, phosphorus-rich passivation layer that significantly reduced corrosion. The IL acted as a mixed-type inhibitor, influencing both anodic and cathodic reactions, thereby reducing the overall corrosion rate. Electrochemical analyses showed that the IL increased the metal's charge transfer resistance, indicating improved barrier properties against the acidic environment. On metal surfaces, ILs can interact with oxides to form more stable compounds that improve surface smoothness and reduce wear. This interaction is particularly beneficial in systems where oxide layers contribute to increased

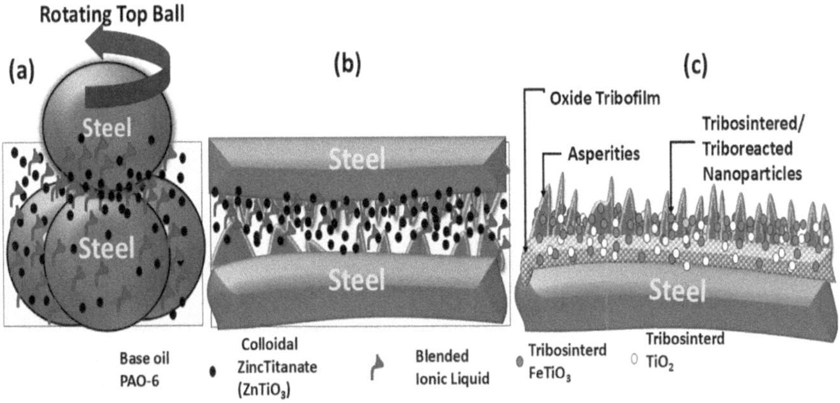

Figure 9.2 Mechanism of tribofilm formation: (a) steel tribopair under lubrication, (b) interaction of additives at the asperity contact, and (c) formation of an oxide tribofilm, primarily composed of phosphate compounds (likely $FePO_4$) along with tribosintered and triboreacted NPs. (Adopted from Ref. [12].)

friction and wear. Due to their high thermal stability, ILs can maintain their lubricating properties even at elevated temperatures where traditional lubricants might degrade. This stability ensures a consistent lubricating film and prevents the breakdown of the lubricant under thermal stress, making ILs suitable for high-temperature applications.

Upendra et al. explored synergy between IL and NP in synthetic base oil [12]. Based on tribological results and surface characterization, they reported a synergistic effect between the IL and nanoparticles (NPs) in film formation (see Figure 9.2). Their XPS spectra have shown the disintegration of zinc titanate under tribostress, leading to the formation of a transfer film on the worn surface, primarily composed of $FeTiO_3$ (binding energy ≈ 464.5 eV). Elemental dot mapping corroborated the presence of phosphorus in significant proportion (1.59%). Additionally, they reported that the phosphonium IL underwent disintegration and tribochemical reaction with the steel surface, forming a likely $FePO_4$ compound, as previously confirmed by XPS spectroscopy (binding energy ≈ 133.5 eV) in an earlier study (Table 9.2).

9.6 COMPARATIVE STUDY OF ILs WITH COMMERCIAL ZINC DIALKYLDITHIOPHOSPHATE (ZDDP)

Jun Qu et al. lubricated different tribopair [steel vs coated surface, i.e., BAM, TiB_2, and diamond-like carbon (DLC)] with ZDDP and phosphonium organophosphate IL ($[P_{66614}]$[DEHP]) to understand tribofilm

Table 9.2 Comparative study of IL with commercial ZDDP

Cation	*Anion*	*Base oil*	*Additives*	*Tribopair*	*Tribo-meter*	*AWA %*	*AFA %*	*Surf. char.*	*References*
P_{8888}	DEHP	GTL 4 cst	ZDDP	CI vs steel	Reciprocating sliding test	<68%	≈ 30%	TEM, EDS, XPS	[65]
P_{66614}	DEHP					<75%			
P_{66614}	BTMPP					<25%	–		
N_{888H}	DEHP					<34%	–		
P_{8888}	DEHP	GTL 4cst and 8 cst blend	ZDDP	CI vs steel	Ball-on-disc reciprocating	90%	–		[66]
P8888	DEHP	PAO 4	ZDDP	CI vs Steel	Ball on disc	84.21	–	SEM EDS XPS	[67]
[P_{66614}]	[DEHP]	SAE 0W-30	ZDDP	Steel vs coated surface (BAM, TiB_2 and DLC)	Ball-on-disc reciprocating	50	36	SEM, AES, FIB, TEM, XPS	[65]

SEM, scanning electron microscopy; TEM, transmission electron microscopy; EDS, energy-dispersive X-ray spectroscopy; XPS, X-ray photoelectron spectroscopy; AES, Auger electron spectroscopy; FIB, focused ion beam.

formation on the coated surface [65]. They reported a nearly similar friction reduction for boride and DLC-coated surface and negligible material loss due to wear, irrespective of lubricant type, while steel balls (counter face surface) showed wear of various degrees. For surfaces lubricated with ILs and ZDDP additives, a direct correlation was observed: for any particular coating, steel ball experience less wear if lubricated with ILs, indicating that ILs performed better than ZDDP. Based on surface characterization, it was reported that the tribofilms of ZDDP and IL were found on both DLC and boride coatings, displaying varying surface coverage and thickness. The boride coatings exhibited a substantial contact area (75%–80%) and film thickness (60–70 nm), whereas DLC had lower coverage (20%–30%) and thinner films (<25 nm). Similarly, ZDDP and IL tribofilms were traced on all three coatings comprising nano-structured composites, including metal phosphates, metal oxides, sulfides, and metallic NPs (derived from worn debris of the metal substrate). Iron compounds in both forms (metallic and oxidized forms) were present in the IL and ZDDP tribofilms on non-metallic coatings. This suggests the participation of substrate wear debris in tribofilm formation, supporting the digestion theory of wear particles proposed by Martin et al. for ZDDP and indicating its applicability to IL tribofilm formation [68]. For boride coatings, oxidation products of TiB_2 were potentially engaged in tribofilm formation, enhancing bonding between the tribofilm and boride surfaces. Conversely, no proof of interactions between hydrogenated DLC and anti-wear additives was obtained, potentially explaining the thicker and more extensive tribofilms on boride coatings compared to DLC. Their study provided unequivocal evidence for two critical processes in tribofilm formation: contact surface reaction and wear debris digestion. Wear debris digestion is believed to be the primary contributor to the bulk of the tribofilm, while contact surface reaction generates robust interface bonding to the substrate.

9.7 CHALLENGES AND LIMITATIONS OF ILs

The environmental impact of ILs is an area of ongoing research. Despite their many advantages, the environmental impact of ILs is not fully understood. Concerns about the potential toxicity and persistence of ILs in the environment have led to calls for more research into their long-term effects. Developing biodegradable and non-toxic ILs is a priority for making this technology more sustainable. While ILs are often touted as green solvents due to their low volatility and recyclability, concerns about their toxicity and biodegradability remain. Some studies have focused on developing biodegradable ILs that maintain the advantageous properties of traditional ILs, while reducing their environmental footprint. One of the main challenges associated with the use of ILs as lubricant additives is their cost. The synthesis of ILs can be expensive, particularly for task-specific ILs designed for specific applications. Additionally, the scalability of IL production for industrial use remains a significant barrier to widespread adoption. While ILs are known for their thermal stability, their performance can be affected by other factors, such as moisture absorption and chemical compatibility with other lubricant components. Research is ongoing to address these stability issues and to develop ILs that are robust across a wider range of operating conditions.

REFERENCES

1. Maurya, U., Vasu, V.: Boehmite nanoparticles for potential enhancement of tribological performance of lubricants. *Wear.* 498–499, 204311 (2022). https://doi.org/10.1016/J.WEAR.2022.204311
2. Spikes, H.A.: Additive-additive and additive-surface interactions in lubrication. *Lubrication Science.* 2, 3–23 (1989). https://doi.org/10.1002/ls.3010020102
3. Fujita, H., Spikes, H.A.: The formation of zinc dithiophosphate antiwear films. *Proceedings of the Institution of Mechanical Engineers, Part J: Journal of Engineering Tribology.* 218, 265–277 (2004). https://doi.org/10.1243/1350650041762677
4. Rizvi, S.Q.A.: *A Comprehensive Review of Lubricant Chemistry, Technology, Selection, and Design* (2009). ASTM International. https://doi.org/10.1520/MNL59-EB
5. Jyothish, P., Maurya, U., Vasu, V.: Advancements in metalworking fluid for sustainable manufacturing. In: Navneet Khanna, Kishor Kumar Gajrani, Khaled Giasin, J. Paulo Davim (Eds.), *Sustainable Materials and Manufacturing Technologies.* pp. 58–81. CRC Press, London (2023).
6. Brinksmeier, E., Meyer, D., Huesmann-Cordes, A.G., Herrmann, C.: Metalworking fluids – Mechanisms and performance. *CIRP Annals – Manufacturing Technology.* 64, 605–628 (2015). https://doi.org/10.1016/j.cirp.2015.05.003
7. Geier, J., Lessmann, H., Schnuch, A., Uter, W.: Contact sensitizations in metalworkers with occupational dermatitis exposed to water-based metalworking fluids: Results of the research project "FaSt." *International Archives of Occupational and Environmental Health.* 77, 543–551 (2004). https://doi.org/10.1007/s00420-004-0539-9
8. IARC monographs programme on the evaluation of carcinogenic risks to humans. *IARC Monographs on the Evaluation of Carcinogenic Risks to Humans.* 73, 9–641 (1999).
9. Skerlos, S.J.: Prevention of metalworking fluid pollution: Environmentally conscious manufacturing at the machine tool. *Environmentally Conscious Manufacturing.* 95–122 (2007). https://doi.org/10.1002/9780470168196
10. Skerlos, S.J., Hayes, K.F., Clarens, A.F., Zhao, F.: Current advances in sustainable metalworking fluids research. *International Journal of Sustainable Manufacturing.* 1, 180–202 (2008). https://doi.org/10.1504/IJSM.2008.019233

11. Holmberg, K., Andersson, P., Erdemir, A.: Tribology international global energy consumption due to friction in passenger cars. *Tribiology International.* 47, 221–234 (2012). https://doi.org/10.1016/j.triboint.2011.11.022
12. Maurya, U., Vasu, V., Saini, V.: Nano-innovations in lubrication: Unveiling the synergistic boundary lubrication potential of zinc titanate with phosphonium ionic liquid. *Tribiology International.* 199, 109987 (2024). https://doi.org/10.1016/J.TRIBOINT.2024.109987
13. Upendra, M., Ajay, K., Vinay, S., Gananath, T., Gulam Asharaf, H.: Unravelling the effect of chain and branch content on viscosity of polyisobutylene-mineral oil blends by modelling and its tribological properties. *Tribology in Industry.* 2, 45–56 (2024). https://doi.org/10.61552/JME.2024.03.005
14. Welton, T.: Room-temperature ionic liquids. Solvents for synthesis and catalysis. *Chemical Reviews.* 99, 2071–2084 (1999). https://doi.org/10.1021/cr980032t
15. Rogers, R.D., Seddon, K.R.: Ionic liquids – Solvents of the future? *Science (1979).* 302, 792–793 (2003). https://doi.org/10.1126/science.1090313
16. Welton, T.: Ionic liquids: A brief history. *Biophysical Reviews.* 10, 691–706 (2018). https://doi.org/10.1007/s12551-018-0419-2
17. Walden, P.: Molecular weights and electrical conductivity of several fused salts. *Bulletin Scientifique / Académie Impériale des Sciences de Saint Petersbourg.* 8, 131–135 (1914).
18. Wells, A.S., Coombe, V.T.: On the freshwater ecotoxicity and biodegradation properties of some common ionic liquids. *Organic Process Research & Development.* 10, 794–798 (2006). https://doi.org/10.1021/op060048i
19. Tavares, A.P., Rodríguez, O., Macedo, E.A.: New generations of ionic liquids applied to enzymatic biocatalysis. In: Jun-ichi Kadokawa (Ed.), *Ionic Liquids – New Aspects for the Future* (167–169). London: IntechOpen (2013). https://doi.org/10.5772/51897
20. Gorke, J., Srienc, F., Kazlauskas, R.: Toward advanced ionic liquids. Polar, enzyme-friendly solvents for biocatalysis. *Biotechnology and Bioprocess Engineering.* 15, 40–53 (2010). https://doi.org/10.1007/s12257-009-3079-z
21. Wilkes, J.S., Levisky, J.A., Wilson, R.A., Hussey, C.L.: Dialkylimidazolium. *Encyclopedia of Lubricants and Lubrication.* 237, 374–374 (2014). https://doi.org/10.1007/978-3-642-22647-2_100172
22. Dong, K., Liu, X., Dong, H., Zhang, X., Zhang, S.: Multiscale studies on ionic liquids. *Chemical Reviews.* 117, 6636–6695 (2017). https://doi.org/10.1021/acs.chemrev.6b00776
23. Cai, M., Yu, Q., Liu, W., Zhou, F.: Ionic liquid lubricants: When chemistry meets tribology. *Chemical Society Reviews.* 49, 7753–7818 (2020). https://doi.org/10.1039/d0cs00126k
24. Plechkova, N. V., Seddon, K.R.: Applications of ionic liquids in the chemical industry. *Chemical Society Reviews.* 37, 123–150 (2008). https://doi.org/10.1039/b006677j
25. Somers, A.E., Biddulph, S.M., Howlett, P.C., Sun, J., MacFarlane, D.R., Forsyth, M.: A comparison of phosphorus and fluorine containing IL lubricants for steel on aluminium. *Physical Chemistry Chemical Physics.* 14, 8224–8231 (2012). https://doi.org/10.1039/c2cp40736a

26. Reeves, C.J., Siddaiah, A., Menezes, P.L.: Ionic liquids: A plausible future of bio-lubricants. *Journal of Bio- and Tribo-Corrosion*. 3, 1–8 (2017). https://doi.org/10.1007/s40735-017-0076-1
27. Zhou, Y., Dyck, J., Graham, T.W., Luo, H., Leonard, D.N., Qu, J.: Ionic liquids composed of phosphonium cations and organophosphate, carboxylate, and sulfonate anions as lubricant antiwear additives. *Langmuir*. 30, 13301–13311 (2014). https://doi.org/10.1021/la5032366
28. Sharma, V., Doerr, N., Aswath, P.B.: Chemical-mechanical properties of tribofilms and their relationship to ionic liquid chemistry. *RSC Advances*. 6, 22341–22356 (2016). https://doi.org/10.1039/c6ra01915c
29. Gusain, R., Singh, R., Sivakumar, K.L.N., Khatri, O.P.: Halogen-free imidazolium/ammonium-bis(salicylato)borate ionic liquids as high performance lubricant additives. *RSC Advances*. 4, 1293–1301 (2014). https://doi.org/10.1039/c3ra43052a
30. Gusain, R., Gupta, P., Saran, S., Khatri, O.P.: Halogen-free bis(imidazolium)/bis(ammonium)-di[bis(salicylato)borate] ionic liquids as energy-efficient and environmentally friendly lubricant additives. *ACS Applied Materials & Interfaces*. 6, 15318–15328 (2014). https://doi.org/10.1021/am503811t
31. Holbrey, J.D., Seddon, K.R.: Tetrafluoroborates: Ionic liquids and ionic liquid crystals. *Journal of the Chemical Society-Dalton Transactions*, (13), 2133–2139 (1999). https://doi.org/10.1039/a902818h
32. Ngo, H.L., LeCompte, K., Hargens, L., McEwen, A.B.: Thermal properties of imidazolium ionic liquids. *Thermochim Acta*. 357–358, 97–102 (2000). https://doi.org/10.1016/S0040-6031(00)00373-7
33. Gordon, C.M., Holbrey, J.D., Kennedy, A.R., Seddon, K.R.: Ionic liquid crystals: Hexafluorophosphate salts. *Journal of Materials Chemistry*. 8, 2627–2636 (1998). https://doi.org/10.1039/a806169f
34. Zheng, W., Mohammed, A., Hines, L.G., Xiao, D., Martinez, O.J., Bartsch, R.A., Simon, S.L., Russina, O., Triolo, A., Quitevis, E.L.: Effect of cation symmetry on the morphology and physicochemical properties of imidazolium ionic liquids. *Journal of Physical Chemistry B*. 115, 6572–6584 (2011). https://doi.org/10.1021/jp1115614
35. Mohammad, A., Inamuddin, D. eds: *Green Solvents II*. Springer Netherlands, Dordrecht (2012).
36. Fredlake, C.P., Crosthwaite, J.M., Hert, D.G., Aki, S.N.V.K., Brennecke, J.F.: Thermophysical properties of imidazolium-based ionic liquids. *Journal of Chemical & Engineering Data*. 49, 954–964 (2004). https://doi.org/10.1021/je034261a
37. Xue, L., Gurung, E., Tamas, G., Koh, Y.P., Shadeck, M., Simon, S.L., Maroncelli, M., Quitevis, E.L.: Effect of alkyl chain branching on physicochemical properties of imidazolium-based ionic liquids. *Journal of Chemical & Engineering Data*. 61, 1078–1091 (2016). https://doi.org/10.1021/acs.jced.5b00658
38. Zeng, Z., Phillips, B.S., Xiao, J.C., Shreeve, J.M.: Polyfluoroalkyl, polyethylene glycol, 1,4-bismethylenebenzene, or 1,4-bismethylene-2,3,5,6-tetrafluorobenzene bridged functionalized dicationic ionic liquids: Synthesis and properties as high temperature lubricants. *Chemistry of Materials*. 20, 2719–2726 (2008). https://doi.org/10.1021/cm703693r

39. Yao, M., Liang, Y., Xia, Y., Zhou, F., Liu, X.: Higherature tribological properties of 2-substituted imidazolium ionic liquids for si 3n 4-steel contacts. *Tribology Letters*. 32, 73–79 (2008). https://doi.org/10.1007/s11249-008-9364-8
40. Maton, C., De Vos, N., Stevens, C. V.: Ionic liquid thermal stabilities: Decomposition mechanisms and analysis tools. *Chemical Society Reviews*. 42, 5963–5977 (2013). https://doi.org/10.1039/c3cs60071h
41. Sowmiah, S., Srinivasadesikan, V., Tseng, M.C., Chu, Y.H.: On the chemical stabilities of ionic liquids. *Molecules*. 14(9), 3780–3813 (2009). https://doi.org/10.3390/molecules14093780
42. Tao, R., Tamas, G., Xue, L., Simon, S.L., Quitevis, E.L.: Thermophysical properties of imidazolium-based ionic liquids: The effect of aliphatic versus aromatic functionality. *Journal of Chemical & Engineering Data*. 59, 2717–2724 (2014). https://doi.org/10.1021/je500185r
43. Shirota, H., Mandai, T., Fukazawa, H., Kato, T.: Comparison between dicationic and monocationic ionic liquids: Liquid density, thermal properties, surface tension, and shear viscosity. *Journal of Chemical & Engineering Data*. 56, 2453–2459 (2011). https://doi.org/10.1021/je2000183
44. Wooster, T.J., Johanson, K.M., Fraser, K.J., MacFarlane, D.R., Scott, J.L.: Thermal degradation of cyano containing ionic liquids. *Green Chemistry*. 8, 691–69 (2006). https://doi.org/10.1039/b606395k
45. Salgado, J., Parajó, J.J., Fernández, J., Villanueva, M.: Long-term thermal stability of some 1-butyl-1-methylpyrrolidinium ionic liquids. *Journal of Chemical Thermodynamics*. 74, 51–57 (2014). https://doi.org/10.1016/j.jct.2014.03.030
46. Salgado, J., Villanueva, M., Parajó, J.J., Fernández, J.: Long-term thermal stability of five imidazolium ionic liquids. *Journal of Chemical Thermodynamics*. 65, 184–190 (2013). https://doi.org/10.1016/j.jct.2013.05.049
47. Minami, I., Kamimura, H., Mori, S.: Thermo-oxidative stability of ionic liquids as lubricating fluids. *Society of Tribologists and Lubrication Engineers – 62nd Annual Meeting of the Society of Tribologists and Lubrication Engineers*. 2, 1038–1040 (2007). https://doi.org/10.1002/jsl.36
48. Ye, C., Liu, W., Chen, Y., Yu, L.: Room-temperature ionic liquids: A novel versatile lubricant. *Chemical Communications*. 21, 2244–2245 (2001). https://doi.org/10.1039/b106935g
49. Iglesias, P., Bermúdez, M.D., Carrión, F.J., Martínez-Nicolás, G.: Friction and wear of aluminium-steel contacts lubricated with ordered fluids-neutral and ionic liquid crystals as oil additives. *Wear*. 256, 386–392 (2004). https://doi.org/10.1016/S0043-1648(03)00442-3
50. Zhou, Y., Qu, J.: Ionic liquids as lubricant additives: A review. *ACS Applied Materials & Interfaces*. 9, 3209–3222 (2017). https://doi.org/10.1021/acsami.6b12489
51. Yu, B., Bansal, D.G., Qu, J., Sun, X., Luo, H., Dai, S., Blau, P.J., Bunting, B.G., Mordukhovich, G., Smolenski, D.J.: Oil-miscible and non-corrosive phosphonium-based ionic liquids as candidate lubricant additives. *Wear*. 289, 58–64 (2012). https://doi.org/10.1016/j.wear.2012.04.015
52. Qu, J., Bansal, D.G., Yu, B., Howe, J.Y., Luo, H., Dai, S., Li, H., Blau, P.J., Bunting, B.G., Mordukhovich, G., Smolenski, D.J.: Antiwear performance and mechanism of an oil-miscible ionic liquid as a lubricant additive. *ACS Applied Materials & Interfaces*. 4, 997–1002 (2012). https://doi.org/10.1021/am201646k

53. Maurya, U., Vasu, V.: Ionic liquids as a potential sustainable green lubricant for machining in the era of industry 4.0. In: Kishor Kumar Gajrani, Arbind Prasad, Ashwani Kumar (Eds.), *Advances in Sustainable Machining and Manufacturing Processes*. pp. 135–155. CRC Press, Boca Raton (2022).
54. Maurya, U., Vasu, V., Kashinath, D.: Three-way compatibility study among nanoparticles, ionic liquid, and dispersant for potential in lubricant formulation. *Materials Today: Proceedings* (2022). https://doi.org/10.1016/J.MATPR.2022.03.329
55. Maurya, U., Vasu, V., Kashinath, D.: Ionic liquid-nanoparticle-based hybrid-nanolubricant additives for potential enhancement of tribological properties of lubricants and their comparative study with ZDDP. *Tribology Letters*. 70, 11 (2022). https://doi.org/10.1007/s11249-021-01551-6
56. Zhang, S., Hu, L., Qiao, D., Feng, D., Wang, H.: Vacuum tribological performance of phosphonium-based ionic liquids as lubricants and lubricant additives of multialkylated cyclopentanes. *Tribology International*. 66, 289–295 (2013). https://doi.org/10.1016/j.triboint.2013.06.012
57. Barnhill, W.C., Qu, J., Luo, H., Meyer, H.M., Ma, C., Chi, M., Papke, B.L.: Phosphonium-organophosphate ionic liquids as lubricant additives: Effects of cation structure on physicochemical and tribological characteristics. *ACS Applied Materials & Interfaces*. 6, 22585–22593 (2014). https://doi.org/10.1021/am506702u
58. Khemchandani, B., Somers, A., Howlett, P., Jaiswal, A.K., Sayanna, E., Forsyth, M.: A biocompatible ionic liquid as an antiwear additive for biodegradable lubricants. *Tribology International*. 77, 171–177 (2014). https://doi.org/10.1016/j.triboint.2014.04.016
59. Westerholt, A., Weschta, M., Bösmann, A., Tremmel, S., Korth, Y., Wolf, M., Schlücker, E., Wehrum, N., Lennert, A., Uerdingen, M., Holweger, W., Wartzack, S., Wasserscheid, P.: Halide-free synthesis and tribological performance of oil-miscible ammonium and phosphonium-based ionic liquids. *ACS Sustainable Chemistry & Engineering*. 3, 797–808 (2015). https://doi.org/10.1021/sc500517n
60. Nasser, K.I., Liñeira del Río, J.M., Mariño, F., López, E.R., Fernández, J.: Double hybrid lubricant additives consisting of a phosphonium ionic liquid and graphene nanoplatelets/hexagonal boron nitride nanoparticles. *Tribology International*. 163, (2021). https://doi.org/10.1016/j.triboint.2021.107189
61. Fan, M., Yang, D., Wang, X., Liu, W., Fu, H.: DOSS- Based QAILs: As both neat lubricants and lubricant additives with excellent tribological properties and good detergency. *Industrial & Engineering Chemistry Research*. 53, 17952–17960 (2014). https://doi.org/10.1021/ie502849w
62. Upendra, M., Vasu, V.: Ionic liquids and its potential in lubricants: A review. *SSRN Electronic Journal* (2018). https://doi.org/10.2139/ssrn.3320367
63. Upendra, M., Vasu, V.: Synergistic effect between phosphonium-based ionic liquid and three oxide nanoparticles as hybrid lubricant additives. *Journal of Tribology* (2020). https://doi.org/10.1115/1.4045769
64. Ashassi-Sorkhabi, H., Es'haghi, M.: Corrosion inhibition of mild steel in acidic media by [BMIm]Br Ionic liquid. *Materials Chemistry and Physics*. 114, 267–271 (2009). https://doi.org/10.1016/J.MATCHEMPHYS.2008.09.019

65. Qu, J., Meyer, H.M., Cai, Z.B., Ma, C., Luo, H.: Characterization of ZDDP and ionic liquid tribofilms on non-metallic coatings providing insights of tribofilm formation mechanisms. *Wear.* 332–333, 1273–1285 (2015). https://doi.org/10.1016/j.wear.2015.01.076
66. Qu, J., Barnhill, W.C., Luo, H., Meyer, H.M., Leonard, D.N., Landauer, A.K., Kheireddin, B., Gao, H., Papke, B.L., Dai, S.: Synergistic effects between phosphonium-alkylphosphate ionic liquids and zinc dialkyldithiophosphate (ZDDP) as lubricant additives. *Advanced Materials.* 27, 4767–4774 (2015). https://doi.org/10.1002/adma.201502037
67. Zhou, Y., Weber, J., Viola, M.B., Qu, J.: Is more always better? Tribofilm evolution and tribological behavior impacted by the concentration of ZDDP, ionic liquid, and ZDDP-Ionic liquid combination. *Wear.* 432–433, 202951 (2019). https://doi.org/10.1016/j.wear.2019.202951
68. Martin, J.M.: Antiwear mechanisms of zinc dithiophosphate: a chemical hardness approach. *Tribology Letters.* 6, 1–8 (1999).

Chapter 10

Exploring the phenomenon of superlubricity

Understanding the principles and mechanisms for current and future applications

Amit Choudhari, Abhishek Kumar, Kailash Arole, Ashish Kumar Gupta, Avinash Kumar, and Ashwani Kumar

10.1 INTRODUCTION

For centuries, humankind has grappled with the ever-present force of friction, its resistance hindering progress and demanding an energetic toll. The dream of frictionless motion, once relegated to the realm of fantasy, now is tantalizingly close to becoming a reality, thanks to the captivating phenomenon of superlubricity [1,2]. Superlubricity is a fascinating phenomenon in physics and tribology (the study of friction) characterized by the near-total absence of friction between two surfaces; instead of experiencing the resistance and energy loss typically associated with friction, surfaces in superlubricity exhibit near-frictionless sliding motion [3]. While a "vanishing" level of friction is impossible to achieve perfectly, a standard definition used to distinguish superlubricity is a kinetic coefficient of friction less than 0.01 [3–5]. It is important to note that this definition is not entirely set in stone, and further discussion and clarification are ongoing.

The journey into superlubricity begins with a departure from traditional notions of friction. While friction has long been considered an inevitable consequence of surface interaction, superlubricity defies this convention, offering a glimpse into a world where friction can be effectively minimized to near-negligible levels [6]. Central to this paradigm shift is the concept of a kinetic coefficient of friction less than 0.01, marking a departure from conventional frictional behavior and opening new avenues for exploration. However, it is essential to recognize that this threshold is not absolute, and ongoing discussions and refinements continue to shape our understanding of superlubricity. At its essence, superlubricity transcends the boundaries of traditional lubrication mechanisms, harnessing unique physical phenomena to achieve frictionless motion [7,8]. One such phenomenon is atomic-scale sliding, where surfaces glide smoothly at the level of individual atoms,

 DOI: 10.1201/9781003642848-10

flattening asperities and irregularities to facilitate near-frictionless contact [8]. This atomic-scale dance, governed by specific crystal structures and alignments, unveils the intricate choreography that underpins superlubricity, offering insights into its elusive nature [9].

Delving deeper into the realm of superlubricity, one can see the profound influence of quantum effects, rooted in quantum mechanics. Here, the inherent strangeness of the subatomic world manifests in phenomena such as tunneling effects and superlubricity-mediated superconductivity, showcasing the fascinating interplay between the macroscopic and microscopic domains. This exploration enriches our understanding of superlubricity and underscores the interconnectedness of physical phenomena across different scales. A pivotal aspect of understanding superlubricity lies in deciphering the mechanisms during sliding interactions [3,10]. One such mechanism involves lattice mismatch, where crystal lattices of two materials create a "moiré pattern" at the interface, preventing interlocking and promoting smooth sliding [11]. Through a meticulous dissection of exemplary systems, such as graphite on graphene and diamond-like carbon (DLC) thin films [12], we unravel the intricate dance of lattices and interface interactions, shedding light on the mechanisms that underpin frictionless motion [13,14].

Looking toward the horizon, superlubricity holds immense promise for various applications across various fields [15–17]. From energy-efficient machinery to next-generation microelectromechanical systems (MEMS) and biomimetic approaches inspired by nature, the potential impact of superlubricity is vast and multifaceted [18]. By charting the course for future applications, we envision a world of frictionless motion and hope to pave the way for transformative advancements that benefit humanity and our planet.

This review chapter explores the groundbreaking field of superlubricity, aiming to uncover its fundamental principles, dissect its underlying mechanisms, and envision its potential applications. Through a synthesis of scientific inquiry and technological innovation, we endeavor to shed light on this captivating phenomenon and inspire future breakthroughs in the quest for frictionless motion. Delving deeper into these areas helps us understand more about superlubricity. Like a key, this understanding unlocks the mysteries of frictionless motion. It paves the way for a future where its transformative potential can be harnessed to benefit humanity and our planet. This chapter explores this groundbreaking field, delving into the fundamental principles and mechanisms that unlock its transformative potential.

10.2 FUNDAMENTAL PRINCIPLES OF FRICTION AND LUBRICATION

Friction and lubrication are two sides of the same coin, forever intertwined in the dance of surfaces in motion. While friction presents resistance and energy loss, lubrication strives to minimize it, paving the way for smoother, more efficient operation. Understanding their fundamental principles is crucial to appreciating the magic of superlubricity.

Friction arises from the microscopic interactions between surface asperities (bumps and valleys) on opposing surfaces. These asperities interlock as they rub against each other, causing resistance and energy dissipation.

10.2.1 Amontons' *laws of friction*

Amontons' laws of friction, formulated by Leonardo da Vinci and further developed by Amontons and Coulomb, established fundamental principles regarding friction. These laws are:

1. The friction force and the applied load have a direct relationship.
2. The apparent contact area has no bearing on the friction force.

This law distinguishes between static friction, the force required to initiate movement between two surfaces initially at rest, and kinetic friction, which acts when two surfaces are in motion relative to each other. Generally, kinetic friction is lower than static friction, but this may vary depending on experimental conditions [19]. Amontons also introduced the concept of surface texture and asperity interaction, which influence friction. Coulomb further investigated factors such as surface area, materials in contact, average load, and the duration of contact, mainly focusing on wooden surfaces [20]. He also proposed the third law of friction: kinetic friction is independent of sliding velocity. While Amontons' laws were initially derived from experiments on wooden surfaces, they have also been found to apply to various other surfaces. However, it is important to note that these laws are empirical, and exceptions have been observed, particularly in nanometric scenarios [21]. Figure 10.1 depicts the wear mechanisms [22] occurring in a ceramic sample, highlighting the distinct types of material degradation under tribological conditions:

10.2.1.1 Abrasive wear

The presence of rigid particles protruding from the sample surface interacts with the mating ring under applied load. This interaction results in the detachment of particles and the formation of grooves and furrows on the sample surface due to frictional forces. Such wear is a consequence of mechanical abrasion caused by hard asperities or embedded debris from the contact surfaces.

10.2.1.2 Adhesive wear

At elevated load and rotational speed, the surfaces of the ceramic sample and the mating ring may experience localized heating, causing material

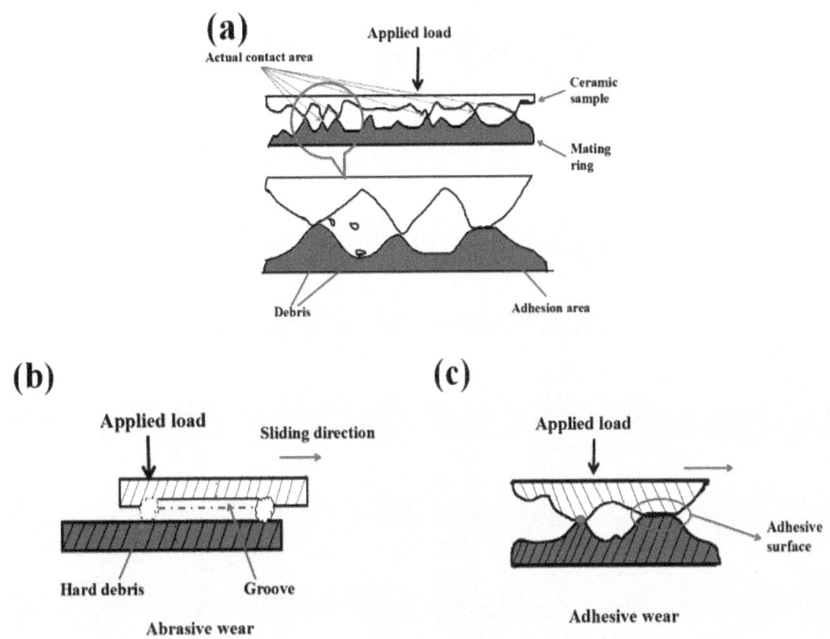

Figure 10.1 Schematic representation of the various wear mechanisms that can occur during the experimental process [22].

bonding between them. This adhesion leads to material transfer and the subsequent detachment of particles, contributing to surface degradation. Adhesive wear is particularly significant in samples exhibiting inferior mechanical properties, as they are more susceptible to such thermal and mechanical effects.

The dominant wear mechanism in the ceramics is influenced by their microstructural features and inherent mechanical properties. Samples with a homogenous dispersion of columnar mullite phases demonstrate superior wear resistance, primarily undergoing mild abrasive wear. Conversely, samples with uneven mullite distribution and reduced mechanical strength are prone to a combination of severe abrasive and adhesive wear.

Table 10.1 presents a comprehensive overview of the standard coefficient of friction values for various material pairs undergoing sliding motion under diverse environmental conditions. Each entry in the table delineates the static (μs) and kinetic (μk) coefficients of friction, showcasing the propensity of different materials to resist motion relative to one another [21,23]. The observations from the table underscore the significant variability in friction coefficients across different materials.

Table 10.1 Standard coefficient of friction values for material 1 sliding across material 2 under varied environmental conditions [21,23]

Material 1	*Material 2*	*Environment*	*Static μs*	*Kinetic μk*
Steel	Steel	Glycerine, 20°C	0.2	–
Gold	Gold	Air	2.8	–
Mild steel	Nylon 6/12	Air	0.24	0.31
Ice	Ice	–70°C	0.5	0.09
Glass tempered	PTFE	Air	0.1	0.1
440C steel	LDPE	Air	–	0.4
Graphite	Graphite	Air	0.1	–
Mild steel	Nylon 6/6	Air	0.2	0.28
440C steel	HDPE	Air	–	0.09
Ice	Ice	0°C	0.1	0.02
Nylon 6/6	Nylon 6/6	Air	0.06	0.07
Polyethylene	Polyethylene	Air	0.2	–
Gold	Silicon	Vacuum (10^{-8} Pa)	1	–
Glass	Glass	Air	0.95	–
Polystyrene	Polystyrene	Air	0.5	–
440C steel	PP	Air	–	0.21
PMMA	PMMA	Air	0.8	–
Steel	Steel	Olive oil, 20°C	0.11	–
Sapphire	Sapphire	Air	0.2	–
Silver	Alumina	Air	–	0.37
Mica	Mica	Air	1	-
Tungsten	Tungsten	Vacuum (10^{-8} Pa)	–	3
PTFE	PTFE	Air	0.08	0.07
Mild steel	Polystyrene	Air	0.28	0.32
Steel	Steel	Air, 20°C	0.58	–
440C steel	PVDF	Air	–	0.31
Tungsten	Tungsten	Vacuum (10^{-8} Pa), exposure to O_2	–	1.3
Diamond	Diamond	Air	0.1	–

Combinations and environmental contexts: Notably, materials like gold exhibit remarkably high friction coefficients, while others, such as polymers, demonstrate relatively lower values. Moreover, environmental factors such as temperature and the presence of lubricants can significantly influence friction behavior. For instance, the coefficients of friction for ice sliding on ice decrease with increase in temperature [24], indicating temperature-dependent frictional properties [25]. Additionally, the impact of vacuum environments on friction is evident, particularly with tungsten pairs. The table highlights the intricate relationship between material properties, environmental conditions, and frictional behavior, emphasizing the necessity of considering these factors in tribological studies and engineering applications. The study of friction in a controlled, efficiently designed system [23,24] may vary in harsh or other environments with normal temperature pressure.

10.2.2 Lubrication

Lubrication, a fundamental principle in tribology, mitigates friction and wear between surfaces in relative motion by introducing an intermediary substance known as a lubricant. At its core, lubrication operates on the principle of interposing a lubricant between interacting surfaces to reduce direct contact and thereby minimize frictional forces [26]. This mechanism is crucial in various mechanical systems, from intricate machinery to everyday applications such as automotive engines and industrial equipment. The lubricant, typically a fluid or solid substance, forms a thin film or boundary layer over the contacting surfaces, effectively separating them and facilitating smoother motion [27]. This separation reduces the surface-to-surface adhesion and frictional resistance, thereby enhancing efficiency and extending the operational lifespan of the components involved. The effectiveness of lubrication hinges on several critical factors, including the properties of the lubricant itself, the nature of the interacting surfaces, and the system's operating conditions [28]. Different lubricants, such as oils, greases, and solid lubricants, offer distinct advantages depending on the application requirements. Additionally, the surface characteristics of the materials in contact, including roughness, hardness, and surface energy, are crucial in determining the success of lubrication. Proper selection and application of lubricants tailored to specific operating conditions are essential for optimal performance and longevity of mechanical systems [29]. Lubrication chemistry plays a pivotal role in achieving superlubricity by optimizing the chemical interactions at the sliding interface, reducing friction through the formation of low-shear boundary layers, and tailoring the molecular composition to withstand extreme operating conditions [30,31].

Furthermore, lubrication principles encompass various regimes, including boundary, mixed, and hydrodynamic lubrication, each dictating the lubrication mechanism based on load, speed, and operating environment [29,32]. In boundary lubrication, the lubricant film is thin, and direct contact between surface asperities may occur, requiring additives to protect against wear under extreme conditions. Mixed lubrication combines boundary and hydrodynamic lubrication, with partial separation of surfaces achieved through a combination of fluid film and boundary lubrication effects. Hydrodynamic lubrication, on the other hand, relies on generating a pressurized fluid film between surfaces in motion, effectively preventing direct contact and minimizing frictional losses [33,34]. Understanding these lubrication regimes and their underlying principles is necessary for improving lubrication and maintaining the reliability and efficiency of mechanical systems across various industries.

10.3 MECHANISMS GOVERNING SUPERLUBRICITY

Superlubricity, characterized by exceptionally low friction and wear, is governed by several vital mechanisms. At the heart of these mechanisms lies the formation of smooth, low-friction interfaces between contacting surfaces [35]. These smooth interfaces can be accomplished by utilizing the nanocomposites with atomically smooth surfaces or applying lubricants like graphite or molybdenum disulfide (MoS_2) [36], which exhibit outstanding lubricating properties. Additionally, the arrangement of atoms in materials plays a crucial role, with layered structures like graphene or hexagonal boron nitride (h-BN) offering favorable conditions for achieving superlubricity [37]. Environmental adaptability is also essential, as certain materials may perform better under specific conditions, such as high temperatures or inert atmospheres. Furthermore, the ability to manipulate interfacial interactions, block chemical interactions, and establish incommensurate contacts are vital aspects that contribute to realizing superlubricity. Overall, a combination of surface smoothness, atomic arrangement, environmental adaptability, and interfacial properties collectively govern the mechanisms underlying superlubricity, enabling smooth sliding with minimal energy dissipation [38,39].

In addition to the previously discussed studies, several other notable investigations have been conducted. Due to the broad spectrum of liquid superlubricity, various mechanisms may operate concurrently depending on different implementation systems. These mechanisms have been categorized into three main types: hydrodynamic effect, electric double-layer effect, and hydration effect. Some instances demonstrate that superlubricity can be achieved solely through one mechanism, while in other cases, a combination of two or all three mechanisms is necessary [40–42]. Figure 10.2 illustrates the primary mechanisms underlying liquid superlubricity. As our understanding of the functional mechanisms of liquid superlubricity advances, efforts are underway to enhance the load-bearing capacity of liquid superlubricious systems [43]. This involves significant improvements in various aspects, including enriching the lubricating fluid system, extending the range of materials used for friction pairs, and enhancing the overall performance of liquid superlubricity. Despite these advancements, two main challenges persist in the practical application of liquid superlubricity: relatively low load-bearing capacity for industrial use and the need to expand the applicable conditions of liquid superlubricity [44]. Superlubricity in Mg and rare earth element alloys is primarily influenced by their unique microstructural characteristics, such as the presence of

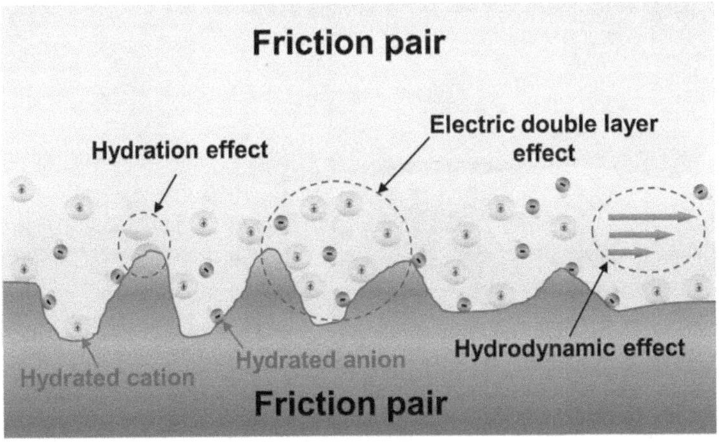

Figure 10.2 Depicting an outline illustrating the primary mechanisms responsible for liquid superlubricity [43].

fine precipitates and grain boundary modifications, which reduce friction and wear by enhancing load-bearing capacity and facilitating low-shear interfacial interactions [45].

The achievement of robust macroscale superlubricity on rough and laser-textured steel substrates, despite their low adhesion force with the steel substrate, was demonstrated by Shi et al. [16], incorporating a typical graphene/MoS_2 nanocomposite coating. In this study, to enhance coating-substrate adhesion strength, polyamide-imide (PAI) was introduced into the graphene/MoS_2 binary coating, resulting in a graphene/MoS_2/PAI ternary coating, which is as shown in Figure 10.3 [46].

Several critical factors must be addressed to achieve macroscale superlubricity [3,47]. Figure 10.3 illustrates the principle behind achieving macroscale superlubricity: achieving extremely low friction coefficients at the macroscale level. Firstly, surface irregularities or segmentation of contact surfaces play a pivotal role in transforming macroscopic contact into micro-contact points. This allows for establishing incommensurate contact, which is vital for reducing friction on a larger scale. Figure 10.3a [46] shows that the macroscopic surface contact is transformed into microscale point contact through micro-asperities on the substrate surface. This microscale point contact facilitates incommensurate contact, which is essential for reducing friction. Moving on to Figure 10.3b [46], a pre-running-in period in the air is introduced to promote the formation of a layer-by-layer slipping structure. This structure is achieved by self-orientating 2D nanomaterials, forming an ordered layer-by-layer slipping structure. Thirdly, as shown in Figure 10.3c, a heterogeneous nanocomposite is strategically designed to manipulate interfacial interactions. Strong covalent-covalent-covalent

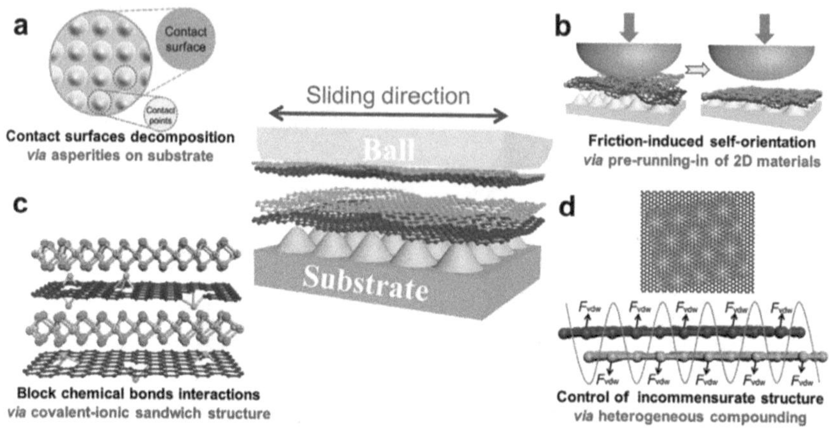

Figure 10.3 The mechanism behind macroscale superlubricity. (a) At the macroscale, contacts are subdivided into microscale point contacts. (b) This leads to the formation of a layered structure. (c) Heterogeneous nanocomposites prevent chemical interactions by blocking them. (d) Incommensurate contact occurs, facilitated by the van der Waals force (*Fvdw*), *enabling low friction* [46].

bonding interactions are obstructed by incorporating a sandwich-like structure of covalent, ionic, and covalent layers [46]. Finally, Figure 10.3d highlights the importance of achieving incommensurate contact. This results in the acquisition of incommensurate contact, further reducing friction and achieving structural superlubricity [46].

10.4 SUPERLUBRICITY: NEAR-ZERO FRICTION REGIME

Superlubricity, as it is a captivating phenomenon in physics and tribology (the study of friction), transcends the limitations of traditional lubrication by achieving near-zero friction between two surfaces. It is as if the surfaces glide effortlessly over each other, defying the ever-present resistance we typically associate with friction [48]. While a perfect "zero friction" state is impossible, the term usually refers to a kinetic coefficient of friction less than 0.01. This is significantly lower than even the best traditional lubricants, indicating a dramatic reduction in resistance. Figure 10.4 illustrates the classification of superlubricity, a state where friction is drastically reduced. Solid, liquid, and mixed solid–liquid superlubricity are the primary categories into which it may be divided. Materials like MoS_2, graphite, graphene, and other 2-D nanomaterials with solid superlubricity—achieved through specific processes—show very little friction in dry environments, especially at the nano/micro scale. However, achieving continuous solid superlubricity

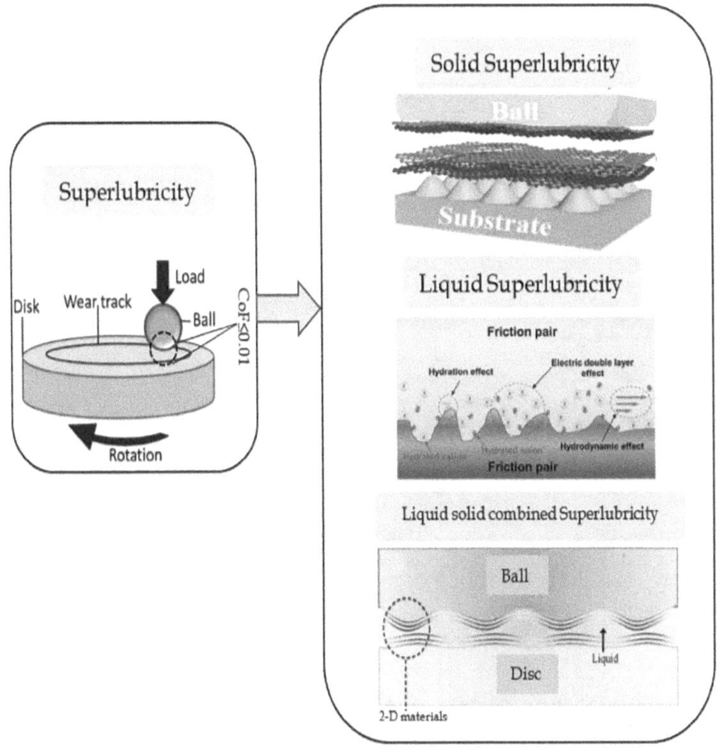

Figure 10.4 Categorization of superlubricity [43,50].

at macroscopic scales remains challenging in ambient environments and typically requires high vacuum or inert gas conditions; hence, designing the vacuum vessel is an essential factor. However, liquid superlubricity relies on suitable liquid lubricants to reduce friction, making it applicable under ambient conditions. Tribological systems with smooth contacts, lower viscosity fluid, good load-bearing capability, and low shear resistance can attain this superlubricity [49]. Low-viscosity lubricants and smooth contacts are the main focus of research in liquid superlubricity. Lastly, adding nanomaterials to liquid lubricants provides a synergistic blend of superlubricity from both solid and liquid components. This approach enhances the load-carrying capacity of the lubricants, enabling superlubricity. Research efforts continue in all these areas [3].

Table 10.2 summarizes several review articles and book chapters published in recent years, briefly discussing the key points.

Materials displaying superlubric behavior exhibit exceptional characteristics, enabling smooth sliding with minimal energy dissipation, as

Table 10.2 List of articles and books related to superlubricity

Ref.	*Title*	*Year*	*Source*	*Key points*	*Additional notes*
[51]	Superlubricity of carbon nanostructures	2020	Carbon	Latest advancements in materials, emerging carbon nanostructures	Focuses on practical applications.
[3]	Emerging superlubricity: a review of the state of the art and perspectives on future research	2018	Applied Physics Reviews	A broad overview of various superlubricity approaches, including theoretical and experimental studies	Discusses potential future research directions.
[47]	Macroscale superlubricity achieved with various liquid molecules: a review	2019	Frontiers in Mechanical Engineering	Progress and prospects of achieving superlubricity at the macroscale using liquid lubricants.	Highlights recent developments and potential applications.
[48]	Nanomaterials in superlubricity	2019	Advanced Functional Materials	A comprehensive exploration of how different nanomaterial sizes (0D–3D) influence superlubricity.	Discusses the impact of nanomaterial properties on friction reduction.
[52]	Superlubricity achieved with two-dimensional nano-additives to liquid lubricants	2020	Friction	Examines the effectiveness of 2D nanomaterials as additives in improving the lubrication qualities of fluids.	Explores potential applications in various tribological systems.
[53]	Superlubricitive engineering: future industry nearly getting rid of wear and frictional energy consumption	2020	Friction	Proposes the concept of "superlubricity engineering" and explores potential applications of superlubricity.	Discusses the broader societal and industrial implications of superlubricity.

(*Continued*)

Table 10.2 (Continued) List of articles and books related to superlubricity

Ref.	*Title*	*Year*	*Source*	*Key points*	*Additional notes*
[54]	Overcoming friction and steps toward superlubricity: a review	2021	Applied Surface Science Advances	Explores the fundamental mechanisms underlying different types of superlubricity (e.g., structural, liquid).	Provides a deeper understanding of the tribological phenomena involved.
[55]	Computational prediction of superlubric layered heterojunctions	2021	ACS Applied Materials & Interfaces	Discusses using computational methods to simulate and predict superlubricity in various materials and systems.	Highlights the role of computational tools in advancing superlubricity research.
[10]	Progress in superlubricity across different media and materialsystems—a review	2022	Frontiers in Mechanical Engineering	Highlights diverse applications of superlubricity across various scales, from nano/micro to meso/macro.	Provides a comprehensive overview of the field.
[43]	Origin of friction and the new frictionless technology: superlubricity: advancements and future outlook	2021	Nano Energy	Discusses the origins of friction and the potential of superlubricity as a "frictionless technology."	Explores potential future applications in various technological fields.

shown in Table 10.3. These materials encompass various compositions and structures, each offering unique advantages. Nanocomposites, with their layered hexagonal arrangement of atoms and atomically smooth surfaces, demonstrate remarkable adaptability to different environmental conditions thanks to their engineered properties tailored for specific applications. Graphite, known for its composite structure and exceptionally low friction coefficient, is popular for dry lubricants due to its smooth and hard surface. Boron nitride (h-BN) [56], characterized by its amorphous, liquid-like arrangement of atoms and excellent thermal and chemical resistance, finds utility as a high-temperature lubricant. Hydrogels [57], composed of cross-linked polymer networks with smooth surfaces, offer self-healing properties and biocompatibility, making them suitable for biomedical applications [58]. DLC materials [59], with their layered trigonal

Table 10.3 A summary of materials displaying superlubric behavior demonstrates exceptionally low friction, allowing smooth sliding with minimal energy dissipation

Type of material	*Arrangement of atoms*	*Surface smoothness*	*Environmental adaptability*	*Additional information*
Nanocomposites	Layered, hexagonal	Atomically smooth	Stable under a wide range of conditions	They can be engineered with specific properties for targeted applications, making them highly versatile.
Graphite	Composite structure	Smooth, hard surface	Relatively stable in various environments	It has an exceptionally low friction coefficient, making it a popular choice for dry lubricants.
h-BN	Amorphous, liquid	Atomically smooth	Excellent thermal and chemical resistance	It is often used as a high-temperature lubricant due to its stability.
Hydrogels	Cross-linked polymer networks	Smooth	Dependent on water content	They can be self-healing and biocompatible, making them suitable for biomedical applications.
DLC	Layered, trigonal	Can be tailored	Depends on matrix and nanoparticles	It offers superior wear resistance and can be customized for specific friction needs.
Molybdenum disulfide (MoS_2)	Layered, trigonal	Varies	Sensitive to moisture and oxidation	Exhibits excellent lubricating properties in dry environments.
Liquid metals	Amorphous, liquid	Atomically smooth	Sensitive to temperature	It can self-contain to rough surfaces but requires careful temperature management.
Tungsten disulfide (WS2)	Layered, trigonal	Smooth	High adaptability	Similar to MoS_2, it offers improved oxidation resistance.

structure and customizable properties, provide superior wear resistance tailored to specific friction requirements. Molybdenum disulfide (MoS_2) [60] and tungsten disulfide (WS2) [61], featuring layered trigonal arrangements, exhibit excellent lubricating properties, albeit with moisture, oxidation, and temperature sensitivities. Liquid metals, characterized by their atomically smooth surfaces, can conform to rough surfaces but require precise temperature control. These materials represent diverse options for achieving superlubric behavior, each offering unique benefits and considerations for various applications.

Xiao et al. [62] investigated the influence of speed on the stability of superlubricity in Si_3N_4/glass tribological pairs lubricated with H_3PO_4 solution by a micro-tribometer. Their study showed that stable superlubricity was attainable within a specific speed range of 0.075 to 0.209 m/s under a load of 3 N, as shown in Figure 10.5a. The failure of superlubricity at both low and high speeds was attributed to the absence of hydrodynamic effects and the inadequacy of the hydrated water film, respectively. Li et al. [63] explored the superlubricity of Si_3N_4/glass tribological pairs lubricated with mixtures of polyhydroxy alcohols and various acid solutions. They found that forming a hydrogen-bond network between polyhydroxy alcohols and water molecules facilitated superlubricity after a running-in period, resulting in a significantly low coefficient of friction ranging from 0.003 to 0.006 under high pressure. Li et al. [64] also investigated the superlubricity of Si_3N_4/glass tribological pairs using mixtures of H_2SO_4 acid and glycerol solution as lubricants. They observed that the coefficient of friction ranged from 0.004 to 0.006 after a running-in period, attributing the superlubricity to the development of a hydrogen-bonded network between glycerol and water molecules on the Stern layer, along with a fluid-hydrated water layer. Furthermore, Li et al. [65] investigated the mechanism behind the superlubricity of H_3PO_4 solution for Si_3N_4/SiO_2 tribological pairs through direct observation methods, observing wear reduction and the formation of a thin film of hydrogen-bond network on the contact surface, contributing to superlubricity as shown in Figure 10.5b. Additionally, Li et al. [66] addressed the superlubricity nature of sapphire/sapphire tribological pairs lubricated with a phosphoric acid solution, achieving a coefficient of friction of 0.004 under high contact pressure, recommending the application of H_2SO_4 acid solution for sapphire bearing lubrication. Barthel et al. [67] studied the origins of superlubricity in boric acid across various environments, noting superlubricity in humidity and acetone vapor conditions. Lastly, Jiang et al. [68] examined the reasoning behind the permanent and temporary failures of liquid superlubricity in Si_3N_4/glass tribological pairs using H_2SO_4 acid solution and found that the higher sliding speeds or loads may lead to permanent or temporary superlubricity failure, respectively.

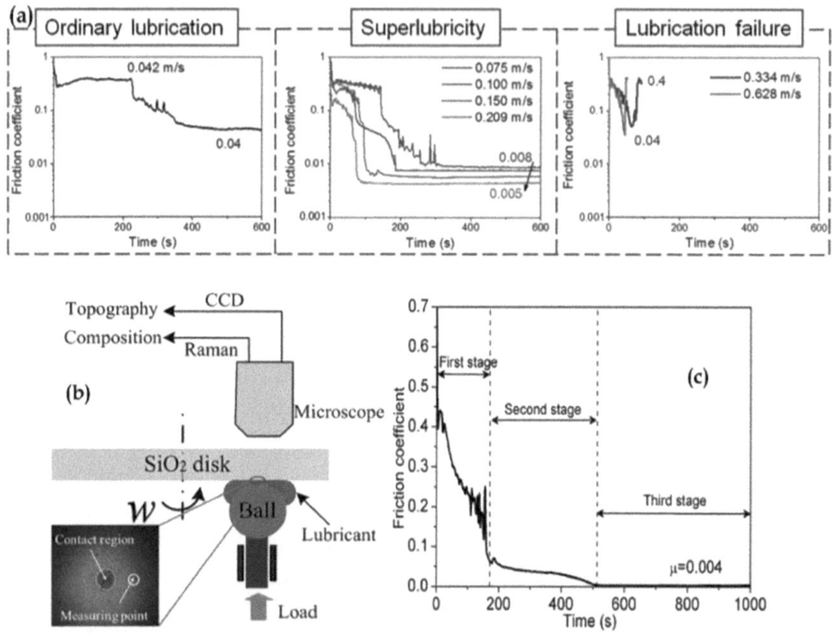

Figure 10.5 (a) The coefficient of friction (CoF) over time for a Si_3N_4/glass tribological pair lubricated with H_3PO_4 solution (pH = 1.5) under an applied load of 3 N. (b) Additionally, it includes a schematic representation of the tribometer used to observe the contact region in situ. (c) Another plot depicts the variation in CoF as a function of time under the lubrication of H_3O_4 solution (pH = 1.5) [50].

10.5 TECHNIQUES USED TO INVESTIGATE SUPERLUBRICITY

Superlubricity, the realm where friction nearly vanishes, continues to fascinate scientists with its potential to revolutionize various fields. However, investigating this elusive phenomenon requires specialized techniques to probe its mechanisms and assess its performance. The quest for achieving superlubricity, characterized by exceptionally low friction and wear, has been a focal point in tribology research due to its potential to revolutionize various engineering applications. Recent studies have delved into diverse lubrication strategies, ranging from solid to liquid lubricants, each offering unique advantages and facing distinct challenges in tribology [69]. Solid superlubricity, often realized through the use of 2D nanomaterials like MXene, h-BN, graphite, graphene, and MoS_2 [70–72], has garnered significant attention for its ability to reduce friction to near-zero levels at the nanoscale [73,74]. However, extending solid superlubricity to macroscopic scales remains a formidable challenge, primarily due to issues such as maintaining the ideal structure of 2D materials during sliding and overcoming changes in the friction mechanism. Despite these challenges, recent advancements, such as the incorporation of polyamide-imide (PAI) into graphene/MoS_2 coatings, have shown promise in improving the adhesion strength between coatings and substrates, thereby enhancing superlubricity even under high contact pressures. Efficient heating, ventilation, and air conditioning (HVAC) systems are crucial for maintaining stable environmental conditions in superlubricity applications. They regulate temperature and humidity, ensuring optimal friction reduction. By minimizing fluctuations, HVAC systems support forming and maintaining super lubricious interfaces, enhancing reliability and reproducibility across diverse conditions.

On the other hand, liquid superlubricity, achieved through suitable liquid lubricants, offers the advantage of adaptability to ambient conditions and the potential for real-world applications [52,75]. Various mechanisms underpin liquid superlubricity, including hydrodynamic effects, electric double-layer effects, and hydration effects, which may operate individually or in combination depending on the specific lubrication system. Despite the progress in understanding these mechanisms and developing adequate lubricating fluids, challenges persist in realizing the full potential of liquid superlubricity for industrial applications [50]. Key obstacles include limitations in load-bearing capacity, particularly for high-pressure scenarios, and the need to expand the range of applicable conditions for liquid superlubricity [48,76]. Recent research has explored the intricate interplay between lubricants and tribological interfaces, revealing insights into liquid superlubricity's speed-dependent stability and lubricant composition's influence on frictional behavior. Studies investigating the lubrication of Si_3N_4/glass tribological pairs with acid solutions and mixtures of polyhydroxy alcohols have highlighted the role of hydrogen-bond networks in facilitating superlubricity [77]. Direct observations of contact regions have provided valuable insights into wear reduction mechanisms and the formation of thin lubricant films crucial for maintaining superlubricity. Despite these advancements, ongoing efforts are needed to address persistent challenges and unlock the full potential of superlubricity for diverse engineering applications, ranging from high-performance bearings to MEMS [50].

Table 10.4 provides a summary of the various techniques used for superlubricity. Understanding the mechanisms behind this phenomenon is critical for unlocking its full potential in multiple areas. To achieve this, scientists employ diverse investigative techniques, each offering unique insights into the intricate world of superlubricity. At the forefront stands atomic force microscopy (AFM). This powerful tool, akin to a tiny, high-precision fingertip, allows researchers to directly probe the forces and friction between surfaces at the nanoscale, where superlubricity often reigns supreme [50]. Scientists can generate critical information on the relationship between

Table 10.4 A summary of the techniques used for investigating the superlubricity

Technique	*Description*	*Information gained*	*Advantages*	*Disadvantages*
Atomic force microscopy (AFM)	It uses a sharp tip to probe the surface forces and friction at the nanoscale.	Friction force vs. standard force curves, surface topography and roughness	High resolution, can measure adhesion and other surface forces	Limited scanning area, requires expertise to interpret data
Tribometers (e.g., pin-on-disk, ball-on-disk)	Measures friction coefficient between two surfaces under controlled conditions (load, speed, environment).	Friction coefficient, wear rate	Versatile, can simulate various conditions	Limited information on mechanisms
SEM	Images surface morphology and composition.	Surface features, chemical composition (with EDX)	High resolution, can visualize wear tracks	Limited information on friction mechanisms
TEM	Images materials at atomic resolution.	Crystal structure, interface characteristics	High resolution, can visualize atomic-scale features	Requires extensive sample preparation
Raman spectroscopy	Analyzes vibrational modes of molecules and materials.	Chemical composition. Presence of lubricants/ tribolayers	Non-destructive, can probe buried interfaces	Limited information on surface interactions
XPS	Analyzes the elemental composition and electronic states of a material's surface.	Chemical composition, chemical bonding state of lubricants/ tribolayers	Provides detailed chemical information	Requires specialized equipment and expertise
Computational modeling	Simulates and predicts friction behavior using computer models.	Insights into underlying mechanisms, Prediction of superlubricity in new materials	A powerful tool for understanding complex phenomena	Requires accurate input data and validation with experimental results

applied load and resulting friction by measuring the minuscule forces involved as the AFM tip interacts with the surface. AFM offers valuable surface topography and roughness insights, allowing researchers to correlate these features with the observed friction behavior [78]. Tribometers, the workhorses of tribology (the study of friction, wear, and lubrication), play a vital role. These instruments simulate real-world scenarios by bringing two surfaces into controlled contact under varying loads, speeds, and environmental conditions. By measuring the resulting friction coefficient, tribometers directly assess a material's or lubricant's ability to achieve superlubricity under specific operating conditions. This information is crucial for evaluating the practical viability of superlubricity solutions in various applications [79,80]. However, the quest for understanding goes beyond just the macroscopic picture. Unveiling the secrets lurking beneath the surface necessitates powerful imaging techniques like scanning electron microscopy (SEM) and transmission electron microscopy (TEM). SEM offers high-resolution surface morphology images and reveals the presence of wear tracks, shedding light on the surface's physical interactions. TEM delves even more deeply, providing atomic-resolution images that can unveil the crystal structure and intricacies of the interface between contacting surfaces. This information is invaluable for understanding the fundamental mechanisms governing superlubricity at the atomic level [81]. Beyond the realm of imaging lies the power of spectroscopy. Techniques like Raman spectroscopy and X-ray photoelectron spectroscopy (XPS) offer valuable insights into the chemical composition of the surfaces and any lubricants or tribolayers present. By analyzing the vibrational modes of molecules (Raman) or the electronic states of atoms (XPS), researchers can identify the presence of specific lubricants, determine their chemical bonding state, and even probe the chemical composition of buried interfaces. This information is crucial for understanding the role of chemical interactions in promoting or hindering superlubricity [82]. Finally, the theoretical realm comes into play with the power of computational modeling. Researchers can gain valuable insights into the fundamental mechanisms governing superlubricity by simulating atomic interactions and friction phenomena using sophisticated computer models. This approach allows scientists to test hypotheses, predict the behavior of new materials, and even guide the design of materials with tailored superlubricity properties [51,83].

10.6 LIMITATIONS AND CHALLENGES AND ADVANCES TO ACHIEVE SUPERLUBRICITY

10.6.1 Limitation

Despite the promising advancements in superlubricity research, several limitations and challenges persist in achieving this highly desirable frictionless state. One significant challenge is maintaining superlubricity under real-world operating conditions, particularly at macroscopic scales and in ambient environments [50]. While superlubricity has been successfully demonstrated in controlled laboratory settings and under specific environmental conditions such as high vacuum or inert gases, extending these conditions to practical applications remains challenging. Furthermore, maintaining consistent and stable superlubricity under diverse conditions of loads, speeds, and temperatures poses an additional challenge [3]. The inherent sensitivity of superlubricious interfaces to external factors like surface roughness, contamination, and temperature fluctuations further complicates efforts to maintain consistent performance. Furthermore, developing reliable and cost-effective methods for producing superlubric materials and coatings at scale is essential for widespread adoption in industrial applications. Addressing these limitations requires interdisciplinary research efforts spanning materials science, surface engineering, tribology, and mechanical engineering to overcome the complex challenges of achieving and sustaining superlubricity in practical settings [2].

10.6.2 Challenges

The quest to achieve superlubricity, a state of near-zero friction, has captivated researchers for decades. While significant progress has been made, recent years have unveiled persistent challenges and exciting advancements in this captivating tribological pursuit. One of the critical challenges lies in achieving robust superlubricity across diverse operating conditions [2]. Most materials exhibiting superlubricity tend to be highly sensitive to temperature, pressure, and environment. Maintaining near-zero friction under real-world operating conditions, which often involve a complex interplay of these factors, remains a significant hurdle. Another challenge lies in scaling superlubricity from the nanoscale to the macroscale. Many promising superlubricity mechanisms observed at the nanoscale, such as those in atomically thin films, struggle to translate effectively to macroscopic devices and systems [10]. Bridging this gap is crucial for realizing the transformative potential of superlubricity in practical applications. However, amidst these challenges, recent advancements offer a glimmer of hope. The exploration of novel materials and their combinations has shown promise. Research into nanocomposites, where different materials are combined at the nanoscale, has yielded promising results, offering enhanced robustness and tunability of superlubricity characteristics. Furthermore, computational modeling and simulation tools are playing an increasingly important role. Researchers can gain deeper insights into the fundamental mechanisms governing superlubricity by simulating material interactions and friction phenomena at the atomic level. This knowledge can be harnessed to design and engineer materials with tailored superlubricity properties. As we move forward, the field of superlubricity research demands a multifaceted approach. Combining advancements in material science, tribology, physics, and computational modeling is critical to overcoming existing challenges and unlocking the full potential of superlubricity. This realm promises to revolutionize various industries, from transportation and energy to manufacturing and healthcare [84], by minimizing friction-related losses and wear, leading to a more sustainable and efficient future.

10.6.3 Recent advancements

Recent advancements in solid superlubricity, particularly with 2-D nanomaterials, have showcased promising potential for reducing friction and wear in industrial applications. Various 2-D nanomaterials, including MoS_2, BN, boric acid, and graphene, have been extensively studied due to their unique physical and chemical properties. These materials exhibit superior tribological properties attributed to weak interlayer interactions, making them capable of achieving a superlubricative state. Researchers have made significant strides in understanding the mechanisms underlying solid superlubricity. For instance, Martin [60] explored the possibility of solid superlubricity with MoS_2 films under high vacuum conditions, reporting exceptionally low friction coefficients. Dienwiebel et al. [85] observed superlubricity between graphite surfaces using a novel frictional force microscope, while Onodera et al. [86] explored the lubrication mechanisms of MoS_2 with computational techniques; research has also broadened to encompass other 2D materials like WS_2 (tungsten disulfide) and MXenes [87]. Despite these advancements, challenges persist in achieving solid superlubricity at macroscopic scales and maintaining ultra-low friction under ambient conditions [88]. The inability to sustain ideal material structures during sliding and significant changes in friction mechanisms are vital obstacles. Efforts are ongoing to address these challenges, but further research is needed to realize the full potential of solid superlubricity in practical applications.

10.7 CONCLUSIONS

In conclusion, superlubricity represents a fascinating frontier in tribology research, offering the tantalizing prospect of achieving near-zero friction under specific conditions. Throughout this review, we have delved into the fundamental principles of friction and lubrication, including Amonton's Laws of Friction, which provide a foundational understanding of the factors influencing frictional behavior. We have explored various lubrication mechanisms and their role in governing superlubricity, ranging from manipulating interfacial forces to forming layered structures and utilizing nanomaterials. Furthermore, our examination of superlubricity has highlighted its potential to usher in a new era of friction reduction, with implications for a wide range of applications, from high-performance machinery to nanoscale devices. We have discussed the techniques employed to investigate superlubricity, including experimental methods such as tribometers and AFM, as well as computational modeling approaches, enabling researchers to gain deeper insights into the underlying mechanisms. Significant challenges and limitations remain despite the remarkable progress in understanding and harnessing superlubricity. These include the difficulty of achieving robust superlubricity at macroscopic scales, the need to overcome changes in friction mechanisms under different operating conditions, and the practical challenges associated with implementing superlubricity in real-world applications. However, amidst these challenges, there have been notable advancements, such as developing novel lubricant formulations, integrating nanomaterials for enhanced lubrication performance, and exploring innovative surface texturing techniques to promote superlubricity. These advancements offer promising avenues for overcoming the existing limitations and realizing the full potential of superlubricity in diverse engineering contexts. This review has provided a comprehensive overview of superlubricity principles, mechanisms, techniques, challenges, and advancements. By deepening our understanding of this phenomenon, we are poised to unlock new opportunities for friction reduction, energy efficiency, and enhanced performance across various technological domains.

REFERENCES

1. Han, T., S. Zhang, and C. Zhang, Unlocking the secrets behind liquid superlubricity: A state-of-the-art review on phenomena and mechanisms. *Friction*, 2022. **10**(8): pp. 1137–1165.
2. Ramezani, M., et al., Superlubricity of materials: Progress, potential, and challenges. *Materials*, 2023. **16**(14): p. 5145.
3. Baykara, M.Z., M.R. Vazirisereshk, and A. Martini, Emerging superlubricity: A review of the state of the art and perspectives on future research. *Applied Physics Reviews*, 2018. **5**(4): p. 041102-(1-18). https://doi.org/10.1063/1.5051445
4. Hirano, M., Superlubricity: A state of vanishing friction. *Wear*, 2003. **254**(10): pp. 932–940.
5. Wang, K., et al., Unraveling the friction evolution mechanism of diamond-like carbon film during nanoscale running-in process toward superlubricity. *Small*, 2021. **17**(1): p. 2005607.
6. Chen, Z., et al., Controllable superlubricity of glycerol solution via environment humidity. *Langmuir*, 2013. **29**(38): pp. 11924–11930.
7. Drexler, K.E., *Radical Abundance: How a Revolution in Nanotechnology Will Change Civilization*. 2013: Public Affairs.
8. Jia, Q., et al., Catalytic superlubricity via in-situ formation of graphene during sliding friction on Au@ aC: H films. *Carbon*, 2022. **186**: pp. 180–192.
9. Kuwahara, T., et al., Mechano-chemical decomposition of organic friction modifiers with multiple reactive centres induces superlubricity of ta-C. *Nature Communications*, 2019. **10**(1): p. 151.

10. Ayyagari, A., et al., Progress in superlubricity across different media and material systems—A review. *Frontiers in Mechanical Engineering*, 2022. **8**: p. 908497.
11. Yang, M., et al., Metal-graphene interfaces in epitaxial and bulk systems: A review. *Progress in Materials Science*, 2020. **110**: p. 100652.
12. Erdemir, A. and C. Donnet, Tribology of diamond-like carbon films: Recent progress and future prospects. *Journal of Physics D: Applied Physics*, 2006. **39**(18): p. R311.
13. Modabberasl, A., et al., Fabrication of DLC thin films with improved diamond-like carbon character by the application of external magnetic field. *Carbon*, 2015. **94**: pp. 485–493.
14. Sánchez, N.A., et al., Characterization of diamond-like carbon (DLC) thin films prepared by rf magnetron sputtering. *Thin Solid Films*, 2000. **373**(1–2): pp. 247–250.
15. Sose, A.T., S.Y. Joshi, L.K. Kunche, F. Wang, and S.A. Deshmukh, A review of recent advances and applications of machine learning in tribology. *Physical Chemistry Chemical Physics*, 2023. 25(6): pp. 4408–4443.
16. Shi, Y., et al., Robust macroscale superlubricity enabled by tribo-induced structure evolution of MoS_2/metal superlattice coating. *Composites Part B: Engineering*, 2023. **250**: p. 110460.
17. Pinto, G.M., et al., From two-dimensional materials to polymer nanocomposites with emerging multifunctional applications: A critical review. *Polymer Composites*, 2023. **44**(3): pp. 1438–1470.
18. Erdemir, A. and O. Eryilmaz, Achieving superlubricity in DLC films by controlling bulk, surface, and tribochemistry. *Friction*, 2014. **2**: pp. 140–155.
19. Kim, H.-J., et al., Static and kinetic friction characteristics of nanowire on different substrates. *Applied Surface Science*, 2016. **379**: pp. 452–461.
20. Barber, J.R., Multiscale surfaces and Amontons' law of friction. *Tribology Letters*, 2013. **49**: pp. 539–543.
21. Faghihnejad, A. and H. Zeng, Fundamentals of surface adhesion, friction, and lubrication, in *Polymer Adhesion, Friction, and Lubrication,* Editor H. Zeng, Chapter 1, 2013: John Wiley & Sons, Inc. pp. 1–57.
22. Huang, Y.-q., et al., Tribological properties of Mullite/3Y-TZP ceramics with different content of mullite fabricated by gel-casting. *Applied Surface Science*, 2019. **476**: pp. 232–241.
23. Bowden, F.P. and D. Tabor, *The Friction and Lubrication of Solids.* Vol. 1. 2001: Oxford University Press.
24. Kennedy, F.E., E.M. Schulson, and D.E. Jones, The friction of ice on ice at low sliding velocities. *Philosophical Magazine A*, 2000. **80**(5): pp. 1093–1110.
25. Schulson, E.M. and A.L. Fortt, Friction of ice on ice. *Journal of Geophysical Research: Solid Earth*, 2012. **117**(B12): pp. 1–15. https://doi.org/10.1029/2012JB009219
26. Khonsari, M.M. and E.R. Booser, *Applied Tribology: Bearing Design and Lubrication*. 2017: John Wiley & Sons.
27. Hutchings, I. and P. Shipway, *Tribology: Friction and Wear of Engineering Materials*. 2017: Butterworth-Heinemann.
28. Hirani, H., *Fundamentals of Engineering Tribology with Applications*. 2016: Cambridge University Press.

29. Mang, T. and W. Dresel, *Lubricants and Lubrication*. 2007: John Wiley & Sons.
30. Choudhari, A., A. Kumar, A.K. Gupta, A. Kumar, and A. Kumar, A comprehensive review on introduction to lubrication chemistry, in *Performance Characterization of Lubricants*, Editors, A. Kumar, A. Kumar, and A. Kumar, Chapter 1, 2024: CRC Press. pp. 1–32. https://doi.org/10.1201/9781003535447-1.
31. Kumar, S., A. Kumar, A. Kumar, A. Kumar, A. Choudhari, A.K. Gupta, and S. Faisal, State of the art in lubricants for extreme cold weather, in *Performance*, in *Characterization of Lubricants*, Editors, A. Kumar, A. Kumar, and A. Kumar, Chapter 12, 2024: CRC Press. pp. 227–261. https://doi.org/10.1201/9781003535447-12
32. Raghuvanshi, N.S., A. Johri, M. Saxena, A.K. Gupta, A. Choudhari, A. Kumar, A. Kumar, and A. Kumar, Thermally stable lubricants, in *Performance Characterization of Lubricants*. 2024: CRC Press. pp. 262–288.
33. Lugt, P.M. and G.E. Morales-Espejel, A review of elasto-hydrodynamic lubrication theory. *Tribology Transactions*, 2011. **54**(3): pp. 470–496.
34. Faisal, S., N. Chowdhury, A. Kumar, A. Kumar, A. Kumar, A. Choudhari, and A.K. Gupta, Lubricant: Soft materials (hydrogel), in *Performance Characterization of Lubricants*, Editors, A. Kumar, A. Kumar and A. Kumar, Chapter 10. 2024: CRC Press. pp. 191–212. https://doi.org/10.1201/9781003535447-10
35. Erdemir, A. and J.-M. Martin, *Superlubricity*. 2007: Elsevier.
36. Cho, M.H., et al., Tribological properties of solid lubricants (graphite, Sb2S3, MoS2) for automotive brake friction materials. *Wear*, 2006. **260**(7–8): pp. 855–860.
37. Song, Y., et al., Robust microscale superlubricity in graphite/hexagonal boron nitride layered heterojunctions. *Nature Materials*, 2018. **17**(10): pp. 894–899.
38. Li, J., et al., Superlubricity of silicone oil achieved between two surfaces by running-in with acid solution. *RSC Advances*, 2015. **5**(39): pp. 30861–30868.
39. Han, T., C. Zhang, and J. Luo, Macroscale superlubricity enabled by hydrated alkali metal ions. *Langmuir*, 2018. **34**(38): pp. 11281–11291.
40. Klein, J., et al., Reduction of frictional forces between solid surfaces bearing polymer brushes. *Nature*, 1994. **370**(6491): pp. 634–636.
41. Ma, L., et al., Origins of hydration lubrication. *Nature Communications*, 2015. **6**(1): p. 6060.
42. Li, J., et al., Investigation of the difference in liquid superlubricity between water-and oil-based lubricants. *RSC Advances*, 2015. **5**(78): pp. 63827–63833.
43. Luo, J., M. Liu, and L. Ma, Origin of friction and the new frictionless technology—Superlubricity: Advancements and future outlook. *Nano Energy*, 2021. **86**: p. 106092.
44. Han, T., et al., Origins of superlubricity promoted by hydrated multivalent ions. *The Journal of Physical Chemistry Letters*, 2019. **11**(1): pp. 184–190.
45. Kumar, A., A. Choudhari, A.K. Gupta, and A. Kumar, Rare-Earth based magnesium alloys as a potential biomaterial for the future. *Journal of Magnesium and Alloys*, 2024. 12(10): pp. 3841–3897. https://doi.org/10.1016/j.jma.2024.10.006

46. Li, P., et al., Toward robust macroscale superlubricity on engineering steel substrate. *Advanced Materials*, 2020. **32**(36): p. 2002039.
47. Ge, X., J. Li, and J. Luo, Macroscale superlubricity achieved with various liquid molecules: A review. *Frontiers in Mechanical Engineering*, 2019. **5**: p. 2.
48. Zhai, W. and K. Zhou, Nanomaterials in superlubricity. *Advanced Functional Materials*, 2019. **29**(28): p. 1806395.
49. Schreiber, P.J. and J. Schneider, Liquid superlubricity obtained for self-mated silicon carbide in nonaqueous low-viscosity fluid. *Tribology International*, 2019. **134**: pp. 7–14.
50. Dhanola, A., N. Khanna, and K.K. Gajrani, A critical review on liquid superlubricitive technology for attaining ultra-low friction. *Renewable and Sustainable Energy Reviews*, 2022. **165**: p. 112626.
51. Chen, X. and J. Li, Superlubricity of carbon nanostructures. *Carbon*, 2020. **158**: pp. 1–23.
52. Wang, H. and Y. Liu, Superlubricity achieved with two-dimensional nano-additives to liquid lubricants. *Friction*, 2020. **8**: pp. 1007–1024.
53. Luo, J. and X. Zhou, Superlubricitive engineering—Future industry nearly getting rid of wear and frictional energy consumption. *Friction*, 2020. **8**: pp. 643–665.
54. Shekhar, H. and R. Dumpala, Overcoming friction and steps towards superlubricity: A review of underlying mechanisms. *Applied Surface Science Advances*, 2021. **6**: p. 100175.
55. Gao, E., et al., Computational prediction of superlubric layered heterojunctions. *ACS Applied Materials & Interfaces*, 2021. **13**(28): pp. 33600–33608.
56. Mandelli, D., et al., Sliding friction of graphene/hexagonal–boron nitride heterojunctions: A route to robust superlubricity. *Scientific Reports*, 2017. 7(1): p. 10851.
57. Chowdhury, N., et al. Chowdhury, N., S. Faisal, A. Kumar, A. Choudhari, A.K. Gupta, A. Kumar, and A. Kumar, Characteristics of hydrogel properties in the advancement of biotribology, in *Application of Biotribology in Biomedical System*, Editors, A. Kumar, A. Kumar, and A. Kumar, Chapter 2, 2024: Springer Nature. pp. 27–78. https://doi.org/10.1007/978-3-031-58327-8_2
58. Gupta, A.K., A. Choudhari, A. Kumar, A. Kumar, A. Gupta, S. Faisal, and A. Kumar, Composites for drug eluting devices: Emerging biomedical applications, in *Application of Biotribology in Biomedical System*, Editors, A. Kumar, A. Kumar, and A. Kumar, Chapter 10, 2024: Springer Nature. pp. 251–311. https://doi.org/10.1007/978-3-031-58327-8_10
59. Bouchet, M.D.B., C. Matta, T. Le-Mogne, J.M. Martin, Q. Zhang, W. Goddard III, ... J. Ye, Superlubricity mechanism of diamond-like carbon with glycerol. Coupling of experimental and simulation studies. *Journal of Physics: Conference Series*, 2007. 89(1): p. 012003 IOP Publishing.
60. Martin, J.-M., Superlubricity of molybdenum disulfide, in *Superlubricity*, Editors, A. Erdemir and J.-M. Martin, Chapter 13, 2007: Elsevier Science B.V. pp. 207–225. https://doi.org/10.1016/B978-044452772-1/50044-5
61. Tian, J., et al., Tribo-induced interfacial material transfer of an atomic force microscopy probe assisting superlubricity in a WS2/graphene heterojunction. *ACS Applied Materials & Interfaces*, 2019. **12**(3): pp. 4031–4040.

62. Xiao, C., et al., Speed dependence of liquid superlubricity stability with H 3 PO 4 solution. *RSC Advances*, 2017. 7(78): pp. 49337–49343.
63. Li, J., C. Zhang, and J. Luo, Superlubricity achieved with mixtures of polyhydroxy alcohols and acids. *Langmuir*, 2013. **29**(17): pp. 5239–5245.
64. Li, J., et al., Superlubricity achieved with mixtures of acids and glycerol. *Langmuir*, 2013. **29**(1): pp. 271–275.
65. Li, J., L. Ma, S. Zhang, C. Zhang, Y. Liu, and J. Luo, Investigations on the mechanism of superlubricity achieved with phosphoric acid solution by direct observation. *Journal of Applied Physics*, 2013. **114**(11): p. 114901.
66. Li, J., et al., Investigations of the superlubricity of sapphire against ruby under phosphoric acid lubrication. *Friction*, 2014. **2**: pp. 164–172.
67. Barthel, A.J., J. Luo, and S.H. Kim, Origin of ultra-low friction of boric acid: Role of vapor adsorption. *Tribology Letters*, 2015. **58**: pp. 1–12.
68. Jiang, Y., et al., Temporary or permanent liquid superlubricity failure depending on shear-induced evolution of surface topography. *Tribology International*, 2021. **161**: p. 107076.
69. Ahmed Abdalglil Mustafa, W., et al., A review on potentials and challenges of nanolubricants as promising lubricants for electric vehicles. *Lubrication Science*, 2022. **34**(1): pp. 1–29.
70. Berman, D., A. Erdemir, and A.V. Sumant, Approaches for achieving superlubricity in two-dimensional materials. *ACS Nano*, 2018. **12**(3): pp. 2122–2137.
71. Gour R., N. Baliyan, A. Pal, A. Kumar, A. Kumar, and A. Kumar, Biotribology, in *Application of Biotribology in Biomedical System*, Editors, A. Kumar, A. Kumar, and A. Kumar, Chapter 8, 2024: Springer Nature. pp. 217–234. https://doi.org/10.1007/978-3-031-58327-8_8
72. Kumar A., A. Kumar, A. Kumar, A. Choudhari, A.K. Gupta, and S. Faisal, Introduction to biotribology: A science of surface interaction, in *Application of Biotribology in Biomedical System*, Editors, A. Kumar, A. Kumar, A. Kumar, Chapter 1, 2024: Springer Nature. pp. 1–26. https://doi.org/10.1007/978-3-031-58327-8_1
73. Swapnalin, J., B. Koneru, R. Boddula, D. Rangappa, and P. Banerjee, 2D nanomaterials as lubricant additives, in *Micro and Nano Technologies, Nanotechnology for Advanced Biofuels*, Editors, A.K. Nadda, T.A. Nguyen, S. Sharma, M. Bilal, and R.K. Gupta, Chapter 6, 2023: Elsevier. pp. 97–112. https://doi.org/10.1016/B978-0-323-91759-9.00012-5
74. Choudhari, A., A.K. Gupta, A. Kumar, A. Kumar, A. Gupta, N. Chowdhury, and A. Kumar, Wear and friction mechanism study in knee and hip rehabilitation: A comprehensive review, in *Application of Biotribology in Biomedical System*, Editors, A. Kumar, A. Kumar, A. Kumar, Chapter 13, 2024: Springer Nature. pp. 345–432. https://doi.org/10.1007/978-3-031-58327-8_13
75. Ma, Q., et al., Achieving macroscale liquid superlubricity using glycerol aqueous solutions. *Tribology International*, 2021. **160**: p. 107006.
76. Liu, Y., Approaching superlubricity under liquid conditions and boundary lubrication—Superlubricity of biomaterials, in *Superlubricity* (Second Edition), Editors, A. Erdemir, J.M. Martin, and J. Luo, Chapter 21, 2021: Elsevier. pp. 405–437. https://doi.org/10.1016/B978-0-444-64313-1.00021-1

77. Du, C., et al., Achieving macroscale superlubricity with ultra-short running-in period by using polyethylene glycol-tannic acid complex green lubricant. *Friction*, 2023. **11**(5): pp. 748–762.
78. Trache, A. and G.A. Meininger, Atomic force microscopy (AFM). *Current Protocols in Microbiology*, 2008. **8**(1): pp. 2C–2.
79. Dhanola, A. and H.C. Garg, Tribological challenges and advancements in wind turbine bearings: A review. *Engineering Failure Analysis*, 2020. **118**: pp. 104885.
80. Chand, N., *An Introduction to Tribology of FRP Materials*. 2000: Allied Publishers.
81. Inkson, B.J., Scanning electron microscopy (SEM) and transmission electron microscopy (TEM) for materials characterization, in *Materials Characterization Using Nondestructive Evaluation (NDE) Methods*, Editors, G. Hübschen, I. Altpeter, R. Tschuncky, and H.-G. Herrmann, 2016: Elsevier. pp. 17–43.
82. Andrade, J.D., X-ray photoelectron spectroscopy (XPS). *Surface and Interfacial Aspects of Biomedical Polymers: Volume 1 Surface Chemistry and Physics*, 1985: pp. 105–195.
83. Wang, H. and Y. Liu, Superlubricity achieved with two-dimensional nano-additives to liquid lubricants. *Friction*, 2020. **8**(6): pp. 1007–1024.
84. Tang, L., et al., Recent advances in superlubricity of liposomes for biomedical applications. *Colloids and Surfaces B: Biointerfaces*, 2022. **218**: p. 112764.
85. Dienwiebel, M., et al., Superlubricity of graphite. *Physical Review Letters*, 2004. **92**(12): p. 126101.
86. Onodera, T., et al., A computational chemistry study on friction of h-MoS2. Part I. Mechanism of single sheet lubrication. *The Journal of Physical Chemistry B*, 2009. **113**(52): pp. 16526–16536.
87. Eftekhari, A., Tungsten dichalcogenides (WS2, WSe2, and WTe2): Materials chemistry and applications. *Journal of Materials Chemistry A*, 2017. **5**(35): pp. 18299–18325.
88. Zheng, X., et al., Robust ultra-low-friction state of graphene via moiré superlattice confinement. *Nature Communications*, 2016. **7**(1): p. 13204.

Chapter 11

Advancements in lubricants toward energy efficiency and sustainability

Adesh Kumar Tomar, Krishnkant Sahu, and Max Marian

11.1 INTRODUCTION

Lubricants are substances made to serve many functions and are indispensable for the proper functioning of tribo-systems. The use of lubricants by humans can be traced back to ancient civilizations when plant oils and animal fats were employed to reduce transport friction. Research into lubricating materials reaches a climax with the mechanical industries and the appearance of special industrial environments such as aerospace [1]. The efficiency and stability of lubrication are important for the successful operation of any machine. As a result, the production of new lubricating materials and the discovery of novel lubrication technologies are critical for enhancing product reliabilities and lifetimes, while reducing natural resources and energy waste [2].

An adequate quantity of lubricants in the tribo-pairs is very important for smooth and reliable operations of machines. The performance of tribological systems greatly depends on the viscosity of a lubricant. The selection/formulation of a lubricant for a specific engineering application is mainly decided by the existing operating parameters, i.e., loading, speed, temperature range, environmental conditions, the material of mating surface to be lubricated, etc. Lubricants may be solids, liquids, or gas films. Generally, solid lubricants are used to decrease boundary or dry friction [3,4]. Solid lubricants like graphite, molybdenum disulfide, and waxes and greases are commonly used for light-duty applications. Further, polymer coatings such as polytetrafluoroethylene (Teflon) and polyethylene can minimize friction and are successfully employed in light-duty applications. However, various industries and the transportation sectors use liquid lubricants more frequently due to their specific advantages as compared to solid lubricants. The essential benefits of liquid lubricants are the formation of hydrodynamic films to support external load, removing dirt particles, and dissipation of the heat generated in the bearing clearance space. Normally, mineral oils are the base ingredient used for synthesizing most of the lubricants. The mineral oil constitutes nearly 90% of the lubricant proportion by volume, and the rest comes from the chemical additives used to impart specific

 DOI: 10.1201/9781003642848-11

properties to the lubricant [5]. The additives used in base oil are generally classified as extreme pressure additives, oxidation inhibitors, viscosity index improvers, detergents, dispersants, etc. [6]. The relative quantity of these additives in the base oil differs from application to application. Another category of lubricants is synthetic oils, which are generally used in the bearing system requiring stability of lubricant viscosity over a considerable range of temperature. Synthetic lubricants are specifically synthesized for severe operating conditions such as extreme temperatures, elevated speeds heavily loaded, etc. Although synthetic lubricants provide good lubricity over a broad range of temperatures, high cost and environmental disposal issues are the main deterrents for their use in common engineering applications.

11.2 PURPOSE OF LUBRICANTS

Lubrication is the method of putting a substance known as a lubricant to reduce wear between two contacting surfaces that are close to each other and moving relative to each other. The applied load in lubricants is supported by either the liquid being pumped under pressure between the surfaces or by the pressure created inside the liquid as a result of the motion of the lubricating fluid between the surfaces. Adequate lubrication allows the equipment to run smoothly and continuously, preventing seizures or excessive stresses on bearings and reducing wear. Lubricants also help to dissipate heat generated during operation by transferring heat away from frictional surfaces. This prevents components from overheating, which can cause damage or failure. Lubricants can also prevent corrosion by establishing a barrier between metal surfaces and moisture or other corrosive chemicals in the environment. This helps to avoid rust and corrosion, which can erode the integrity of components over time. Further, lubricants can also help to decrease noise and vibrations produced during operation by reducing friction and metal-to-metal contact.

11.3 CLASSIFICATIONS OF LUBRICANTS BASED ON DIFFERENT CRITERIA

The various types of lubricants classified based on different criteria are shown in Figure 11.1. In general, lubricants can further be classified as Newtonian and non-Newtonian lubricants based on their mathematical relations between the shear stress and shear strain rate. Further, the non-Newtonian lubricants are described by rheopectic, thixotropic, pseudoplastic, and Bingham lubricant models. Nowadays, many rheological models, such as the continuous Bingham model (used in ER / MR lubricants), cubic shear stress, couple stress lubricant model, micropolar

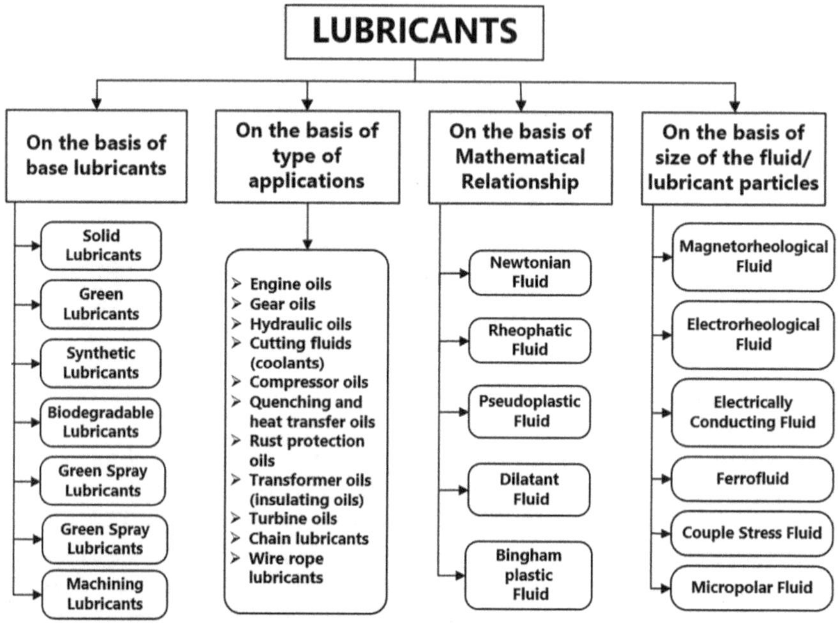

Figure 11.1 Types of lubricants based on different criteria.

lubricant model, power-law model, magnetohydrodynamic model, etc., are typically used to empirically describe the behavior of non-Newtonian lubricants [7–13].

11.3.1 Newtonian lubricant

The classical Newtonian theory neglects the size of fluid particles and assumes that a Newtonian lubricant is assumed to possess only 3° of translational freedom. If a lubricant exhibits a linear relationship between shear stress and shear strain rate, then it is called a Newtonian lubricant. Various lubricants such as air, water, mineral oil, and gasoline behave as Newtonian lubricants under normal conditions. The relationship between shear rate and shear stress from the Newtonian law of viscosity is given as follows [14]:

$$\tau = \mu\left(\frac{\mathrm{d}u}{\mathrm{d}y}\right),$$

where τ is the shear stress, μ is the viscosity, and $\mathrm{d}u/\mathrm{d}y$ is the shear rate.

11.3.2 Non-Newtonian lubricant

The blending of mineral oil with different additives makes the lubricant behavior non-Newtonian [15,16]. Power-law and cubic shear stress law models are used to theoretically categorize various lubricants as dilatant (shear-thickening), piezo-viscous, and pseudoplastic (shear-thinning) lubricants [8,17,18]. These non-Newtonian lubricant models exhibit a non-linear relationship between shear stress and shear strain rate. Many researchers [19,20] have used a couple stress and micropolar lubricants, assuming them to be non-Newtonian lubricants in many engineering and scientific research applications. In such lubricants, the additives may undergo micro-translation, micro-rotation, and even deformation between the bearing surfaces. Based on the mathematical relationship, the non-Newtonian lubricants can be described by the following relationship between shear stress and strain rate:

$$\tau = A\left(\frac{\mathrm{d}u}{\mathrm{d}y}\right)^n + B,$$

where A and B are constants depending upon the type of lubricants and condition of flow, and n is the flow behavior index. The non-Newtonian lubricant models like couple stress lubricant and micropolar lubricant are also discussed in the following section.

11.3.2.1 Couple stress lubricants

Mineral oil and synthetic lubricants are often blended with different packages of additives to achieve the desired tribological performance. The right type and the right quantity of additives used to blend the lubricants are decided based on the operating conditions of the bearings. Long-chain additives are frequently added in lubricants used in fluid film bearings running at high speeds and high loads to stabilize the viscosity variation with temperature. In the classical Newtonian theory, the fluid particles are considered to be point masses that interact with neighboring particles/additives only in terms of force.

In 1966, Stokes [21] proposed that the size of additives has a significant influence on the flow behavior of non-Newtonian lubricants. Owing to the finite size of the particle, a fluid particle/additive interacts with neighboring fluid particles in terms of force and couple [19,20,22]. This assumption induced couple stress in the momentum conservation equation of the lubricant, making the stress tensor anti-symmetric. This postulated theory was termed as couple stress theory. Due to the finite size of particles, the couple stress fluids are in micro-motion (translation and rotation). However, Stokes summarized that the rotational field of couple stress lubricant can be described in terms of velocity field. Some common lubrication applications, where couple stress theory has been used to model fluids/ lubricants are lubrication of synovial joints, synthetic lubricants, bio-fluids, polymer-thickened mineral oils, colloidal suspensions, extrusion of polymeric fluids, exotic lubrication, etc.

11.3.2.2 Micropolar lubricants

The micromorphic continuum theory [23–26] is the most generalized theory used to discuss the behavior of lubricants featuring microstructures. The microstructure [27] in many lubricants has a finite size and possesses material orientations. These microstructures undergo independent micro-motion (i.e. translation, rotation, and deformation) compared to the bulk of the lubricant. A lubricant consisting of microstructure [28] consists of 9 degrees of freedom (three translations, three rotations, and three stretches). However, when the microstructures in the lubricant are assumed to be rigid, the motion of fluid particles reduces to micro-translation and micro-rotation. With these assumptions, the micromorphic continuum theory is termed as Eringen's micropolar theory [29], and such fluids are classified as micropolar fluids. The behavior of polymeric fluids, bio-fluids, and fluids consisting of randomly oriented rigid microstructures can be suitably explained by using Eringen's micropolar theory. Also, the behavior of lubricant contaminated with dirt and metal particles during the normal operation of the fluid film bearing [30] can be approximated by using the micropolar theory. The analysis of micropolar lubricant requires the simultaneous solution of linear and angular momentum conservation equations along with mass conservation equation.

11.4 MINIMUM QUANTITY LUBRICATION (MQL) FOR SUSTAINABILITY

The implementation of MQL system could result in significant cost reductions in lowering fluid consumption. MQL, also known as "micro-lubrication" [31] and "near-dry machining" [32], is a new approach for supplying metal cutting lubricants to the tool-work interface. The tool performance can be greatly increased with a small volume of fluid using this technology. There are two fundamental types of MQL systems: through-tool and external spray. The exterior spray system is made up of tubes with one or more nozzles connected to a coolant reservoir. It works well for practically all machining operations and is portable and cheap. There are two configurations of through-tool MQL systems, depending on how the air-oil mist is produced. The external mixing, or one-channel system, is the first. Here, air and oil are combined externally before being pumped to the cutting zone via the tool and spindle. These systems have the benefits of simplicity and affordability; their tool coolant capability allows them to be easily retrofitted to the machines with high pressure. The dispersion and separation of oil mist during its passage from the nozzle is a drawback. The second type of system is the internal mixing or two-channel systems. The spindle is designed to accommodate two parallel tubes that carry air and oil to an external mixing mechanism located next to the tool holder, where the mist is produced. A spindle that is specifically made for this purpose is needed. The lag time between tool changes and the oil delivery rate during a cut is reduced in these systems. However, these systems are more challenging to maintain due to the housing of crucial components inside the spindle.

11.5 SMART LUBRICANTS

Over the last few decades, tribologists and lubrication practicing engineers have directed their research efforts in the area of formulating a new class of lubricants, whose properties and thus lubricating performance can be controlled as per the operating conditions design constraints. These classes of lubricants are referred to as smart lubricants. Smart lubricants behave in a Newtonian way under the normal operating condition, but once an external electric or magnetic field is applied, they become activated and offer adapted lubricating properties. These fluids can be activated and deactivated in a very short duration of time (nano-seconds) upon the application and removal of an electric or magnetic field respectively. The last few decades have witnessed an increasing use of smart lubricants in various engineering applications such as fluid film bearings, power transmissions, magnetic damping systems, etc. There are three main types of fluid/field interactions:

- Ferrohydrodynamics (FHD) deals with magnetic lubricants in the presence of magnetic fields, such as ferrofluids and magneto-rheological (MR) lubricants.
- Magneto-hydrodynamics (MHD) deals with the interactions between magnetic fields and electrical conductors (such as electrically conducting lubricants).
- Electro-hydrodynamics (EHD) deals with the effects of electrical forces [33].

In the following sub-sections, some insights on smart lubricants where the presence of magnetic/electric fields have considerable influence on the flow characteristics and performance of the lubricants are provided.

11.5.1 MR lubricants

MR lubricant technology is becoming a very interesting area of research for tribologists. MR lubricants are among the smart and intelligent lubricants due to their reversible and controllable characteristics [34–37]. MR lubricants are suspensions of micron-sized magnetic particles in carrier fluids, i.e., hydrocarbons, water, mineral oil, silicon oil, polyesters, polyethers, and so on [12,38]. Examples of some commercially available MR lubricants are MRF122-2ED, MRF-122EG, MRF-132DG, MRX-126PD, MRX-140ND, MRX-242AS, MRX336AG, etc. The application of magnetic field causes the magnetic particles to align in one direction, forming a chain-like structure. Consequently, the lubricant has a yield shear stress that is dependent on the applied magnetic field [39–42]. However, the transformation of the lubricant from a liquid state to a jelly state is completely reversible, as upon removal of a magnetic field, the lubricant instantly recovers to a liquid state. The magnitude of transformation (viscosity change) can be controlled by the magnetic field strength applied in the journal bearing. The transformation of the MR lubricant under an applied magnetic field is depicted in Figure 11.2a–d. At first, the magnetic particles in a carrier fluid remain in the colloidal suspended condition when there is no magnetic field (Figure 11.2a). When a magnetic field is applied across parallel plates (Figure 11.2b), these particles reorganize themselves toward the field's direction, creating chain-like microstructures that resemble

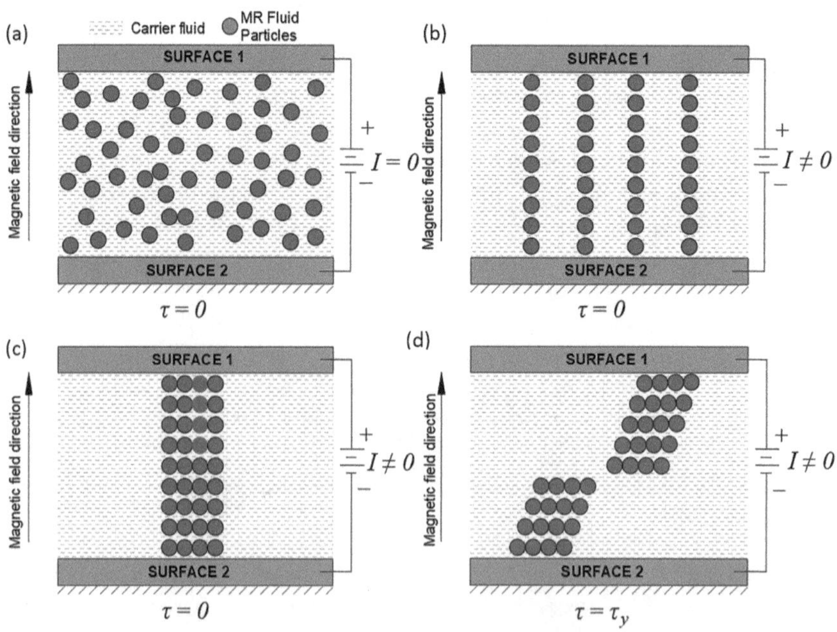

Figure 11.2 Transformation of the MR lubricant under the application of a magnetic field. (a) The magnetic particles remain in the colloidally suspended condition in a carrier fluid under zero magnetic field. (b) The magnetic particle rearranges in the direction of the applied magnetic field to form chain-like microstructures. (c) The agglomeration of the microstructure chains occurs between the surfaces for a high magnetic field. (d) The particle chains break for a typical value of shear stress and MR lubricant flows as a liquid.

chains. Further, an agglomeration of microstructure chains between the surfaces occurs when the strength of the magnetic field increases (Figure 11.2c). The viscosity of the lubricant is increased by multiple orders of magnitude depending on the strength of the applied field due to the creation of an aligned microstructure between the surfaces. The lubricant flow is induced by a shear force, the strength of which depends on the applied magnetic field. The particle chains break for a specific value of shear stress, and the MR lubricant flows as a fluid. This value of shear stress is termed as static yield stress $\left(\tau_y\right)$ (Figure 11.2d). The additives are mandatory to prevent aggregation and sedimentation, as well as to provide additional lubricating properties. These include polymers, surfactants, and thixotropic agents. The micron-size particles in MR lubricant have a field-induced magnetic moment, and their Brownian motion is insignificant.

The particle aggregation processes in MR lubricants are reversible and quite extensive, and they are induced by the applied field (magnetic), which is one of their most important feature in developing field-controlled flow characteristics. The rheological properties of the magnetic fluid depend on the magnetic field intensity, shear rate, additives, carrier fluid properties, concentration, shape, size, and, type of the MR lubricant particles [39,43–47]. MR lubricants have several technological advantages over equivalent electro-rheological (ER) lubricants including higher control effect, better stability, and lower power requirements [40]. Typical applications of MR lubricants include fluid film journal bearing [34,36], artificial joints [48], control valves [49], brakes [12], clutches [50], shock absorbers [51], and seismic vibration dampers [52]. Numerous different applications include the use of magnetic field to control thermal energy transfer, sound propagation, precision polishing, chemical sensing and biomedical applications, and isothermal magnetic advection [39].

11.5.2 Electrically conducting lubricants

The operating speeds of machines have increased tremendously over the years to achieve higher production rates. Lubrication engineers have undertaken numerous efforts to synthesize new lubricants for efficient and reliable operation of the machines, which are not only capable of sustaining high speeds and heavy loads, but also offer excellent lubricity and stable viscosity over a wider range of operating temperatures. As such, fluid film bearings used in aerospace and nuclear applications are made to operate under extreme temperature environments. The viscosity of commercially available lubricants falls drastically under such elevated temperature conditions. Therefore, commercially available lubricants are not a good choice in engineering applications as the lubricant operating in such extreme environments needs to have high viscosity. Electrically conducting lubricants based upon MHD are one of the examples of smart lubricants that are specifically formulated for the abovementioned applications [53,54]. These lubricants possess high thermal and electrical conductivity, which makes them an excellent choice as a lubricant under such environments. When an electrically conducting lubricant flows through a transverse magnetic field, it will experience a body force known as the Lorentz force, which tends to reduce the flow of the lubricant in the clearance space among the bearing surfaces. This leads to the accumulation of a higher quantity of lubricant between the bearing clearance surfaces, which ultimately leads to a higher fluid film pressure or fluid film reaction to support a heavier external load [55,56]. In other words, the presence of a magnetic field enhances the apparent viscosity of the lubricant. The reasonably higher external load can be supported by the use of electrically conducting lubricant in fluid film journal bearings [9,10,54]. Liquid metal-based alloys of sodium, mercury, and gallium [9,57–59] are a few examples of electrically conducting lubricants. MobilTM DTE 932GT is another example of electrically conducting lubricants, which are manufactured commercially by Mobils, for lubricating high-speed turbomachinery. Electrically conducting lubricants are also widely used in industrial and R&D activities, such as purification of crude oil, power generation in the steam/gas power plant, MHD accelerators, and propulsion units to drive marine and space ships propulsion, etc. [60,61].

11.5.3 Ferrofluid lubricants

Ferrofluids (FFs) or magnetic fluids (MFs) or ferromagnetic lubricants are stable colloidal suspensions of magnetic particles (diameter range 5–20 nm) in a carrier liquid. Such magnetic particles are usually formed from magnetic materials, namely, cobalt ferrite ($Co \cdot Fe_2O_4$), magnetite ($FeO \cdot Fe_2O_3$), or maghemite ($\gamma \cdot Fe_2O_3$) [62,63]. These are smart materials that combine normal liquid behavior with superparamagnetic properties. The dynamic capabilities (stiffness, damping, and stability) and self-sealing properties of journal bearings can be improved using FFs [62]. The magnetic particles are coated with a suitable material shell to avoid agglomeration. FFs are classified into two groups based on the coating: ionic if the coating is an electric shell and surfacted if the coating is a molecule. In ionic FFs [64,65], nanoparticles are electrically charged to keep the colloidal system stable. FFs have recently been used in many commercials and industrial applications. When the FFs interact with the magnetic field, magnetic forces are generated and create pressure inside the fluid [66]. As a result, the load-carrying capacity of the journal-bearing system is improved. For stability, the necessary criterion is the size of the particles. They must be small enough to prevent settling or sedimentation of the FF in a magnetic field due to thermal agitation, which is caused by the Brownian motion of the particles [67]. Since the formation of agglomerates in FFs is restricted by the thermal motion of nanoparticles, field-induced changes in flow behavior are usually insignificant. The FF remains in a liquid state in the presence of a magnetic field, even when magnetized to saturation [67–69].

11.5.4 ER lubricants

ER lubricants are examples of smart lubricants, which have found extensive application in fluid film journal bearings. The use of ER lubricants offers excellent control over the rotor dynamic performance of the journal bearings [7,45,70–72]. The fluid film stiffness and damping coefficients of the bearings can be tailor-made as per the requirement by varying the strength of an electric field. The use of these ER lubricants not only improves the tribological performance of bearings but also enhances the service life of the bearing. An ER lubricant essentially consists of dielectric particles suspended in an insulating viscous medium, usually oil. When an electric field is applied across an ER lubricant, it instantly transforms into a jelly-type substance. However, the transformation of the lubricant from liquid to jelly state is completely reversible, as upon removal of the electric field, the lubricant recovers to the liquid state within milliseconds. The magnitude of transformation (viscosity change) can be controlled by the strength of the electric field/potential difference applied across the bearing surfaces. The ability to quickly change their material/lubricating properties has made ER lubricants one of the most versatile materials for the development of smart machines/structures. Although the phenomenon of electrorheology was first reported in the 18th century [73], the first patent for such fluids was filed by Winslow [74] in 1947. The transformation of the ER lubricant structure under the influence of an electric field is the same as for the MR lubricant, as indicated in Figure 11.2a–d. Theoretically, an ER lubricant is described by a continuous Bingham model. The behavior of lubricant in this model is characterized by yield shear stress and viscosity. The flow in an ER lubricant can only be induced, once a typical shear stress $\left(\tau_y\right)$ has been applied. Apart from fluid film bearings, the ER lubricant has been extensively used in engineering applications associated with power transmissions (clutches, brakes, torque converters, etc.), hydraulics (ER valves), vibration control (ER dampers) [75–79], etc. In all of these engineering applications/devices, the ER lubricant operates with either of the three flow modes, namely, flow mode, squeeze mode, and shear mode. These three modes of operations are associated with the design of ER lubricant devices, which utilize the post-yield regime of the lubricant behavior as per the required mode of applications. In a flow mode configuration, the ER lubricant is forced to flow between two electrodes and electric potential is used to control pressure drop across the device. The flow mode of operation is generally used in ER dampers and ER shock absorbers. In the shear mode, one of the electrodes is free for translation or rotation and the electric potential controls the force (torque)/velocity relationship. The shear mode of operation is commonly used in ER clutches and ER brakes, in places where rotational motion must be controlled. The third mode of operation is called the squeeze-flow mode in which the ER lubricant is squeezed between two electrodes which are free to move, contracting is expanding the electrode gap. The squeeze-flow mode of operation is mostly used for applications for controlling small movement but large force involvement.

11.6 CONCLUSIONS

Rapid developments in lubricant technology are accelerating progress toward meeting sustainability goals in a variety of applications. The transition from traditional petroleum-based lubricants to smart lubricant alternatives is more than a trend; it is a necessity driven by environmental, economic, and social considerations. The implementation of new technologies, such as MQL and smart lubricants, could be advantageous for lubricant sustainability. Researchers, engineers, policymakers, and industry stakeholders are focusing on developing creative solutions that reduce environmental impact, optimize resource utilization, and improve overall efficiency.

ACKNOWLEDGMENTS

Adesh Kumar Tomar and M. Marian greatly acknowledge the postdoctoral fellowship from the School of Engineering of Pontificia Universidad Católica de Chile (PD-DII23–1251) as well as the financial support given by ANID-Chile within the project Fondecyt de Postdoctorado No. 3240005. M. Marian greatly acknowledges the financial support from Programa de Inserción Académica (PIA) 2022 from the Vicerrectoría Académic y Prorectoría of Pontificia Universidad Católica de Chile within the and the Schaeffler FAG Foundation within the Future Technology Award 2022.

REFERENCES

1. Gong, H., Yu, C., Zhang, L., Xie, G., Guo, D., Luo, J. Intelligent lubricating materials: a review. *Composites Part B: Engineering* 202: 108450 (2020).
2. Johansson, J., Devlin, M.T., Prakash, B. Lubricant additives for improved pitting performance through a reduction of thin-film friction. *Tribology International* 80:122–130 (2014).
3. Kumar, V., Gautam, R.K., Tyagi, R. Tribological behavior of Al-based self-lubricating composites. *Composite Interfaces* 23:481–492 (2016).
4. Oh, D.-S., Kang, K.-H., Kim, H.-J., Kim, J.-K., Won, M.-S., Kim, D.-E. Tribological characteristics of micro-ball bearing with V-shaped grooves coated with ultra-thin protective layers. *Tribology International* 119:481–490 (2018).
5. Bhushan, B. *Introduction to Tribology*. John Wiley & Sons (2013).

6. Scott, W., Suntiwattana, P. Effect of oil additives on the performance of a wet friction clutch material. *Wear* 181:850–855 (1995).
7. Nikolakopoulos, P., Papadopoulos, C. Controllable high speed journal bearings, lubricated with electro-rheological fluids. An analytical and experimental approach. *Tribology International* 31:225–234 (1998).
8. Kumar, P., Khonsari, M. On the role of lubricant rheology and piezo-viscous properties in line and point contact EHL. *Tribology International* 42:1522–1530 (2009).
9. Sasada, T., Kurosaki, Y., Kamijo, K. MHD journal bearing in a magnetic field perpendicular to its axis: 2nd report, experiment with mercury. *Bulletin of JSME* 17:1652–1656 (1974).
10. Kamiyama, S. Magnetohydrodynamic Journal Bearing (Report 1). *ASME Journal of Lubrication Technology* 91(3): 380–386 (1969). https://doi.org/10.1115/1.3554944
11. Carlson, J.D. MR fluids and devices in the real world. *International Journal of Modern Physics B* 19:1463–1470 (2005).
12. Varela-Jiménez, M., Luna, J.V., Cortés-Ramírez, J., Song, G. Constitutive model for shear yield stress of magnetorheological fluid based on the concept of state transition. *Smart Materials and Structures* 24:045039 (2015).
13. Mang, T., Dresel, W. *Lubricants and Lubrication*. John Wiley & Sons (2007).
14. Rao, B.N., Sekhar, A.S. Analysis of magneto rheological fluid Journal Bearing. *Applied Mechanics and Materials* 895:152–157 (2019). https://doi.org/10.4028/www.scientific.net/amm.895.152
15. Wu, Z., Dareing, D.W. Non-Newtonian effects of powder-lubricant slurries in hydrostatic and squeeze-film bearings. *Tribology Transactions* 37:836–842 (1994).
16. Spikes, H. The behaviour of lubricants in contacts: current understanding and future possibilities. *Proceedings of the Institution of Mechanical Engineers, Part J: Journal of Engineering Tribology* 208:3–15 (1994).
17. Tomar, A.K., Sharma, S.C. Non-Newtonian lubrication of hybrid multi-recess spherical journal bearings with different geometric shapes of recess. *Tribology International* 171:107579 (2022).
18. Tomar, A.K., Sharma, S.C. Study on surface roughness and piezo-viscous shear thinning lubricant effects on the performance of hole-entry hybrid spherical journal bearing. *Tribology International* 168:107349 (2022).
19. Ahmad, N., Singh, J. A model for couple-stress fluid film mechanism with reference to human joints. *Proceedings of the Institution of Mechanical Engineers, Part J: Journal of Engineering Tribology* 221:755–759 (2007).
20. Khatri, C.B., Sharma, S.C. Influence of couple stress lubricant on the performance of textured two-lobe slot-entry hybrid journal bearing system. *Proceedings of the Institution of Mechanical Engineers, Part J: Journal of Engineering Tribology* 231:366–384 (2017).
21. Stokes, V.K. Couple stress in fluids. *Physics of Fluids* 9:1709–1715 (1966).
22. Stokes, V.K. Couple stresses in fluids. In: *Theories of Fluids with Microstructure*. Berlin, Heidelberg: Springer (1984). https://doi.org/10.1007/978-3-642-82351-0_4
23. Eringen, A.C. Mechanics of micromorphic materials. In *Applied Mechanics*, pp. 131–138. Berlin, Heidelberg: Springer (1966).

24. Ariman, T., Turk, M., Sylvester, N. Microcontinuum fluid mechanics—a review. *International Journal of Engineering Science* 11:905–930 (1973).
25. Ariman, T., Turk, M., Sylvester, N. Applications of microcontinuum fluid mechanics. *International Journal of Engineering Science* 12:273–293 (1974).
26. Eringen, A.C. Simple microfluids. *International Journal of Engineering Science* 2:205–217 (1964).
27. Eringen, A.C., Okada, K. A lubrication theory for fluids with microstructure. *International Journal of Engineering Science* 33:2297–2308 (1995).
28. Eringen, A.C. Micropolar fluids with stretch. *International Journal of Engineering Science* 7:115–127 (1969).
29. Eringen, A.C. Theory of micropolar fluids. *Journal of Mathematics and Mechanics* 16(1): 1–18 (1966).
30. Khonsari, M.M., Brewe, D.E. On the performance of finite journal bearings lubricated with micropolar fluids. *Tribology Transactions* 32:155–160 (1989).
31. Boubekri, N., Shaikh, V. Machining using minimum quantity lubrication: a technology for sustainability. *International Journal of Applied Science and Technology* 2:111–115 (2012).
32. Klocke, F., Eisenblätter, G. Dry cutting. *CIRP Annals* 46:519–526 (1997).
33. Rosensweig, R.E. *Ferrohydrodynamics.* Courier Corporation (2013).
34. Urreta, H., Leicht, Z., Sanchez, A., Agirre, A., Kuzhir, P., Magnac, G. Hydrodynamic bearing lubricated with magnetic fluids. *Journal of Intelligent Material Systems and Structures* 21:1491–1499 (2010).
35. Laukiavich, C., Braun, M.J., Chandy, A.J. A comparison between the performance of ferro-and magnetorheological fluids in a hydrodynamic bearing. *Proceedings of the Institution of Mechanical Engineers, Part J: Journal of Engineering Tribology* 228:649–666 (2014).
36. Bompos, D.A., Nikolakopoulos, P.G. CFD simulation of magnetorheological fluid journal bearings. *Simulation Modelling Practice and Theory* 19:1035–1060 (2011).
37. Sharma, S.C., Tomar, A.K. Study on MR fluid hybrid hole-entry spherical journal bearing with micro-grooves. *International Journal of Mechanical Sciences* 202–203:106504 (2021).
38. Data, T. MRF-132DG magneto-rheological fluid. *Lord Prod Sel Guide Lord Magnetorheol Fluids* 54:11 (2011).
39. Jolly, M.R., Bender, J.W., Carlson, J.D. Properties and applications of commercial magnetorheological fluids. *Journal of Intelligent Material Systems and Structures* 10:5–13 (1999).
40. Wang, J., Meng, G. Magnetorheological fluid devices: principles, characteristics and applications in mechanical engineering. *Proceedings of the Institution of Mechanical Engineers, Part L: Journal of Materials: Design and Applications* 215:165–174 (2001).
41. Data, L.T. *MRF-122-2ED Magneto-Rheological Fluid.* LORD Corporation World Headquarters (2002).
42. Tomar, A.K., Sharma, S.C., Sahu, K. Performance analysis of textured spherical hybrid journal bearings operated with magnetorheological fluid. *Journal of Tribology* 146:014602 (13 pages) (2023).

43. Phule, P.P. Magnetorheological (MR) fluids: principles and applications. *Smart Materials Bulletin* 2001:7–10 (2001).
44. Odenbach, S. Recent progress in magnetic fluid research. *Journal of Physics: Condensed Matter* 16:R1135 (2004).
45. Peng, J., Zhu, K.-Q. Effects of electric field on hydrodynamic characteristics of finite-length ER journal bearings. *Tribology International* 39:533–540 (2006).
46. Jolly, M.R., Carlson, J.D., Munoz, B.C. A model of the behaviour of magnetorheological materials. *Smart Materials and Structures* 5:607 (1996).
47. Sahu, K., Sharma, S.C., Tomar, A.K. Analysis of MR fluid lubricated slot entry hybrid conical journal bearing with texturing arrangements. *Tribology International* 188:108788 (2023).
48. Carlson, J.D., Matthis, W., Toscano, J.R. *Smart Prosthetics Based on Magnetorheological Fluids. Smart Structures and Materials 2001: Industrial and Commercial Applications of Smart Structures Technologies*, pp. 308–316. International Society for Optics and Photonics (2001).
49. Leicht, Z., Urreta, H., Sanchez, A., Agirre, A., Kuzhir, P., Magnac, G. Theoretical and experimental analysis of MR valve. *Journal of Physics: Conference Series*, pp. 012070. IOP Publishing (2009).
50. Jacob, R. Magnetic Fluid Torque and Force Transmitting Device. Google Patents (1951).
51. Gołdasz, J., Sapiński, B. *Insight into Magnetorheological Shock Absorbers.* Springer (2015).
52. Forte, P., Paternò, M., Rustighi, E. A magnetorheological fluid damper for rotor applications. *International Journal of Rotating Machinery* 10:175–182 (2004).
53. Roberts, P.H. *An Introduction to Magnetohydrodynamics.* Longmans London (1967).
54. Sasada, T., Kurosaki, Y., Honda, K., Kamijo, K. MHD journal bearing in a magnetic field perpendicular to its axis: 1st report, analysis of an infinitely long bearing. *Bulletin of JSME* 17:1645–1651 (1974).
55. Lin, J.R. Magneto-hydrodynamic squeeze film characteristics between annular disks. *Industrial Lubrication and Tribology* 53(2):66–71 (2001).
56. Dudzinsky, S., Young, F.J., Hughes, W.F. On the load capacity of the MHD Journal Bearing. *ASME Journal of Lubrication Technology* 90(1):139–144 (1968).
57. Roberts, W. Tribology in nuclear power generation. *Tribology International* 14:17–28 (1981).
58. Gerkema, J. Gallium-based liquid-metal full-film lubricated journal bearings. *ASLE Transactions* 28:47–53 (1985).
59. Maurya, U., Vasu, V., Saini, V. Nano-innovations in lubrication: unveiling the synergistic boundary lubrication potential of zinc titanate with phosphonium ionic liquid. *Tribology International* 199:109987 (2024). https://doi.org/10.1016/J.TRIBOINT.2024.109987.
60. Makinde, O.D., Chinyoka, T. MHD transient flows and heat transfer of dusty fluid in a channel with variable physical properties and Navier slip condition. *Computers & Mathematics with Applications* 60:660–669 (2010).

61. Roy, S. Wingless Hovering of Micro Air Vehicle. Google Patents (2013).
62. Kuzhir, P. Free boundary of lubricant film in ferrofluid journal bearings. *Tribology International* 41:256–268 (2008).
63. Torres-Díaz, I., Rinaldi, C. Recent progress in ferrofluids research: novel applications of magnetically controllable and tunable fluids. *Soft Matter* 10:8584–8602 (2014).
64. Massart, R. Preparation of aqueous magnetic liquids in alkaline and acidic media. *IEEE Transactions on Magnetics* 17:1247–1248 (1981).
65. Depeyrot, J., Da Silva, G., Alves, C.R., Sousa, E.C., Magalhães, M., Figueiredo Neto, A.M., et al. Static magneto optical birefringence of new electric double layered magnetic fluids. *Brazilian Journal of Physics* 31:390–397 (2001).
66. Neuringer, J.L., Rosensweig, R.E. Ferrohydrodynamics. *The Physics of Fluids* 7:1927–1937 (1964).
67. Scherer, C., Figueiredo Neto, A.M. Ferrofluids: properties and applications. *Brazilian Journal of Physics* 35:718–727 (2005).
68. Joseph, A., Mathew, S. Ferrofluids: synthetic strategies, stabilization, physicochemical features, characterization, and applications. *ChemPlusChem* 79:1382–1420 (2014).
69. Susan-Resiga, D., Vékás, L. Yield stress and flow behavior of concentrated ferrofluid-based magnetorheological fluids: the influence of composition. *Rheologica Acta* 53:645–653 (2014).
70. Dimarogonas, A.D., Kollias, A. Electrorheological fluid-controlled "smart" journal bearings. *Tribology Transactions* 35:611–618 (1992).
71. Jang, S., Tichy, J.A. Internal damper characteristics of rotor system with submerged ER fluid journal bearing. *International Journal of Rotating Machinery* 3:61–71 (1997).
72. Tomar, A.K., Sharma, S.C. A study of hole-entry grooved surface hybrid spherical journal bearing operating with electrorheological lubricant. *Journal of Tribology* 142:111802 (19 pages) (2020).
73. Duff, A.W. The viscosity of polarized dielectrics. *Physical Review (Series I)* 4:23 (1896).
74. Winslow, W. *Methods and Means of Translating Electrical Impulses in to Mechanical Forces*. US Patent (1947).
75. Carlson, J.D., Duclos, T.G. Electrorheological Fluid Force Transmission and Conversion Device. Google Patents (1990).
76. Papadopoulos, C.A. Brakes and clutches using ER fluids. *Mechatronics* 8:719–726 (1998).
77. Stangroom, J.E. Electrorheological fluids. *Physics in Technology* 14:290 (1983).
78. Stevens, N., Sproston, J., Stanway, R. Experimental evaluation of a simple electroviscous damper. *Journal of Electrostatics* 15:275–283 (1984).
79. Stanway, R., Sproston, J., El-Wahed, A. Applications of electro-rheological fluids in vibration control: a survey. *Smart Materials and Structures* 5:464 (1996).

Chapter 12

Used oil disposal

Biswadeep Pal

12.1 INTRODUCTION

Globally, a substantial volume of used lubricating oils is generated from various sources like railway workshops, industrial facilities, ship garages, defense machinery, and automobile workshops. Over time, these oils have become polluted and unable to produce satisfactory efficiencies due to change in chemical parameters. So, it is necessary to regular removal of dust and carbon residue to manage the daily work. Discarding these used lubricating oils raised environmental concerns and regulatory challenges. Conventional methods such as burning for energy generation, landfilling, greasing, and indiscriminate dumping have proven to be neither cost-effective nor environmentally friendly (Okeke et al., 2022). Recycling provides a more suitable and cost-effective alternative to treat the used oil. Various techniques like heating and filtration, re-refining within crude oil processes, acid/clay treatment, and solvent extraction with clay or hydrotreating are available for refining these oils (Sarkar et al., 2023). While environmental concerns are associated with most methods, the distillation process offers a cleaner option but is typically more expensive.

The unsuitability of lubricating oil for further use is primarily due to the accumulation of impurities within the oil, which alters its composition. The main impurities consist of combustion by-products containing water, leading to sludge formation and rust. Additionally, soot and carbon cause the oil to darken, and lead, most commonly tetraethyl lead, is found in the range varied from 2 to 1200 ppm due to bearing wear (Yang, 2008; Mekonnen & Ababa, 2014). The limited chemical constituents present in the used lubricant oil are reported in Table 12.1. Unburnt gasoline or diesel fuel, especially during startup, also contributes to oil contamination. Moreover, abrasive road dust particles that enter the engine through the air cleaner further contaminate the lubricant oil.

The amount of water used in a lubricating oil varies based on the type of vehicle that is used. During running conditions of any engine, it will be possible to get a small amount of water; more precisely, moisture formation is observed in an oil cooler and engine cooling system and vaporized

DOI: 10.1201/9781003642848-12

Table 12.1 Typical levels of contaminants found in used oils

Categories and components	*Automotive used oil concentration*	*Industrial used oil contamination*
Metals		
Arsenic	5–25 ppm	NG
Barium	50–500 ppm	NG
Cadmium	2 ppm	NG
Chromium	3–30 ppm	NG
Lead	100–1200 ppm	NG
Zinc	100–1200 ppm	NG
Chlorinated hydrocarbons		
Dichloro difluoromethane		NG
Trichloro trifluoroethane	1000–4000 ppm	
Tetrachloroethylene		1000–6000 ppm
1,1,1-Trichloroethane		
Trichloroethylene		
Other organic compounds		
Benzene	100–300 ppm	100–300 ppm
Toluene	500–5000 ppm	500–5000 ppm
Xylene	500–5000 ppm	500–5000 ppm
Benzo(a)anthracene	10–50 ppm	NG
Benzo(a)pyrene	5–20 ppm	NG
Naphthalene	100–1400 ppm	NG
PCBs	NG* -20ppm	100–1000 ppm

Source: Taken from Yang (2008).

to the atmosphere (Rincon et al., 2005). The presence of excess water has provided several adverse effects, including viscosity alteration and emulsion formation, potentially causing issues with sprocket or bearings (Mensah-Brown, 2015).

Sometimes, used oil contains a huge amount of water, which needs to be eradicated before reusing. As a result, it is always recommended in most of the cases to remove the free water first just by keeping it settled for 5–6 hours and then heating it up to 110°C for an hour to remove both free water and emulsified water. It is important to note that since the boiling point of water is 100°C in the selective treatment process (STP), using temperatures above this point is typically effective (Vecchione et al., 2016). However, when temperatures exceed 200°C, there is a risk of losing lighter hydrocarbon fractions, and the water becomes contaminated, requiring comprehensive treatment. Sometimes if tight emulsion is formed, then chemical demulsification is also recommended to remove the microemulsified water. In general, two types of emulsion were formed: microemulsion and macroemulsion. Macroemulsion is easily separated due to gravity settling, but microemulsion is tough to separate due to bond formation between oil and water with impurities (Umar et al., 2018).

Used oil typically contains 1%–15% water in the form of free water or emulsions, same percentage of light hydrocarbons, a recycling rate ranging from 60% to 90%, and 7%–20% additives and pollutants (Audibert, 2011). Lubricant oils are produced from petroleum and nonpetroleum-based feedstock. They consist of heavier petroleum hydrocarbon base stock with hydrocarbons in the carbon range of 18 to 34, blended with additives to enhance specific properties. One crucial property of lubricant oil is that determines its ability to maintain its viscosity. Viscosity is defined as resistance of oil to flow between two moving parts so that oil can spread in all the possible parts throughout the engine, but enough care must be taken to check its lubricity. As oil viscosity depends on temperature, it is essential to know the value of the viscosity index. A low viscosity index means the viscosity changes more with temperature and vice versa (Shahnazar et al., 2016). Other important properties of lubricant oil include specific density, pour point, refractive index, total acid number, cardoon residual, and flash point.

12.2 DUMPING BURNED LUBE OIL CAUSES ECOLOGICAL IMBALANCE

Burning any material can lead to the release of its basic elements and compounds into the air, which can pose potential dangers to both human well-being and the natural world when substantial emissions take place. These risks can have direct impacts, such as leading people to develop respiratory problems when they breathe these tiny particles, or indirect effects, as these particles, often carrying contaminants like heavy metals, settle on crops and make their way into the food chain. It is important to note, however, that specific oils can be burned safely if strict conditions are met (Nolan, 2014). These conditions encompass maintaining precise control over the composition of material, using filters and scrubbers to remove emitted particles and chemicals, designing chimneys that effectively disperse emissions, ensuring efficient combustion by regulating factors like temperature and residence time, and establishing clear guidelines for ash containment and disposal to minimize environmental harm.

12.3 REFINING

The choice of regeneration technologies for used lubricant oils is influenced by several factors, including the technology itself (operating conditions, recovered amount, product quality, and progression phase), fiscal considerations (water and energy charges, equipment costs, and scale), and environmental protection aspects [removing polychlorinated biphenyls (PCBs), managing acidic sludge, handling residual oil sludge, and minimizing hazardous chemical use (Boyde, 2002). Various regeneration methods are available to produce reusable lubricating oil (Pirro et al., 2016). Regardless of the specific technology, it is generally followed a four-step process. Firstly, dewater and defuel the used oil to separate foreign components, like moisture and light fuel compounds, which may have infiltrated during use. Secondly, a de-asphalting step removes solid foreign elements, such as metal particles from wear and additives. This is achieved by exploiting differences in physical properties like solubility and boiling points. Thirdly, fractionation classifies materials based on boiling points to select or eliminate specific components. Finally, finishing treatments are employed to eliminate remaining foreign elements, such as chlorine, nitrogen, oxygen, and sulfur, which could not be removed in the earlier steps. Different technologies may vary in the de-asphalting and finishing procedures. The most recent technology, thin film evaporation (TFE), offers three processing trends based on various finishing methods. Physicochemical techniques for lubricant regeneration include vacuum distillation with clay, chemical material finishing (involving thermal, chemical materials, solvent, clay, or integrated approaches), and vacuum distillation with hydrofinishing. The choice of distillation process depends on the type of process, pollutant concentrations in used lubricants, and the desired product quality. While TFE is prevalent in America and Europe, acid/clay technology is commonly used in developing countries. Re-refining is employed for used lubricants with high contaminant levels and yields by-products used as fuel. The specific re-refining process significantly impacts pollutant concentrations and final product quality.

12.4 ADDITIVES

Over the past century, there has been a growing utilization of additives in lubricating oils to enhance their chemical stability, improve performance, and reinforce their physiochemical properties. These additives serve various purposes, including boosting ignition and combustion effectiveness, stabilizing heat content, safeguarding the motor against friction, carbon formation, corrosion, abrasion, and deposition, as well as reducing the dissemination of pollutants, among other benefits. More recently, incorporating nanomaterials into oils and lubricants to reduce friction and extend their durability has gained significant attention due to their ability to form protective tribofilms, enhance thermal stability, and improve load-carrying capacity, ultimately leading to increased efficiency and longer service life of mechanical components. (Ali et al., 2019; Ghalme & Bhalerao, 2017; Gulzar et al., 2016).

The primary source of oil contamination during usage is the chemical breakdown of additives and their interactions, leading to the creation of corrosive and undesirable components. A particular concern is the presence of polycyclic aromatic hydrocarbons (PAHs) due to their well-established carcinogenic properties (Boehm, 1964). However, the levels of pollutants vary based on different constituent factors, including the detergent types, added diluents, storage system, and management and usage practices.

For instance, a reduction in the use of leaded gasoline has led to a significant decrease in lead concentrations in used oil, subsequently lowering the required levels of bromine and chlorine additives. This, in turn, diminishes the presence of halogenated hydrocarbons in used oil.

12.5 THE CLASSIFICATION OF USED LUBRICANTS

Different categories of used oil exist based on their cleanliness and composition. Clean used oil, which includes cutting and hydraulic oils, can be processed through re-refining and cleaning methods. Mixed used oil, combined with clean oil, like hydraulic and diathermic oils without chlorine, can undergo regeneration or re-refining. Mixed used oil, which encompasses inseparable oil during production, can be processed through gasification using plasma jets and incinerators to produce gases for industrial use. Total used oil, comprising unseparated oils and heavy base oil, can be processed through intensive, soft processing, or direct incineration methods. Finally, synthetic oils fall under the category of other used oil and are primarily processed through thermal cracking.

12.6 CONVENTIONAL METHODS FOR TREATING USED LUBE OIL

12.6.1 Acid-clay method

Historically, the most effective method for treating used oils was the acid-clay process (Hamawand et al., 2013). While this process could yield high-quality lubrication stocks, it also generated significant quantities of petroleum-contaminated acid-clay sludge (Jagaba et al., 2022). Acid-clay sludge as hazardous waste, necessitating proper management and disposal (Hu et al., 2013). Consequently, in the acid-clay process, which is shown in Figure 12.1, became uneconomical due to the high costs associated with handling its residues.

In this process, used lubricant oil undergoes dehydration and distillation, followed by acid treatment using sulfuric acid and impurities are eliminated by clay. However, this process poses negligible environmental security, primarily because it generates substantial volumes of acidic sludge. The nature of the processing, whether reprocessing or regenerative, depends on factors

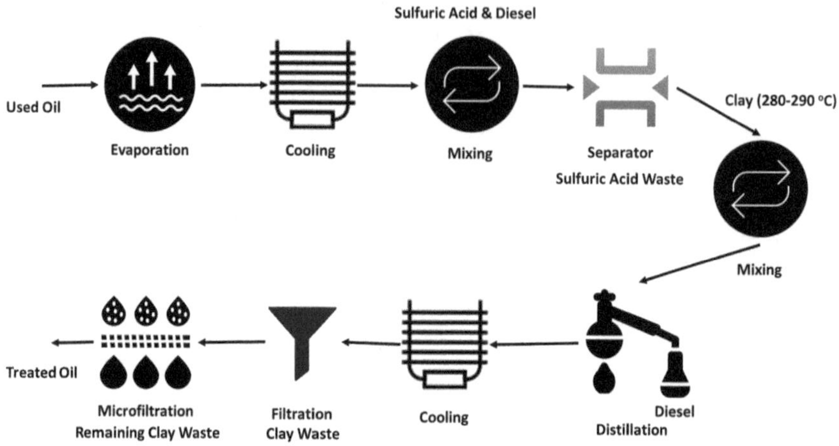

Figure 12.1 Typical process diagram for acid-clay treatment (taken from Hamad et al., 2004).

such as the concentration of impurities, chemical constituents of lubricant oil, and the desired value of the regenerated oil.

Hook et al. (2009) reported that acid-clay treatment is the most adapted technology worldwide for used oil recycling. As a result, acidic sludge was produced, which was identified as potentially hazardous. An eco-friendly method is needed to reduce the volume of acidic sludge, although there is insufficient scientific information regarding this (Unnisa, 2017).

There are alternative management practices for acidic sludge, such as incineration, which can also harness the energy generated from combustion. The heating value of acidic sludge is approximately 4000 kcal/kg during incineration (Liu et al., 2017). Acidic sludge primarily consists of unsaturated compounds, which are polar and asphaltene-like in composition. Acidic sludge can be modified and neutralized to mitigate environmental and health hazards, and it can be used as a raw material for bitumen production, often involving physical modification with additives like bentonite, polymers, and nano-composites. The findings demonstrate that products derived from acidic sludge can be used in construction, road-building, and other applications based on specific bitumen criteria and characteristics.

Jafari and Hassanpour (2014) reported that the acidic sludge has potential to be transformed into valuable products, such as organic fertilizers, explosives materials, paint, ink, chemical fibers, and industrial detergents. The processing of acidic sludge provides a range of commercial products which includes surfactants, sulfuric acid, light hydrocarbons, coke, activated carbon, furnace fuels, various types of bitumen and asphalt, thermal insulators, and more. Acidic sludge also serves as an additive in bitumen materials and the production of carbon rods. Bitumen materials used as raw material or additives in various industrial applications, from automotive fan belts to polyvinyl chloride (PVC) flooring.

Converting acidic sludge from used oil refineries into intermediate sludge suitable for producing soft, unoxidized, or hard, oxidized asphalt. This acidic sludge, produced in oil refineries, is non-biodegradable, highly acidic, and toxic. Several industrial processes combine acidic sludge with other compounds to produce asphalt, often incorporating various additives to modify its characteristics and quality (Nazem & Tavakoli, 2017).

It was suggested that biodiesel can be economically viable at various plant capacities, while sometimes breakeven prices are seen for biodiesel plants of different sizes. It was noted that acidic sludge can be converted into valuable products and raw materials for various applications. It can also yield commercial products, including surfactants, sulfuric acid, light hydrocarbons, and more, making it a versatile resource. The transformation of acidic sludge has significant potential, and its value can be harnessed through various processes and applications. These studies collectively shed light on the economic and environmental aspects of managing acidic sludge from used motor oil reprocessing.

12.6.2 Extraction

Solvent extraction is a method known for producing high-quality base oils while minimizing environmental pollution. However, it operates under elevated pressure and demands the expertise of skilled personnel and advanced systems. Various solvents have been utilized in the solvent extraction process, which includes 2-propanol, 1-butanol, MEK, ethanol, toluene, and acetone. Several studies were conducted and demonstrated that MEK exhibited superior performance, characterized by minimal oil loss and effective removal of sludge. On the other hand, butan-1-ol yielded the highest sludge removal rate (Nezhdbahadori et al., 2018; Taiwo & Otolorin, 2009; Al-Doury, 2019).

Oladimeji et al. (2018) studied on both composite and single solvents, such as MEK and propan-2-ol, were employed in used lubricant oil. The results favored MEK, which yielded the highest oil output, which is around 96%. Notably, the ratio of solvent to oil had a more significant influence on the properties of oil compared to temperature. Additional research has confirmed the effectiveness of solvent extraction and adsorption in recycling waste lubricating oils. It was investigated the use of blends of multiple solvents and activated alumina adsorbent, concluding that higher solvent-oil ratios led to increased product recovery (Martins, 1997).

Mohammed et al. (2013) explored the use of various adsorbent materials, including almond shell, eggshell, walnut shell, and locally prepared acid-activated clay. Ben Hammouda et al. (2021) recommended the use of magnesite as a sorbent due to its cost-effectiveness and superior adsorption performance.

Hydrocarbon solvents, such as stabilized and liquefied petroleum gas (LPG) condensate, have been employed to reduce asphaltene content, ash content, and carbon residue in lubricating oil (Al Zubaidi & Tamimi, 2018; Zare-Shahabadi et al., 2010)). Kamal and Khan (2009) introduced a solvent selection method based on Burrel's classification, categorizing alcohols as high-capacity solvents, ketones as moderate-capacity solvents, and hydrocarbons as low-capacity solvents due to their ability to form hydrogen bonds. Over time, solvent extraction gradually replaced the acid treatment method, as it improved the stability of base oils against oxidation and their viscosity-temperature characteristics. This process dissolves undesirable aromatic components, leaving behind desirable saturated hydrocarbon components, particularly alkanes (Prince, 1992; Thorat et al., 2021)).

Although the oil produced through solvent extraction is comparable to that obtained via the acid-clay method, the extraction process results in significant solvent losses and requires highly skilled maintenance. Furthermore, extraction is conducted at higher pressures (exceeding 10 atm), necessitating specialized high-pressure sealing systems, which makes the construction of solvent processing unit costly (Henderson, 2022; Eshaghi & Hamrang, 2021).

Subsequently, propane extraction emerged as a viable alternative. Propane has the capacity to dissolve waxy and paraffinic materials, as well as partially dissolve oxygenated substances. However, heavy aromatic compounds like asphaltenes are not soluble in liquid propane. Despite its potential, propane extraction is considered a hazardous and highly flammable process (Huang et al., 2022).

12.6.3 Membrane technology

Membrane technology represents another method used for regenerating lubricating oils, making use of three specific types of polymer fiber membranes: polyethersulfone (PES), polyvinylidene fluoride (PVDF), and polyacrylonitrile (PAN). It is a continuous process designed to remove metal particles or dust from used engine oil, ultimately improving the flow characteristics and flash point of the recovered oils (Dmitrieva et al., 2022; Englehardt, 1993)). Despite these acknowledged benefits, the expensive nature of these membranes may make them vulnerable to damage and fouling by larger particles, as noted by AlSawaftah et al. (2021).

12.6.4 Acid washing

Acid washing like glacial acetic acid and formic acid, introduced later, has been found to be a valuable technique for recycling used lubricant oils due to its remarkable ability to dissolve and attract polar contaminants and metal particles present in used oil (Cooney et al., 2011). This method offers cost advantages over conventional approaches, primarily owing to the cost-effectiveness and ready availability of acetic acid. The recycled oil product generated through this method has demonstrated the potential for reuse as an oil blend for engine lubricants, as supported by Rincon et al. (2005).

Research from Widodo et al. (2018) indicates that the acetic acid extraction process for used/waste engine oil can yield up to 90% (depending on the method employed) of base lubricating oil. In essence, if the recycling of used lubricant oil is handled carefully with the appropriate technology, a significant amount of base oil can be recovered with a relatively cost-effective process, producing high yield and high-quality lubricating oil base stock that rivals virgin lube oil base stock. Various studies have shown that for dry used lubricant oil, optimal purification is done using acetic acid, depending on the initial contaminant levels.

Consequently, the capacity of acetic acid to separate the base oil from the contaminated used oil has been explored through a series of experiments that aim to create two distinct layers: a clear reddish lubricant oil layer and sludge at the bottom of the measuring cylinder. Weighing both the lube oil and sludge produced in each test allows for the determination of the optimal acid-oil ratio for a specific oil sample (Gaber et al., 2011).

12.6.5 Vacuum distillation

Several studies have been dedicated to vacuum distillation (VD) applied to used oil. The key stages involved in the VD process encompass an initial pretreatment phase, focusing on the removal of impurities that may lead to fouling and equipment corrosion. Following this, the procedure advances to the distillation phase, which is responsible for separating water and light hydrocarbons from the oil, a concept elucidated by Pinheiro et al. (2019). Subsequently, vacuum distillation is conducted, utilizing either a falling film apparatus or a conventional VD column.

12.6.6 Hydrotreating

The VD is then succeeded by hydrotreating, an operation conducted under high pressure and temperature in the presence of a catalyst. The primary objective of hydrotreating is to eliminate chlorine, sulfur, nitrogen, and organic constituents from the distillate. This hydrotreated oil is subsequently further fractionated under high vacuum conditions, resulting in the production of distinct components suitable for engineering, hydraulic, and diesel-powered oil applications (Schmidt & Eugene, 1986). The tar formation after VD treatment was further applied road construction as well as coating in roof.

12.6.7 Hydrogenation

Initially, used oil is heated and combined with hot hydrogen within a pressurized mixing chamber (Kupareva et al., 2013). This blended mixture is then conveyed to a separator, which directs it to a residue stripper. The mixture subsequently undergoes a catalytic reactor process, with the primary goal of eliminating soluble metals, and then passes through a hydrofinishing reactor. In the hydrofinishing reactor, essential processes such as dechlorination and desulfurization are carried out, among others. These treatments lead to notable improvements in the odor, chemical properties, and color of the treated hydrocarbons (Boyde, 2002).

The residual product generated by this process consists of a high-boiling-range hydrocarbon, which can be further fractionated into neutral oil products characterized by varying viscosities. These neutral oil products can also be effectively employed as components for blending in the production of lubricating oils. A summary of various conventional methods and their advantages and disadvantages are shown in Table 12.2.

12.6.8 Combined technologies

Advanced technologies for recycling used oil employ a combination of methods recommended due to the complex nature of the process, sometimes conventional methods not achieved the desired yield. Different companies have developed specific techniques for reprocessing waste oils which required advanced equipment and procedures. Here is an overview of some of these technologies:

- KTI process: This process was introduced in Greece in year of 1992, where this KTI process was operated with the combination of VD and hydrofinishing. Initially, this process begins with atmospheric distillation operation to eradicate moisture and lighter components of the waste oil and the heavier component residues fractionated with the help of VD unit at the temperature range with its boiling range, which is approximately 250°C (Kajdas, 2000). Subsequently, hydrogenation is also useful in the application to remove heavy metals like sulfur, nitrogen, oxygenated compounds, etc. (Mang, 2017).
- STP: Another sophisticated method, STP, also combines VD and hydrofinishing. This procedure results in reduced emissions, making it environmentally friendly. The key steps include dehydration, VD, separation, and hydrofinishing to separate lube oil from the rest (Madanhire et al., 2016).
- Vaxon process: This process is widely adopted in various countries, including Denmark, Germany, Saudi Arabia, and Spain; the Vaxon process incorporates a distinctive Vacuum Cyclone Flash Evaporator in the VD step to minimize oil cracking (Kupareva et al., 2013). The process starts with chemical treatment using alkali to remove impurities. Later, aromatic content was separated with the help of different solvents like dimethyl formamide (DMF) and n-methyl-2-pyrrolidone. In the final stage, VD was incorporated to separate the raffinate into different heavier grade compounds.

Table 12.2 Different recycling methods and their advantages and limitations

Methods	*Advantages*	*Limitation*
Acid-clay (Udonne, 2011; Mekonnen & Ababa, 2014)	• This method sees frequent usage • Production costs are reduced • The process and operation are simple and not complicated, requiring skilled operators	• It results in waste generation, contributing to pollution • It induces equipment corrosion and shortens its lifespan • Achieves low oil recovery rates
Vacuum distillation (Hamawand et al., 2013)	• Appropriate for large-scale facilities with high capacity • It is an environmentally friendly process with no pollution • Demands advanced equipment • Yields high-quality base oils	• Demands a substantial capital investment • Necessitates highly skilled operators due to the use of advanced equipment
Hydrotreating (Kajdas, 2000)	• Eliminate chlorine, sulfur, nitrogen from used oil • Hazardous waste is less generated	• Regeneration of the catalyst in the reactor is required
Acid washing (Fonseca et al., 2019)	• Remove heavy metals • High yield oil recovery • Low-cost process	• High corrosion • Low catalyst recovery • Purification difficult
Solvent extraction (Hall et al., 1990; Khan et al., 2023)	• The solvent is reusable • It does not lead to any pollution • Achieves efficient recovery of high-quality base oils	• Economically viable only for large-scale facilities • Demands operators with advanced skills and expertise
Membrane technologies (Cao et al., 2009)	• Continuous operation, • Low energy consumption, easy adaptability, versatility	• Overburden with dirt, metals and degraded the quality
Hydrogenation (Pelemo et al., 2020)	• Reducing the metal concentration • Improve odor, chemical properties, and color	• Process can work effectively in low hydrocarbon but clogging occurs during work with higher molecular hydrocarbon

- Revivoil technology: This method employs high-pressure catalytic hydrogenation and comprises three main steps: dehydration to eliminate water and light hydrocarbons, de-asphalting to remove asphaltic and bituminous products, and hydrofinishing to eliminate chlorine, sulfur, and other contaminants.

- IFP technology/Snamprogetti: Combining VD and the hydrogenation process, this technology involves atmospheric distillation to remove light hydrocarbons and water, followed by VD and extraction of the oil-containing portion using propane and lastly hydrogenation for separation of propane, asphaltic compounds, oxidized hydrocarbons, and solids.
- Proterra recycling technology: The Proterra recycling technology involves two principal processes one is solvent extraction and another is VD. Light oils, vacuum residues, extracts, residual vapor, and effluent water are found from this technology.
- Recently, a different combination of chemical methods is applied to recover the base oil from the used oil. Ayeni et al. (2021) applied the extraction process with the help of methyl-ethyl ketone for extraction purposes, followed by adsorption of carbon, soot, and metal compounds with the help of activated carbon.

12.6.9 Current technologies

Recycling used oil is of paramount importance due to its high content of potentially harmful pollutants and heavy metals, which pose environmental risks. Therefore, the development of eco-friendly, sustainable, and cost-efficient recycling methods is imperious. Various contemporary approaches to recycling used oil are outlined:

- Pyrolytic distillation method (DLF): Arpa et al. (2010) introduced a technique to produce a fuel referred to as diesel-like fuel through pyrolytic distillation. A typical process flow diagram of the Pyrolytic distillation unit is shown in Figure 12.2. This process entails filtering the waste engine oil using a qualitative filter. Subsequently, three additives, Na_2CO_3, zeolite, and CaO, are blended with the purified oil at different ratios and subjected to thermal and pyrolytic treatments to convert it into fuel. The properties of fuel closely resemble those of conventional diesel fuel.
- Pyrolysis: Pyrolysis has emerged as a substitute method for the efficient conversion of used lubricants into refined products. It is a thermal decomposition process that fractionate substances at higher temperatures (approximately 200°C–1000°C range) in an inert or less oxygenate atmosphere. While not yet widely adopted, pyrolysis has gained attention due to its potential to produce high-energy products from waste materials (Demirbas et al., 2015). Microwave pyrolysis is a thermochemical process used for converting waste materials into valuable products with specific electrical power input and flow rate (Lam et al., 2017).

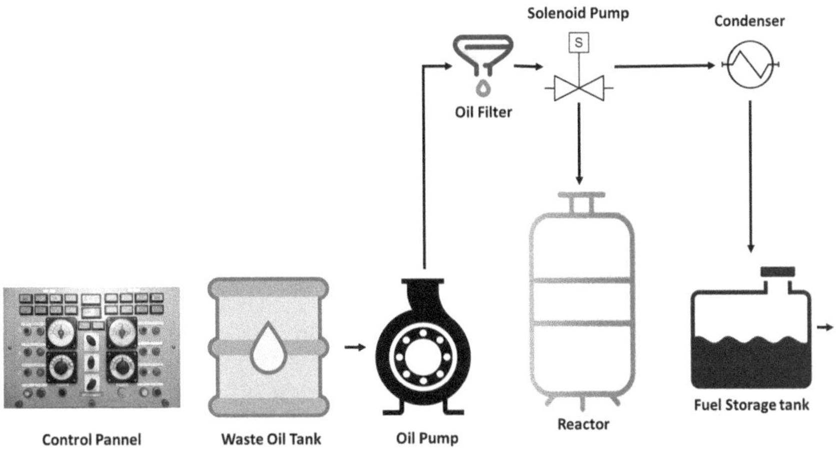

Figure 12.2 Pyrolytic distillation unit (taken from Arpa et al., 2010).

12.7 MANAGEMENT OF BY-PRODUCTS AFTER THE PROCESS

It is important to note that dumping and burning used oil are not included in the list of recommended oil management practices, as they are not environmentally responsible methods for oil disposal. Instead, the following eco-friendly options are recommended for the disposal of waste oils:

- Regeneration of new base oil: Various technologies have been developed to regenerate used lubricating oil, producing different fractions of base oil, fuel by-products, and heavier residue streams. The regeneration process results in fuel by-products (the lighter components) that can be used as fuel, while the obtained base oil can be used as lubricating oil after blending with additives.
- Thermal cracking: Thermal cracking is a process that breaks down used oil into valuable products by subjecting it to high temperatures in the absence of oxygen. This method can yield various hydrocarbon fractions suitable for various applications.
- Incineration: Waste oil can be incinerated or used as fuel, allowing for the extraction of energy from the oil. This approach is more environmentally friendly than simple burning, as it can be controlled and harnessed for energy generation.

The heavier residual stream, containing carbonaceous substances, can be utilized as a blending component in the bitumen processing industry. This residue is typically incorporated into construction products such as road surfaces. These recommended options are environmentally responsible and contribute to reducing the adverse impact of waste oil disposal on the environment.

12.8 EXPERIMENTAL ANALYSIS

12.8.1 Calorific value

The calorific value represents the energy obtained by burning 1 kg of fuel. It is providing information on engine efficiency, energy output estimation, and fuel blending potential. An adiabatic bomb calorimeter was used to measure this value. A bomb was placed in a water-filled copper vessel containing a heat-preserving stirring mechanism. Lube oil was introduced into the bomb, and a nickel-chrome wire connected to the calorimeter made contact with the oil. The instrument was pressurized to initiate the calorimeter, and a thermometer was used to monitor water temperature until stabilization was reached.

12.8.2 Flash point

The flash point is the temperature at which a flammable liquid will release enough vapor to create a combustible mixture with air. Under specific conditions, when a small flame is introduced, this mixture should briefly ignite. Since it signifies the temperature at which flammable vapor is generated, the flash point is typically the most valuable single indicator of the potential fire hazard.

12.8.3 Viscosity

Viscosity measures a resistance of fluid to gradual deformation under shear stress. It has two important parameters: dynamic viscosity, which signifies an internal resistance by fluid and is the force required to move one plane horizontally in the fluid separated by a unit distance; and kinematic viscosity, which is calculated by dividing dynamic viscosity by the mass density of fluid.

Viscosity is influenced by temperature, pressure, and density and inversely varies with temperature. Viscosity tests help detect contamination in used engine oil. Increased oil viscosity can result from dissolved or suspended oxidized or polymerized substances, while decreased engine oil viscosity may indicate fuel contamination (Hu et al., 2013).

12.8.4 Viscosity index

The viscosity index is an empirical number reflecting how viscosity changes with temperature. A high viscosity index suggests minimal viscosity variation with temperature, providing better engine protection across a wide temperature range. It also indicates good thermal stability and low-temperature flow characteristics, often linked to the absence of aromatic and volatile compounds.

12.8.5 Pour point

The pour point of an engine oil is the lowest temperature at which the oil remains in a liquid state. Engine base oils can contain waxes and paraffins that solidify in cold temperatures, and oils with a higher wax and paraffin content tend to have a higher pour point. The pour point of lubricating oil is significantly affected by its viscosity; oils with higher viscosity values typically have higher pour points. In cold weather, particularly during engine startup, the pour point of the oil is crucial. It must flow smoothly into the oil pump, even at low temperatures, to ensure proper engine lubrication.

12.8.6 Ash content

The ash content of the extracted base oil sample was analyzed using the ASTM D482 standard. Understanding the level of ash-forming material in a product can help determine its suitability for a specific application. Ash can originate from metallic compounds that are soluble in oil or water, or it can come from foreign solid particles like dirt and rust.

12.8.7 Heavy metals

To determine the metal content (specifically, iron, copper, magnesium, zinc, and manganese) in new, used, and recycled lubricating oils. The ash collected in the crucible was dissolved in a solution containing 50 mL of aqua regia, which is the combination between HNO_3 and HCl at a ratio of 1:3. The solution was heated on a hot plate until its volume was reduced by half. Distilled water was added to the solution, and this process of dilution was repeated three times. Finally, the solution was brought up to the desired volume using distilled water and stored in a 250-mL volumetric flask. The next step involved determining the metal content of this solution, which contains the dissolved ash. This was done using a flame atomic absorption spectrophotometer. The percentage of metals in these solutions was then calculated. These percentage values were subsequently converted to represent the metal content in the corresponding oils, which includes new, used, and recycled lubricating oils. This procedure allows for the quantitative analysis of the metal content in the different types of lubricating oils, providing important information for quality control and assessment of their condition.

12.8.8 Sulfur content

Sulfur content is another most important parameter for conversion of good quality lube oil. In most of the cases, sulfur creates the corrosion problems in the processing equipment as well as engines. High-quality lube oils, especially those designed for modern engines, are formulated to have very low sulfur content, often well below the 2–500 ppm range (Hoque, 2019; Mikulski et al., 2022). On the other hand, used lube oils having sulfur content 1000 ppm or more due to long run time.

12.8.9 Specific gravity

Specific gravity is defined as the ratio between the mass of a fixed volume of a substance and the mass of the same volume of water (Welty et al., 2020). The value is substance-specific and varies depending on the temperature. The specific gravity of a substance is significantly influenced by its chemical composition, particularly in the context of lubricating oil. Specifically, an increase in the concentration of cyclic compounds in lubricating oil leads to a higher specific gravity, while an increase in saturated compounds results in a lower specific gravity.

12.8.10 Carbon residue

Carbon residue refers to the amount of coke forming residue that remains after the thermal disintegration of engine oil in an insufficient amount of air (Kasar et al., 2020). The carbon residue test is used to assess the engine oil's propensity to deposit carbonaceous material in internal combustion engines. It is worth noting that the carbon residue value of engine/motor lubricating oil provides an indication of the amount of carbon deposits the oil may form in an engine's combustion chamber. For instance, the inclusion of ash-forming detergent additives in engine oil can increase the carbon residue value and generally reduce the oil's tendency to create deposits (Jafari et al., 2014).

12.9 SOCIOECONOMIC FACTOR

The disposal of waste oil has both advantages and disadvantages. Despite the challenges of managing hazardous waste, recycling waste oil offers numerous benefits for individuals, communities, and the environment. While residues from recycling processes can be hazardous, they can be effectively managed by employing appropriate handling methods, with secure landfill disposal as a last resort.

Interestingly, underdeveloped countries have a unique advantage when it comes to waste oil recycling. They can reduce energy intensity (specific energy consumption for recycling) by using manual labor for various operations. These manual tasks include management barrels and containers, handling cargo, labeling, book keeping, and other organizational measures, which account a significant portion of the energy used in recycling processes. Additionally, running filter presses are adjusted by manual labors to certain modifications of fabricated designs.

Overall, recycling operations create a wide range of employment opportunities, significantly benefiting local economies and the workforce. This highlights the advantages of recycling waste oil, particularly in underdeveloped countries (Hassanpour, 2020).

12.10 EMPLOYMENT GENERATION

The re-refining industry presents significant potential for job creation and plays a crucial role in conserving rapidly depleting natural petroleum resources. Governments, especially in developing countries, should actively encourage and support this industry. They can provide financial assistance through measures such as subsidies, tax rebates, or other fiscal incentives to promote environmentally friendly waste oil management (Qadir et al., 2021; Dulal et al., 2015). In many developing countries, environmental rules and regulations related to pollution control, covering aspects like air, water, and soil, are already in place.

Local institutional and regulatory requirements are vital factors in selecting the most suitable technology for waste oil recycling. When choosing a specific technology, a careful assessment of its advantages and disadvantages must be conducted, considering all relevant aspects. Regardless of the complexity of the technologies employed, a significant portion of recycling activities is common across all methods. These common activities include:

- Procurement of used oil
- Transport through the recycling plant
- Storage of used oil before reprocessing
- Stage-wise transport to the processing equipment
- Periodic maintenance of reactors, pipelines, motors, and pumps.

To illustrate the significance of waste oil management, consider India used oil import from Singapore, Malaysia, and Germany. That used oil is blended into different oils for further usage. This highlights the significant potential for waste oil recycling and the role it can play in the sustainable management of resources.

12.11 CONCLUSION

The arrangement and layout of operational units in an industry depend on the quality of the product oil and the level of contaminants in used oil. Recycling technologies for used lubricants can be expensive but offer several advantages, such as reducing hazardous waste, achieving oil quality similar to crude oil, and producing minimal by-products and pollutants. Emerging technologies like plasma technology show promise in removing pollutants and reducing the toxicity of by-products, contributing to sustainable development.

Combined technology processes are common in developed countries but less accessible in developing nations. While they produce lower pollutant levels, they require advanced and costly equipment. Conventional treatment methods have improved but still pose environmental risks. Solvent extraction is an environmentally controllable method but has higher costs and lower yields compared to conventional methods. VD and hydrogenation techniques also face cost challenges, making the quest for an efficient, eco-friendly, and cost-effective treatment technique a significant challenge.

An analysis of used lube oil shows that many key parameters (calorific value, pour point, kinematic viscosity, and flash point) are need to consider before further processing. This suggests that lube oil can be blended with additives at specific ratios, depending on its intended use as a substitute. The study highlights the following key research:

- Improper disposal of used lubricating oil can lead to soil and water pollution, potentially harming the ecosystem.
- Recycling of used lube oil can effectively replace base oil that offering economic value.
- Used lube oil is a cost-effective source of energy compared to other alternatives.
- However, this practice is ideal because it generates a lot of employment opportunities, which reduce the high demand of lube oil generation in the refinery.

REFERENCES

Al Zubaidi, Isam, and Adil Al Tamimi. "Soil remediation from waste lubricating oil." *Environmental Technology & Innovation* 9 (2018): 151–159.

Al-Doury, Muzher Mahdi Ibrahem. "Treatment of oily sludge using solvent extraction." *Petroleum Science and Technology* 37, no. 2 (2019): 190–196.

Ali, Imran, Anastasia Kucherova, Nariman Memetov, Tatiana Pasko, Kirill Ovchinnikov, Vladimir Pershin, Denis Kuznetsov, Evgeny Galunin, Vladimir Grachev, and Alexey Tkachev. "Advances in carbon nanomaterials as lubricants modifiers." *Journal of Molecular Liquids* 279 (2019): 251–266.

AlSawaftah, Nour, Waad Abuwatfa, Naif Darwish, and Ghaleb Husseini. "A comprehensive review on membrane fouling: Mathematical modelling, prediction, diagnosis, and mitigation." *Water* 13, no. 9 (2021): 1327.

Arpa, Orhan, Recep Yumrutas, and Ayhan Demirbas. "Production of diesel-like fuel from waste engine oil by pyrolitic distillation." *Applied Energy* 87, no. 1 (2010): 122–127.

Audibert, Francois. *Waste Engine Oils: Rerefining and Energy Recovery*. Netherlands: Elsevier (2011).

Ayeni, Augustine O., Sarah Tagwai, Deima T. Dick, Oluranti Agboola, Ayodeji A. Ayoola, Damilola E. Babatunde, and Babalola A. Oni. "A two-stage coupling process for the recovery of base oils from used lubricating oils." In *IOP Conference Series: Earth and Environmental Science* (vol. 655, no. 1, p. 012029). United Kingdom, Covenant University Ota Ogun State, Nigeria: IOP Publishing (2021).

ben Hammouda, Samia, Zhi Chen, Chunjiang An, and Kenneth Lee. "Recent advances in developing cellulosic sorbent materials for oil spill cleanup: A state-of-the-art review." *Journal of Cleaner Production* 311 (2021): 127630.

Boehm, Paul D. "(pp. 313–337). United States: Academic Press (1964).

Boyde, Steve. "Green lubricants. Environmental benefits and impacts of lubrication." *Green Chemistry* 4, no. 4 (2002): 293–307.

Cao, Yuhe, Feng Yan, Jianxin Li, Xiaoping Liang, and Benqiao He. "Used lubricating oil recycling using a membrane filtration: Analysis of efficiency, structural and composing." *Desalination and Water Treatment* 11, no. 1–3 (2009): 73–80.

Cooney, T. I., F. Cardona, and T. Tran-Cong. "Kinetics of in situ epoxidation of hemp oil under heterogeneous reaction conditions: An overview with preliminary results." In *Proceedings of the 1st International Postgraduate Conference on Engineering, Designing and Developing the Built Environment for Sustainable Wellbeing (eddBE 2011)*. University of Southern Queensland, Australia (2011).

Demirbas, Ayhen, Mohammed A. Baluabaid, Mohamad Kabli, and Waqar Ahmad. "Diesel fuel from waste lubricating oil by pyrolitic distillation." *Petroleum Science and Technology* 33, no. 2 (2015): 129–138.

Dmitrieva, Evgenia S., Tatyana S. Anokhina, Eduard G. Novitsky, Vladimir V. Volkov, Ilya L. Borisov, and Alexey V. Volkov. "Polymeric membranes for oil-water separation: A review." *Polymers* 14, no. 5 (2022): 980.

Dulal, Hari Bansha, Rajendra Dulal, and Pramod Kumar Yadav. "Delivering green economy in Asia: The role of fiscal instruments." *Futures* 73 (2015): 61–77.

Englehardt, James D. "Pollution prevention technologies: A review and classification." *Journal of Hazardous Materials* 35, no. 1 (1993): 119–150.

Eshaghi, Soroush, and Farzad Hamrang. "An innovative techno-economic analysis for the selection of an integrated ejector system in the flare gas recovery of a refinery plant." *Energy* 228 (2021): 120594.

Fonseca, Jhessica Marchini, Joel Gustavo Teleken, Vitor de Cinque Almeida, and Camila da Silva. "Biodiesel from waste frying oils: Methods of production and purification." *Energy Conversion and Management* 184 (2019): 205–218.

Gaber, Seleem E., Mahmoud S. Rizk, and Mohamed M. Yehia. "Extraction of certain heavy metals from sewage sludge using different types of acids." *Biokemistri* 23, no. 1 (2011): 41–48.

Ghalme, Sachin, and Yogesh J. Bhalerao. "Application of nanoparticles as additive for lubricant nano-materials in tribology." *Recent Patents on Materials Science* 10, no. 2 (2017): 88–96.

Gulzar, Mubashir, Haji Hassan Masjuki, Md Abul Kalam, Mahendran Varman, Nurin Wahidah Mohd Zulkifli, Riaz Ahmad Mufti, and Rehan Zahid. "Tribological performance of nanoparticles as lubricating oil additives." *Journal of Nanoparticle Research* 18 (2016): 1–25.

Hall, Dorothy W., Joseph A. Sandrin, and Rhonda E. McBride. "An overview of solvent extraction treatment technologies." *Environmental Progress* 9, no. 2 (1990): 98–105.

Hamad, Ahmad, Essam Al-Zubaidy, and Muhammad E. Fayed. "Assessment of used motor oil recycling opportunities in the United Arab Emirates." *Journal of King Saud University-Engineering Sciences* 16, no. 2 (2004): 215–227.

Hamawand, Ihsan, Talal Yusaf, and Sardasht Rafat. "Recycling of waste engine oils using a new washing agent." *Energies* 6, no. 2 (2013): 1023–1049.

Hassanpour, Malek. "Techno-economic assessment of recycling acidic sludge project of reprocessing industries to value-added gaseous products using Plasmatron." *Environmental Health Engineering and Management Journal* 7, no. 3 (2020): 183–192.

Henderson, Marc Simon. *The Development of a Supercritical Extraction Method to Extract Zero Valent Gold.* PhD diss., School of Chemical and Metallurgical Engineering, Faculty of Engineering and the Built Environment, University of the Witwatersrand, Johannesburg, South Africa (2022).

Höök, Mikael, Robert Hirsch, and Kjell Aleklett. "Giant oil field decline rates and their influence on world oil production." *Energy Policy* 37, no. 6 (2009): 2262–2272.

Hoque, Rezaul. *Background Analysis and Development of a Lubricating Oil Management System for Bangladesh.* Bangladesh University of Engineering and Technology, Central Library, Bangladesh (2019).

Hu, Guangji, Jianbing Li, and Guangming Zeng. "Recent development in the treatment of oily sludge from petroleum industry: A review." *Journal of Hazardous Materials* 261 (2013): 470–490.

Huang, Lijuan, Yu Wang, Liang Zhang, Yuliang Su, Zhe Zhang, and Shaoran Ren. "Influence of pressure on the flammability limits and explosion pressure of ethane/propane-air mixtures in a cylinder vessel." *Journal of Loss Prevention in the Process Industries* 74 (2022): 104638.

Jafari, Ahmad Jonidi, and Malek Hassanpour. "Analysis and comparison of used lubricants, regenerative technologies in the world." *Resources, Conservation and Recycling* 103 (2015): 179–191.

Jafari, Ahmad Jonidi, Mitra Gholam, and Mehdi Farzadkia. "A novel method for recovery of acidic sludge of used-motor oil reprocessing industries to bitumen using bentonite and SBS." *Iranian Journal of Health, Safety and Environment* 1, no. 2 (2014): 59–66.

Jagaba, Ahmad Hussaini, Shamsul Rahman Mohamed Kutty, Ibrahim Mohammed Lawal, Abdullahi Haruna Birniwa, Augustine Chioma Affam, Nura Shehu Aliyu Yaro, Abdullahi Kilaco Usman et al. "Circular economy potential and contributions of petroleum industry sludge utilization to environmental sustainability through engineered processes: A review." *Cleaner and Circular Bioeconomy* 3(2022): 100029.

Kajdas, Czeslaw. "Major pathways for used oil disposal and recycling. Part 1." *Tribotest* 7, no. 1 (2000): 61–74.

Kamal, Ashraf, and Fasihullah Khan. "Effect of extraction and adsorption on re-refining of used lubricating oil." *Oil & Gas Science and Technology-Revue de l'IFP* 64, no. 2 (2009): 191–197.

Kasar, Pamreishang, Durlubh Kumar Sharma, and Mohammed Ahmaruzzaman. "Thermal and catalytic decomposition of waste plastics and its co-processing with petroleum residue through pyrolysis process." *Journal of Cleaner Production* 265 (2020): 121639.

Khan, Aftab Hasnat, Nadia Batool, and Abdul Aziz. "Recovery of base oil from used oil through solvent extraction followed by adsorption: A review." *Environmental Protection Research* 3, no. 2 (2023): 319–340.

Kupareva, Antonina, Päivi Mäki-Arvela, and Dmitry Yu Murzin. "Technology for rerefining used lube oils applied in Europe: A review." *Journal of Chemical Technology & Biotechnology* 88, no. 10 (2013): 1780–1793.

Lam, Su Shiung, Wan Adibah Wan Mahari, Ahmad Jusoh, Cheng Tung Chong, Chern Leing Lee, and Howard A. Chase. "Pyrolysis using microwave absorbents as reaction bed: An improved approach to transform used frying oil into biofuel product with desirable properties." *Journal of Cleaner Production* 147 (2017): 263–272.

Liu, Hongbo, Hang Xiao, Bo Fu, and He Liu. "Feasibility of sludge deep-dewatering with sawdust conditioning for incineration disposal without energy input." *Chemical Engineering Journal* 313 (2017): 655–662.

Madanhire, Ignatio, and Charles Mbohwa. "Recycling of used oil." *Mitigating Environmental Impact of Petroleum Lubricants.* Germany: Springer International Publishing (2016): 127–148.

Mang, Theo. "Disposal of used lubricating oils." *Lubricants and Lubrication.* Germany: Wiley (2017): 237–248.

Martins, João Poças "The extraction– flocculation re-refining lubricating oil process using ternary organic solvents." *Industrial & Engineering Chemistry Research* 36, no. 9 (1997): 3854–3858.

Mekonnen, Hayalu Andargachew, and Addis Ababa. *Recycling of Used Lubricating Oil Using Acid-Clay Treatment Process.* Ethiopia: Addis Ababa (2014).

Mensah-Brown, Henry. Re-refining and recycling of used lubricating oil." ARPN Journal of Engineering and Applied Sciences 10, no. 2 (2015): 797–801.

Mikulski, Maciej, Jacek Hunicz, Kamil Duda, Paweł Kazimierski, Tomasz Suchocki, and Arkadiusz Rybak. "Tyre pyrolytic oil fuel blends in a modern compression ignition engine: A comprehensive combustion and emissions analysis." *Fuel* 320 (2022): 123869.

Mohammed, Rafie R., Inaam AR Ibrahim, Alladdin H. Taha, and Gordon McKay. "Waste lubricating oil treatment by extraction and adsorption." *Chemical Engineering Journal* 220 (2013): 343–351.

Nazem, Mohammad Ali, and Omid Tavakoli. "Bio-oil production from refinery oily sludge using hydrothermal liquefaction technology." *The Journal of Supercritical Fluids* 127 (2017): 33–40.

Nezhdbahadori, Farzad, Mohamad Ali Abdoli, Majid Baghdadi, and Fereydoun Ghazban. "A comparative study on the efficiency of polar and non-polar solvents in oil sludge recovery using solvent extraction." *Environmental Monitoring and Assessment* 190 (2018): 1–9.

Nolan, Dennis P. *Handbook of Fire and Explosion Protection Engineering Principles: For Oil, Gas, Chemical and Related Facilities.* William Andrew (2014).

Okeke, Emmanuel Sunday, Charles Obinwanne Okoye, Timothy Prince Chidike Ezeorba, Guanghua Mao, Yao Chen, Hai Xu, Chang Song, Weiwei Feng, and Xiangyang Wu "Emerging bio-dispersant and bioremediation technologies as environmentally friendly management responses toward marine oil spill: A comprehensive review." *Journal of Environmental Management* 322 (2022): 116123.

Oladimeji, Temitayo E., Jacob A. Sonibare, James A. Omoleye, Abiola A. Adegbola, and Hilary I. Okagbue. "Data on the treatment of used lubricating oil from two different sources using solvent extraction and adsorption." *Data in Brief* 19 (2018): 2240–2252.

Pelemo, Josiah, Freddie L. Inambao, and Emmanuel Idoko Onuh. "Potential of used cooking oil as feedstock for hydroprocessing into hydrogenation derived renewable diesel: A review." *IJERT* 13 (2020): 500–519.

Pinheiro Pires, Anamaria Paiva, Jesus Arauzo, Isabel Fonts, Marcelo E. Domine, Alberto Fernandez Arroyo, Marta Estrella Garcia-Perez, Jorge Montoya, Farid Chejne, Peter Pfromm, and Manuel Garcia-Perez. "Challenges and opportunities for bio-oil refining: A review." *Energy & Fuels* 33, no. 6 (2019): 4683–4720.

Pirro, Don M., Martin Webster, and Ekkehard Daschner. *Lubrication Fundamentals, Revised and Expanded*. United States: CRC Press (2016).

Prince, Ruth. J. "Base oils from petroleum." In *Chemistry and Technology of Lubricants* (pp. 1–31). Boston, MA: Springer US (1992).

Qadir, Sikandar Abdul, Hessah Al-Motairi, Furqan Tahir, and Luluwah Al-Fagih. "Incentives and strategies for financing the renewable energy transition: A review." *Energy Reports* 7 (2021): 3590–3606.

Rincon, Jesusa, Pablo Canizares, and Maria Teresa Garcia. "Waste oil recycling using mixtures of polar solvents." *Industrial & Engineering Chemistry Research* 44, no. 20 (2005): 7854–7859.

Sarkar, Sayantan, Deepshikha Datta, K. S. Deepak, Bikash Kumar Mondal, and Bimal Das. "Comprehensive investigation of various re-refining technologies of used lubricating oil: A review." *Journal of Material Cycles and Waste Management* 25, no. 4 (2023): 1–31.

Schmidt, Gerald K., and Eugene J. Forster. "Modern refining for today's fuels and lubricants, Part 2. No. 861176." SAE Technical Paper (1986).

Shahnazar, Sheida, Samira Bagheri, and Sharifah Bee Abd Hamid. "Enhancing lubricant properties by nanoparticle additives." *International Journal of Hydrogen Energy* 41, no. 4 (2016): 3153–3170.

Taiwo, Elijah A., and John A. Otolorin. "Oil recovery from petroleum sludge by solvent extraction." *Petroleum Science and Technology* 27, no. 8 (2009): 836–844.

Thorat, Saket, Jidnesh Arekar, and Sonali Dhokpande. "Re-refining of used lubricating oil and to determine rheological characteristics: A literature review." *International Research Journal of Engineering and Technology* 8, no. 4 (2021): 3567–3572.

Udonne, Joseph D. "A comparative study of recycling of used lubrication oils using distillation, acid and activated charcoal with clay methods." *Journal of Petroleum and Gas Engineering* 2, no. 2 (2011): 12–19.

Umar, Abubakar, Ismail Bin Mohd Saaid, Aliyu Adebayo Sulaimon, and Rashidah Bint Mohd Pilus. "A review of petroleum emulsions and recent progress on water-in-crude oil emulsions stabilized by natural surfactants and solids." *Journal of Petroleum Science and Engineering* 165 (2018): 673–690.

Unnisa, Syeda Azeem, and Malek Hassanpour. "Development circumstances of four recycling industries (used motor oil, acidic sludge, plastic wastes and blown bitumen) in the world." *Renewable and Sustainable Energy Reviews* 72 (2017): 605–624.

Vecchione, Raffaele, Vincenzo Quagliariello, Dominic Calabria, Vincenzo Calcagno, Elisa De Luca, Rosario V. Iaffaioli, and Paolo A. Netti. "Curcumin bioavailability from oil in water nano-emulsions: In vitro and in vivo study on the dimensional, compositional and interactional dependence." *Journal of Controlled Release* 233 (2016): 88–100.

Welty, James, Gregory L. Rorrer, and David G. Foster. *Fundamentals of Momentum, Heat, and Mass Transfer*. United States: John Wiley & Sons (2020).

Widodo, Setyo, Danu Ariono, Khoiruddin, Ahmad N. Hakim, and I. Gede Wenten. "Recent advances in waste lube oils processing technologies." *Environmental Progress & Sustainable Energy* 37, no. 6 (2018): 1867–1881.

Yang, Chia Yu. *The Feasibility Studies on Sonochemical Processes for Treating Used Oil: Toxin Reduction for Eliminating Recycle Interference*. California, USA: University of Southern California-Faculty of the Graduate School (2008).

Zare-Shahabadi, Abolfazl, Ali Shokuhfar, and Salman Ebrahimi-Nejad. "Preparation and rheological characterization of asphalt binders reinforced with layered silicate nanoparticles." *Construction and Building Materials* 24, no. 7 (2010): 1239–1244.

Chapter 13

New tendencies in lubrication – selected case studies

Andreas Rosenkranz and Upendra Maurya

13.1 INTRODUCTION

Friction and wear are daily-life phenomena that notably impact many branches of our society including mechanical aspects of everyday objects (cars, watches, and brakes, among many others) [1–3] and more sophisticated areas such as medical devices (dental implants or artificial joints) [4]. Based on very fundamental laws of physics, these phenomena govern the interaction between surfaces in contact considering an applied normal load and a relative movement.

Friction, which can be defined as the resistance against the sliding motion, is a familiar concept, which we all experience in our daily lives. Whether it is the grip between shoe soles and pavement, which allows us the properly walk, or the friction between moving mechanical parts, its effects are often crucial for the overall functionality of many mechanical systems. While friction is often perceived as negative characteristics, which should be reduced or even avoided, friction is essential in enabling motion and stability. These attributes become observable when thinking about the traction experienced between the tires and the road or the precision of gears meshing in a mechanical watch [1,2,5].

When considering wear, this aspect typically refers to the gradual loss of material from the surfaces of interacting bodies due to frictional interaction and rubbing. Many different wear types/mechanisms are known such as including abrasion, adhesion, erosion, and oxidative wear. Wear is generally considered as something detrimental, which negatively affects the performance of mechanical systems and leads to catastrophic failures over time. Although often referred to mechanical systems, wear can be found in many different areas such as the erosion of coastal landscapes by ocean currents, or the downgraded absorption efficiency of solar cells by erosive degradation due to air contamination [5].

The detailed understanding of the involved processes and mechanisms in friction and wear is essential to develop solutions to precisely manipulate and tailor friction and wear. In this regard, it can be desirable to

 DOI: 10.1201/9781003642848-13

reduce or increase friction depending on the respective application, while the optimization of wear is generally considered to decrease the material loss and to suppress the main wear mechanisms. Through advances in materials technology, surface engineering, and lubrication strategies, researchers continually strive to optimize friction and wear, thereby enhancing the performance, durability, and efficiency of mechanical systems [1,2,4,6–8].

Historically, the easiest way to reduce friction and wear is to use a liquid lubricant (oil or grease). When two rubbing surfaces are in contact with each other under relative motion, the liquid is placed in the tribological contact with the overall purpose to separate them. This, in turn, avoids direct contact between the involved rubbing surfaces, which reduces friction and wear [9]. In this regard, insufficient lubrication can cause severe damages, thus increasing friction and the resulting energy consumption as well as leading to accelerated wear, which finally results in the failure of the entire component [2,4]. The proper function of the liquid lubricants (oils or greases) depends on several operational parameters such as temperature, viscosity, applied load, and quantity of the available lubricant and often requires complex additive systems to improve their performance under more severe conditions (higher loads and lower velocities) [2].

Although liquid lubricants are generally used in many industrial branches, they have a limited effective working range. The involved additive systems make use of environmentally critical elements (Zn, P, S, and Cr, among others), which has led to legal restrictions regarding their toxicity and difficulties, where conditions such as load, speed, temperature, viscosity, and design of the lubricating system are important for their best performance [2].

Consequently, the tribological community always looks for alternative and more sustainable solutions to optimize the resulting friction and wear performances. In this regard, environmentally friendly bio-lubricants [water, ethanol, glycerol, and ionic liquids (ILs), among others] come into play. Moreover, solid lubrication has become an interesting choice from the perspective of efficiency/sustainability/performance since this approach offers an excellent tribological properties (low friction and excellent wear resistance) under severe conditions such as high temperature and pressures, radiation, vacuum, or ultra-clean conditions, for which traditional lubricants such as oils and greases generally tend to fail [1,4,6,10].

Based on this short introduction and motivation, it becomes evident that the tribological community is urgently looking for alternative and more eco-friendly lubrication concepts. To shed some light on this exciting and quickly developing topic, some promising case studies have been carefully selected to present the latest trends in the scientific community.

13.2 EMERGING TRENDS IN SOLID LUBRICATION

13.2.1 Single-component coatings

With respect to solid lubrication, different material classes can be used to pursue this overall objective. In this regard, soft metals such as In, Au, and Ag, among others, are suitable due to their low hardness and shear resistance. As a polymeric counterpart, polytetrafluoroethylene (PTFE) has been shown to induce beneficial tribological properties due to its capability to form tribofilms. In the last two decades since the discovery of graphene, 2D materials have moved into the focus of applied tribological research including solid lubrication [7,10,11].

From the chemical and structural point of view, 2D materials can be classified into several sub-categories such as carbon-based materials (graphite, graphene, and its derivatives), transition metal dichalcogenides (transition metal dichalcogenides (TMDCs): MoS_2, WS_2, $MoSe_2$, and WSe_2), MXenes (most studied family member connects to $Ti_3C_2T_x$), nitrides (h-BN and g-C_3N_4), black phosphorus (BP), and metal and covalent organic frameworks (MOFs and coefficient of friction (COFs)) [1,6,10,12–14]. Since the discovery of graphene and TMDCs, 2D materials have moved toward the focus of solid lubrication owing to their outstanding physical, chemical, and mechanical properties [1,6,10,12–14].

For example, graphene significantly lowers friction and enhances wear performance because of its two-dimensional layered structure, which has low shear resistance due to the influence of van der Waals interactions and the possible formation of sacrificial tribofilms [15–18]. Additionally, defects and oxygen-containing surface groups in graphene oxide (GO) can create structural disorder and boost interfacial hydrogen bonding energy, thereby improving friction [15–18].

TMDCs such as MoS_2 have been widely studied under dry sliding conditions [19]. The enhanced frictional performance is induced by the easy shearing of basal planes due to their week interlayer bonding characteristics and lamellar structure. Moreover, the ionic bond between Mo and S provides a rather high strength, which implies that this material can partially resist deformation and mechanical rubbing [7,19]. However, the tribological properties of TMDCs notably depend on the operational environment. For instance, under humid (>50%) and elevated temperature (>300°C) conditions, MoS_2 tends to show a rather poor friction and wear performance due to oxidation [7,19]. In contrast, GO and hexagonal boron nitride coatings show decreased friction with increasing humidity because of a saturation of dangling bonds, thus improving their resulting tribological performance. Moreover, in N_2- or O_2-rich environments, both h-BN and MoS_2 coatings tend to lose their tribological properties due to oxidation, while graphene and graphite loses lubricating properties in vacuum condition owing to water desorption [15–18,20].

In the last 5 years, MXene nano-sheets have become the focus of applied tribological research. In this regard, MXenes, discovered in 2011, are layered, 2D carbides, nitrides, and carbonitrides of early transition metals with nanoscopic thicknesses [21,22]. MXenes can be abbreviated by the chemical formula $M_{n+1}X_nT_x$ (n=1 to 4), which describes alternating layers of early transition metals (M: groups 3–6 of the periodic table) interspersed with layers of carbon/nitrogen (X) with the outer surfaces of transition metals [21–25]. The crystal structure and chemical formula of MXenes are derived from their 3D crystalline MAX precursors [24,26–28]. MAX phases are chemically denoted by $M_{n+1}AX_n$ (n=1 to 4), for which the M-X layers found in the MXenes are interspersed by layers of group A elements (mainly group 13–16 of the periodic table) [21,22,29,30]. MXenes are synthesized by selectively etching the layers of group A elements. After their removal, the surface terminations (T_x) occupy the binding sites in the M layers previously attached to the group A elements [27,28] (Figure 13.1).

In the last couple of years, researchers have started to explore MXenes' tribological properties due to their theoretically predicted low shear strength between adjacent layers owing to rather weak interlayer bonding characteristics [14,31,32]. In this sense, MXenes' use as lubricant additives or reinforcement phase in polymeric and/or metallic composites has been explored [14,31,32]. Regarding the MXene family, $Ti_3C_2T_x$ has shown to be a promising solid lubricant. In 2019 [33,34], the first studies on the solid lubrication performance of $Ti_3C_2T_x$ have been published. Very recently, multilayer $Ti_3C_2T_x$ coatings with a thickness of 100 nm have shown a 6-fold friction reduction combined with an outstanding wear resistance. The authors reported an extraordinary performance of MXene coatings with low COFs for 100000 sliding cycles, which was traced back to the formation of a beneficial tribolayers [35].

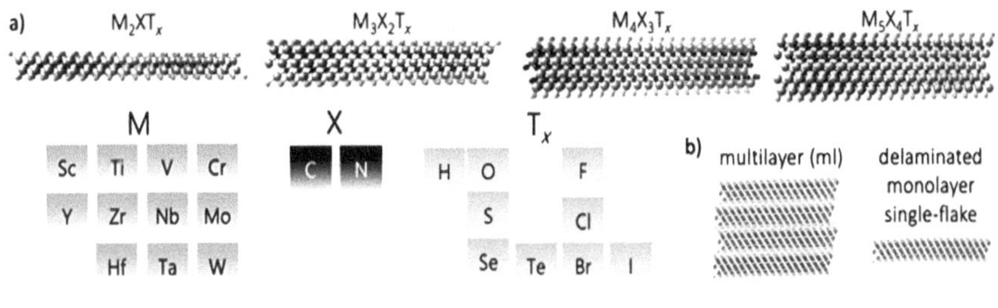

Figure 13.1 Chemical composition and structure of synthesized MXenes to date. (a) Variations in the monolayer thicknesses of MXenes (the top row illustrates representative structures of M_2XT_x, $M_3X_2T_x$, M_4X_3Tx, and M_5X_4Tx, each with different monolayer thicknesses where n ranges from 1 to 4), along with their elemental makeup concerning M, X, and T (bottom row). (b) Illustrations of multilayer and delaminated MXenes (also referred to as single-flake or monolayer MXenes). (Reproduced from Ref. [31].)

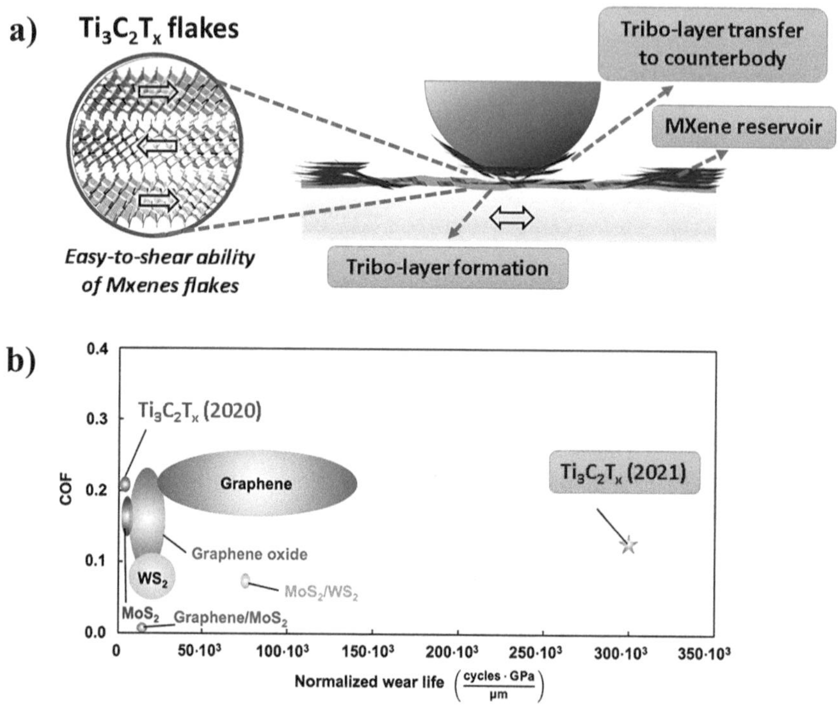

Figure 13.2 (a) Schematic of the underlying wear mechanism between the MXene coating and the counterpart. (b) Normalized service life and COF of MXenes compared to state-of-the-art 2D materials when used as solid lubricants under comparable conditions. (Adapted from Ref. [35].)

The useful coating's lifetime was compared with other state-of-the-art 2D nanomaterials (graphene and MoS_2), thus underscoring that MXene coatings are capable to outperform all state-of-the art solid lubricants regarding the useful lifetime, which significantly contributes toward their **durability** and **longevity** [35].

As can be seen in Figure 13.2, MXenes are highly prospective for their wear resistance, but their major drawback until now relates to their rather high coefficients of friction, which requires further experimental exploration and optimization.

13.2.2 Synergy between single-component coatings and surface textures

Irrespective of the used nanomaterial, single-component coatings always come with some material-specific shortcomings and drawbacks, which may relate to a strong dependency of the observed frictional performance on the tribological testing parameters and adjusted environmental conditions. In this regard, graphene and MoS_2 work best regarding the frictional performance for very different humidity levels. Moreover, these materials come with a rather limited wear resistance, since upon the first sign of wear, coatings made of graphene- and MoS_2-based materials tend to wear off quickly, thus leading to catastrophic wear. Although MXene nano-sheets help to overcome the problems with the existing wear resistance, this material class shows deficits related to the resulting COFs, which asks for further optimization and improvement. This implies that all classes of 2D materials come with their material-specific problems. Therefore, the research community is continuously looking for alternative solutions to further optimize the respective friction and wear performance at the same time. In this regard, the combination of solid lubricant coatings based on 2D materials and surface texturing appears to be a very promising pathway since this approach is expected to induce synergistic effect with respect to friction and wear at the same time.

In this regard, surface texturing connects with the intentional modification of the resulting surface topography (metals, alloys, polymers, and even ceramics can be precisely modified) to generate multi-functional surfaces. From a tribological point of view, surface texturing is beneficial in many different ways. Designed surface textures can create efficient lubricant reservoirs, which help to store the lubricating material (liquid or solid state), thus providing the possibility to replenish the tribological contact with fresh, lubricious material. Moreover, surface textures induce the ability to trap wear particles, thus removing abrasive particles from the tribological interface. Both processes are highly efficient to enhance the resulting tribological performance, as they reduce friction and wear, thereby potentially extending their lifespan.

In this regard, several mechanisms have been identified to explain how surface textures contribute to an improved tribological behavior. On the one hand, they help to enhance the load-bearing capacity by increasing the hydrodynamic pressure under lubricated conditions, aiding in separating and supporting applied loads. On the other hand, surface textures help to intentionally reduce the real area of contact, which inherently reduces the resulting frictional force according to the theory of Bowden and Tabor. Apart from that, the lubricant reservoir effect and the trapping of wear debris are essential in minimizing wear under dry conditions. Upon reducing wear, the abrasive component of the COF can be reduced as well, which goes hand in hand with reduced friction.

In dry contacts, surface textures additionally aid in mitigating friction or wear by decreasing adhesion and the effective contact area, while also trapping wear debris well as the reduction of static friction by reducing the real area of contact. Regardless of the tribological conditions, the efficacy of surface textures and their impact on the tribological performance largely hinge on the testing conditions and alterations in material microstructure and surface chemistry caused during texture fabrication and rubbing.

Surface textures come also with specific drawbacks and shortcomings. Especially under high load conditions, they tend to wear away quickly. Once starting to initiate wear processes, they typically wear off, completely thus leading to catastrophic wear. Furthermore, wear can exacerbate issues by generating abrasive wear debris, thereby accelerating wear and associated degradation processes. These aspects and processes can negatively impact the resulting friction and wear performance, thus asking for further optimization and improvement.

A potential solution to overcome these drawbacks relates to the combination of surface textures with solid lubricant coatings, which represents a promising strategy to further enhance tribological behavior under more severe contact conditions. This combination offers the advantage of providing low friction and low wear at the same time, significantly contributing to the durability and efficiency of mechanical components.

From the experimental point of view, many successful examples of combining surface textures with solid lubricants can be found in the literature [36]. In this regard, several combinations such as graphene and texturing as well as carbon nanotubes have attracted much scientific attention, demonstrating the notable potential of the overall approach.

With regard to MXene nano-sheets, as potentially wear-resistant coatings to be combined with surface texturing, research activities have been recently initiated. Xue et al. pioneered the integration of surface textures and MXenes in both lubricated and dry conditions [37]. They created grooves with a density of 33.3% on CSS-42L bearing steel disks using electrical discharge machining. Subsequently, they infused complex Sn-Ag-Cu or Sn-Ag-Cu/$Ti_3C_2T_x$ powder mixtures into the grooves via high-temperature vacuum infiltration. The frictional behavior, vibration, and noise characteristics were assessed using a reciprocating Si_3N_4 ball-on-disk setup. Under dry conditions at room temperature, the Sn-Ag-Cu/$Ti_3C_2T_x$ composite coating on the grooved surface exhibited reduced noise emission, while the lowest COF was around 0.33. This reduction was attributed to a wave-damping effect of the Sn-Ag-Cu/$Ti_3C_2T_x$ coating, shifting the wear mechanisms toward low tribochemical wear. This was facilitated by the immiscibility of Sn-Ag-Cu with the tin-based alloy substrate and the formation of oxides and intermetallic compounds, providing self-healing, high load-bearing capacity, and wear resistance. Additionally, $Ti_3C_2T_x$ induced crystallinity and a “nano-nail effect”, reducing residual strains on worn surfaces. The combination of grooved textures and Sn-Ag-Cu/$Ti_3C_2T_x$ demonstrated synergistic effects, forming uniform profiles with excellent load-bearing capacity, thereby minimizing self-excited vibration and friction-induced noise.

In rotational dry sliding experiments, Xue et al. observed that the tribological performance was further enhanced at temperatures up to 450°C. This enhancement was attributed to improved crystallinity, microhardness, residual strain, and grain size of the Ag-Cu/ $Ti_3C_2T_x$, which facilitated the formation of enamel layers. Under oil-lubricated conditions, Xue et al. found that surface textures introduced a micro-hydrodynamic effect, resulting in a slight reduction in COF (0.11) compared to untextured references (0.22). When combined with Sn-Ag-Cu/ $Ti_3C_2T_x$ coatings, synergistic effects lowered the resulting COFs to around 0.06, while simultaneously reducing frictional scattering. This was attributed to the good wettability of Sn-Ag-Cu/ $Ti_3C_2T_x$, promoting oil adsorption and fluid film formation during tribological testing. Subsequent studies confirmed similar results and mechanisms in conformal oil-lubricated contacts, simulating journal bearings. The friction-reducing ability was particularly evident at lower rotational speeds and higher loads, where the Sn-Ag-Cu/MXene coating prevented oil film rupture.

Peng et al. utilized a design of experiments and reciprocating Si_3N_4 ball-on-disk trials under both dry and oil-lubricated conditions [38]. They employed the response surface method to optimize micro-channels on M50 bearing steel, depositing a Sn-Ag-Cu/MXene composite coating. These bioinspired channels facilitated the extrusion of solid lubricant to the worn surface during tribological testing, preventing direct contact between rubbing surfaces. The addition of MXenes reinforced the shear resistance of the Sn-Ag-Cu solid lubricant film, synergizing with oil and surface textures to reduce friction and protect against wear.

In a study related to other MXene types, Huang et al. illustrated the synergistic effects of Sn-Ag-Cu/Nb2C composite coatings when combined with bioinspired rhombic textures on M50 bearing steel [39]. This combination facilitated the spreading of the solid lubricant film into a uniform and dense layer, resulting in lower and more stable COFs and significantly reduced wear compared to samples that were solely textured, coated with Sn-Ag-Cu, or left uncoated and untextured.

On the other hand, Zhang et al. investigated how textured surfaces could synergize with a lithium-based grease lubrication system reinforced with MXenes ($Ti_3C_2T_x$) to reduce friction and enhance lubrication efficiency [40]. They found that modifying the shape, morphology, and roughness of the textured surface led to a significant improvement in efficiency, with the multi-stage bionic texture inspired by the nacre of the shell proving most effective compared to single-stage linear groove surfaces. Additionally, when reinforcing the lithium-based grease with MXene, they observed that the lubrication mechanism persisted even after grease degradation, attributable to the action of the solid lubricant MXene. This combination of surface textures, grease, and $Ti_3C_2T_x$ demonstrated synergistic behavior, offering greater adaptability and reliability, resulting in prolonged lifespan and reduced wear rate and coefficient of friction compared to conventional non-textured or reinforced systems.

13.2.3 Multiple-component coatings

An effective strategy to overcome the aforementioned shortcomings and deficiencies of 2D materials is to combine two or more 2D nanomaterials with each other, thus forming a hybrid nanocomposite. This may maximize their structural compatibility and induce synergistic effects, thus resulting in an enhanced tribological performance, which is supposed to outperform the performance of the individual components [41]. In this context, GO has been combined with MoS_2 to form nanocomposites conferring extraordinary lubricating properties with reduced COFs (~0.09) and increased wear resistance due to the formation of protective tribofilms [42,43]. Moreover, when MoS_2 is combined with Ti and studied under humid or dry N_2 conditions, low friction, low particle generation, and high wear resistance were achieved as a consequence of tribo-chemical reactions [44]. Hybrid composite coatings have gained considerable interest in the scientific community due to their outstanding tribological performance under severe conditions, which opens up new strategies to design and optimize novel solid lubricants based upon 2D layered nanomaterials with self-lubrication ability and excellent wear resistance [10,20,45–47]. These studies have incorporated the formation of graphene-based hybrids with other 2D materials such as h-BN, MXenes, and MoS_2 into a wide range of matrices to improve the mechanical performance.

13.3 NANOLUBRICANTS

Nanolubricants have emerged as a promising advancement in tribology, offering enhanced lubrication properties that traditional lubricants cannot match. These lubricants are formulated by dispersing nanoparticles into conventional lubricant bases, resulting in improved performance in reducing friction, wear, and temperature rise in mechanical systems. The application of nanolubricants spans various industries, including automotive, aerospace, and manufacturing, making them a critical area of research in the field of tribology.

Recent studies have emphasized the significant role that nanolubricants play in enhancing thermal conductivity and heat dissipation in mechanical systems. Researchers have observed that the inclusion of nanoparticles such as graphene, carbon nanotubes (CNTs), and metal oxides (e.g., Al_2O_3 and CuO) in lubricant oils leads to a substantial increase in thermal conductivity [48]. For instance, graphene-based nanolubricants have shown a 30%–50% improvement in thermal conductivity compared to conventional lubricants. This enhancement is particularly crucial in high-temperature applications where efficient heat dissipation is required to prevent system failure.

One of the most significant advantages of nanolubricants is their ability to reduce friction and wear, which extends the lifespan of mechanical components. Recent experiments by Upendra et al. demonstrated that nanolubricants containing boehmite (AlOOH) and zinctitanate ($ZnTiO_3$) nanoparticles reduced the coefficient of friction by up to 18% compared to standard lubricants [49,50]. Additionally, the wear rate of metallic surfaces was reduced by nearly 28%. These improvements are attributed to the formation of a protective tribofilm on the contacting surfaces, which reduces direct metal-to-metal contact and minimizes wear.

Nanolubricants have been tested under extreme conditions, such as high pressure, high temperature, and high load scenarios, to evaluate their performance limits. Recent studies by Saini et al. highlighted that nanolubricants with titanium (Ti) nanoparticles exhibit superior performance under such conditions [51]. These nanoparticles provide excellent load-carrying capacity and stability at elevated temperatures, making them suitable for applications in heavy machinery and aerospace where extreme operating conditions are common. With the growing emphasis on sustainability, research has also focused on developing eco-friendly nanolubricants [52]. These green nanolubricants showed promising results in terms of biodegradability, reducing environmental impact while maintaining excellent tribological performance. This trend aligns with the global push toward sustainable industrial practices and offers a viable solution for industries looking to reduce their carbon footprint. The future of nanolubricants will likely see the development and incorporation of advanced nanomaterials with tailored properties. Materials such as MXenes, 2D nanomaterials, and hybrid nanoparticles are expected to gain attention due to their unique properties, such as high surface area, electrical conductivity, and tunable chemical reactivity. These materials could offer unprecedented control over lubrication properties, leading to even greater reductions in friction and wear. Another emerging trend is the development of smart nanolubricants that can adapt to changing operating conditions [53]. Researchers are exploring the use of stimuli-responsive nanoparticles that can alter their behavior based on temperature, pressure, or pH changes. For example, a nanolubricant that thickens under high load conditions could provide enhanced protection against wear, while thinning out during low-load operations to reduce energy consumption. This adaptability could revolutionize lubrication in dynamic environments.

The integration of artificial intelligence (AI) and machine learning (ML) into nanolubricant research is expected to play a crucial role in the future [54,55]. AI-driven models can predict the performance of nanolubricants under various conditions, optimizing formulations for specific applications. Comprehensive research is required to demonstrate the use of ML algorithms to design nanolubricants with optimized nanoparticle concentrations and base oil combinations, achieving optimal tribological performance. This approach not only accelerates the development process but also reduces the need for extensive trial-and-error experimentation. As with any emerging technology, the widespread adoption of nanolubricants will require careful consideration of regulatory and safety aspects. The potential environmental and health impacts of nanoparticles in lubricants are still being studied. Future research will need to address these concerns by developing guidelines for safe use and disposal of nanolubricants. Additionally, standardized testing protocols for evaluating the performance and safety of nanolubricants will be essential for their acceptance in various industries.

13.4 IONIC LIQUIDS

ILs are salts in a liquid state at relatively low temperatures, typically below 100°C. Composed of anions and cations, these unique fluids possess remarkable properties, such as low volatility, high thermal stability, tunable viscosity, and excellent solubility for a variety of materials. These attributes make ILs highly attractive as lubricants or lubricant additives, offering potential improvements in reducing friction, wear, and environmental impact [56,57]. The integration of ILs into lubrication systems is an emerging area of research, with applications spanning from automotive engines to space exploration.

One of the most significant observations in recent studies is the superior lubrication performance of ILs compared to conventional lubricants. Several studies has demonstrated that ILs, when used as lubricants or additives, significantly reduce the coefficient of friction and wear rate between contacting surfaces [58,59]. For example, 1-butyl-3-methylimidazolium hexafluorophosphate ([BMIM][PF6]) has been shown to reduce friction by up to 40% and wear by 50% in steel-on-steel contacts [58,59]. The unique molecular structure of ILs, particularly their ability to form stable boundary films on metal surfaces, is credited with these improvements. ILs exhibit remarkable thermal stability, making them ideal for high-temperature applications where conventional lubricants may degrade. Recent studies have shown that ILs such as tetrafluoroborate-based ILs can maintain their lubricating properties at temperatures exceeding 300°C [58]. This makes them suitable for use in environments like gas turbines and advanced automotive engines, where extreme temperatures are common. The thermal stability of ILs also translates to longer lubricant life, reducing the need for frequent replacements and lowering maintenance costs. Another notable benefit of ILs is their ability to protect metal surfaces from corrosion. Research indicates that ILs with specific anions, such as phosphate or sulfonate, provide excellent corrosion protection in harsh environments [58,59]. These ILs form a protective layer on metal surfaces, preventing the ingress of corrosive agents like water and oxygen. This property is particularly valuable in marine and offshore applications, where equipment is constantly exposed to corrosive conditions. Recent research has also focused on the compatibility of ILs with conventional lubricant additives. Studies have shown that ILs can be blended with anti-wear, anti-oxidant, and anti-corrosion additives without compromising their performance [50,58–61]. In fact, some ILs can enhance the effectiveness of these additives, leading to synergistic effects that further improve lubrication performance. This compatibility opens up opportunities for formulating hybrid lubricants that combine the best properties of ILs and traditional additives. The environmental benefits of ILs are another area of growing interest. Unlike many conventional lubricants, which are derived from petroleum and can be harmful to the environment, many ILs are biodegradable and have low toxicity. Research explored the use of bio-based ILs derived from renewable resources such as cellulose and fatty acids [62,63]. These ILs may show good lubrication properties while being environmentally benign, making them suitable for use in applications where sustainability is a priority, such as in agricultural machinery or eco-friendly vehicles.

The future of ILs in lubrication will likely see the development of tailor-made ILs with properties specifically engineered for particular applications [58,59]. Researchers are exploring the synthesis of ILs with customized cation–anion pairs to achieve desired characteristics, such as enhanced thermal stability, lower friction, or improved solubility in specific base oils. This level of customization could lead to ILs that outperform conventional lubricants in even the most demanding environments, such as space exploration or deep-sea drilling. Another emerging trend is the development of hybrid lubricants that combine ILs with other advanced materials, such as nanoparticles or polymers [50,57,60,61]. The synergy between ILs and these materials can result in lubricants with superior properties, such as enhanced load-carrying capacity, reduced friction, and improved wear resistance. Research has already demonstrated promising results with such hybrid lubricants, indicating a potential direction for future lubricant formulations.

ACKNOWLEDGMENTS

A. Rosenkranz gratefully acknowledges the financial support of ANID-Chile within the projects Fondecyt Regular 1220331 and Fondequip EQM190057 as well as and the Millennium Science Initiative Program (NCN2023_007).

REFERENCES

1. Marian, M., Berman, D., Rota, A., Jackson, R.L., Rosenkranz, A.: Layered 2D nanomaterials to tailor friction and wear in machine elements—A review. *Adv Mater Interfaces.* 9 (2022). https://doi.org/10.1002/admi.202101622
2. Holmberg, K., Erdemir, A.: The impact of tribology on energy use and CO_2 emission globally and in combustion engine and electric cars. *Tribol Int.* 135, 389–396 (2019). https://doi.org/10.1016/J.TRIBOINT.2019.03.024

3. Maurya, U., Vasu, V.: Ionic liquids as a potential sustainable green lubricant for machining in the era of industry 4.0. In: Kishor Kumar Gajrani, Arbind Prasad, Ashwani Kumar (eds.), *Advances in Sustainable Machining and Manufacturing Processes.* pp. 135–155. CRC Press, Boca Raton (2022).
4. Marian, M., Berman, D., Nečas, D., Emami, N., Ruggiero, A., Rosenkranz, A.: Roadmap for 2D materials in biotribological/biomedical applications – A review. *Adv Colloid Interface Sci.* 307, 102747 (2022). https://doi.org/10.1016/J.CIS.2022.102747
5. Costa, H.L., Schille, J., Rosenkranz, A.: Tailored surface textures to increase friction—A review. *Friction.* 10, 1285–1304 (2022). https://doi.org/10.1007/s40544-021-0589-y
6. Zhang, S., Ma, T., Erdemir, A., Li, Q.: Tribology of two-dimensional materials: From mechanisms to modulating strategies. *Mater Today.* 26, 67–86 (2019). https://doi.org/10.1016/J.MATTOD.2018.12.002
7. Rosenkranz, A., Costa, H.L., Baykara, M.Z., Martini, A.: Synergetic effects of surface texturing and solid lubricants to tailor friction and wear – A review. *Tribol Int.* 155, 106792 (2021). https://doi.org/10.1016/J.TRIBOINT.2020.106792
8. Holmberg, K., Andersson, P., Erdemir, A.: Global energy consumption due to friction in passenger cars. *Tribol Int.* 47, 221–234 (2012). https://doi.org/10.1016/j.triboint.2011.11.022
9. Spikes, H.A., Olver, A. V.: Basics of mixed lubrication. *Lubr Sci.* 16, 1–28 (2003). https://doi.org/10.1002/ls.3010160102
10. Rosenkranz, A., Liu, Y., Yang, L., Chen, L.: 2D nano-materials beyond graphene: From synthesis to tribological studies. *Appl Nanosci.* 10, 3353–3388 (2020). https://doi.org/10.1007/s13204-020-01466-z
11. Scharf, T.W., Prasad, S. V.: Solid lubricants: A review. *J Mater Sci.* 48, 511–531 (2013). https://doi.org/10.1007/s10853-012-7038-2
12. Zhang, H., Cheng, H.-M., Ye, P.: 2D nanomaterials: Beyond graphene and transition metal dichalcogenides. *Chem Soc Rev.* 47, 6009–6012 (2018). https://doi.org/10.1039/C8CS90084A
13. Spear, J.C., Ewers, B.W., Batteas, J.D.: 2D-nanomaterials for controlling friction and wear at interfaces. *Nano Today.* 10, 301–314 (2015). https://doi.org/10.1016/J.NANTOD.2015.04.003
14. Malaki, M., Varma, R.S.: Mechanotribological aspects of MXene-reinforced nanocomposites. *Adv Mater.* 32 (2020). https://doi.org/10.1002/adma.202003154
15. Penkov, O., Kim, H.-J., Kim, H.-J., Kim, D.-E.: Tribology of graphene: A review. *Int J Precis Eng Manuf.* 15, 577–585 (2014). https://doi.org/10.1007/s12541-014-0373-2
16. Liu, Y., Ge, X., Li, J.: Graphene lubrication. *Appl Mater Today.* 20, 100662 (2020). https://doi.org/10.1016/J.APMT.2020.100662
17. Geim, A.K.: Graphene: Status and prospects. *Science (1979).* 324, 1530–1534 (2009). https://doi.org/10.1126/science.1158877
18. Berman, D., Erdemir, A., Sumant, A. V.: Graphene: A new emerging lubricant. *Mater Today.* 17, 31–42 (2014). https://doi.org/10.1016/J.MATTOD.2013.12.003
19. Vazirisereshk, M.R., Martini, A., Strubbe, D.A., Baykara, M.Z.: Solid lubrication with MoS2: A review. *Lubricants.* 7, 57 (2019). https://doi.org/10.3390/lubricants7070057

20. Liu, L., Zhou, M., Jin, L., Li, L., Mo, Y., Su, G., Li, X., Zhu, H., Tian, Y.: Recent advances in friction and lubrication of graphene and other 2D materials: Mechanisms and applications. *Friction.* 7, 199–216 (2019). https://doi.org/10.1007/s40544-019-0268-4
21. Naguib, M., Kurtoglu, M., Presser, V., Lu, J., Niu, J., Heon, M., Hultman, L., Gogotsi, Y., Barsoum, M.W.: Two-dimensional nanocrystals produced by exfoliation of Ti_3 AlC_2. *Adv Mater.* 23, 4248–4253 (2011). https://doi.org/10.1002/adma.201102306
22. Naguib, M., Barsoum, M.W., Gogotsi, Y.: Ten years of progress in the synthesis and development of MXenes. *Adv Mater.* 33 (2021). https://doi.org/10.1002/adma.202103393
23. Gogotsi, Y., Anasori, B.: The rise of MXenes. *ACS Nano.* 13, 8491–8494 (2019). https://doi.org/10.1021/acsnano.9b06394
24. Anasori, B., Lukatskaya, M.R., Gogotsi, Y.: 2D metal carbides and nitrides (MXenes) for energy storage. *Nat Rev Mater.* 2, 16098 (2017). https://doi.org/10.1038/natrevmats.2016.98
25. Naguib, M., Mochalin, V.N., Barsoum, M.W., Gogotsi, Y.: 25th anniversary article: MXenes: A new family of two-dimensional materials. *Adv Mater.* 26, 992–1005 (2014). https://doi.org/10.1002/adma.201304138
26. VahidMohammadi, A., Rosen, J., Gogotsi, Y.: The world of two-dimensional carbides and nitrides (MXenes). *Science (1979).* 372 (2021). https://doi.org/10.1126/science.abf1581
27. Lim, K.R.G., Shekhirev, M., Wyatt, B.C., Anasori, B., Gogotsi, Y., Seh, Z.W.: Fundamentals of MXene synthesis. *Nat Synth.* 1, 601–614 (2022). https://doi.org/10.1038/s44160-022-00104-6
28. Alhabeb, M., Maleski, K., Anasori, B., Lelyukh, P., Clark, L., Sin, S., Gogotsi, Y.: Guidelines for synthesis and processing of two-dimensional titanium carbide (Ti_3 C_2 T_x MXene). *Chem Mater.* 29, 7633–7644 (2017). https://doi.org/10.1021/acs.chemmater.7b02847
29. Li, T., Yao, L., Liu, Q., Gu, J., Luo, R., Li, J., Yan, X., Wang, W., Liu, P., Chen, B., Zhang, W., Abbas, W., Naz, R., Zhang, D.: Fluorine-free synthesis of high-purity Ti_3 C_2 T_x (T=OH, O) via alkali treatment. *Angew Chem Int Ed.* 57, 6115–6119 (2018). https://doi.org/10.1002/anie.201800887
30. Alhabeb, M., Maleski, K., Mathis, T.S., Sarycheva, A., Hatter, C.B., Uzun, S., Levitt, A., Gogotsi, Y.: Selective etching of silicon from Ti_3 SiC_2 (MAX) to obtain 2D titanium carbide (MXene). *Angew Chem Int Ed.* 57, 5444–5448 (2018). https://doi.org/10.1002/anie.201802232
31. Rosenkranz, A., Righi, M.C., Sumant, A. V., Anasori, B., Mochalin, V.N.: Perspectives of 2D MXene tribology. *Adv Mater.* 35 (2023). https://doi.org/10.1002/adma.202207757
32. Wyatt, B.C., Rosenkranz, A., Anasori, B.: 2D MXenes: Tunable mechanical and tribological properties. *Adv Mater.* 33 (2021). https://doi.org/10.1002/adma.202007973
33. Lian, W., Mai, Y., Liu, C., Zhang, L., Li, S., Jie, X.: Two-dimensional Ti_3C_2 coating as an emerging protective solid-lubricant for tribology. *Ceram Int.* 44, 20154–20162 (2018). https://doi.org/10.1016/J.CERAMINT.2018.07.309

34. Rosenkranz, A., Grützmacher, P.G., Espinoza, R., Fuenzalida, V.M., Blanco, E., Escalona, N., Gracia, F.J., Villarroel, R., Guo, L., Kang, R., Mücklich, F., Suarez, S., Zhang, Z.: Multi-layer $Ti_3C_2T_x$-nanoparticles (MXenes) as solid lubricants – Role of surface terminations and intercalated water. *Appl Surf Sci.* 494, 13–21 (2019). https://doi.org/10.1016/J.APSUSC.2019.07.171
35. Grützmacher, P.G., Suarez, S., Tolosa, A., Gachot, C., Song, G., Wang, B., Presser, V., Mücklich, F., Anasori, B., Rosenkranz, A.: Superior wear-resistance of Ti_3 C_2 T_x multilayer coatings. *ACS Nano.* 15, 8216–8224 (2021). https://doi.org/10.1021/acsnano.1c01555
36. Takadoum, J., Houmid Bennani, H.: Influence of substrate roughness and coating thickness on adhesion, friction and wear of TiN films. *Surf Coat Technol.* 96, 272–282 (1997). https://doi.org/10.1016/S0257-8972(97)00182-5
37. Xue, Y., Wu, C., Shi, X., Zhang, K., Huang, Q.: High temperature tribological behavior of textured CSS-42L bearing steel filled with Sn-Ag-Cu-Ti3C2. *Tribol Int.* 164, 107205 (2021). https://doi.org/10.1016/J.TRIBOINT.2021.107205
38. Peng, Y., Wang, Y., Ding, J., Shi, Y., Jin, T., Zhang, W.: Balling behavior and diamond graphitization of metal-matrix diamond composites by selective laser melting: A quantitative investigation. *Int J Refract Metals Hard Mater.* 124, 106830 (2024). https://doi.org/10.1016/J.IJRMHM.2024.106830
39. Huang, Y., Ma, H., Meng, Y., Mao, Y.: Synergistic lubrication effects of Sn–Ag–Cu and MXene–Ti_3 C_2 to improve tribological properties of M50 bearing steel with microporous channels. *Ind Lubr Tribol.* 76, 327–336 (2024). https://doi.org/10.1108/ILT-12-2023-0381
40. Zhang, J., Jiang, D., Liao, L., Cui, L., Zheng, R., Liu, J.: $Ti_3C_2T_x$ MXene based hybrid electrodes for wearable supercapacitors with varied deformation capabilities. *Chem Eng J.* 429, 132232 (2022). https://doi.org/10.1016/J.CEJ.2021.132232
41. Xin, Y., Li, T., Gong, D., Xu, F., Wang, M.: Preparation and tribological properties of graphene oxide/nano-MoS2 hybrid as multidimensional assembly used in the polyimide nanocomposites. *RSC Adv.* 7, 6323–6335 (2017). https://doi.org/10.1039/c6ra27108a
42. Hou, K., Wang, J., Yang, Z., Ma, L., Wang, Z., Yang, S.: One-pot synthesis of reduced graphene oxide/molybdenum disulfide heterostructures with intrinsic incommensurateness for enhanced lubricating properties. *Carbon N Y.* 115, 83–94 (2017). https://doi.org/10.1016/j.carbon.2016.12.089
43. Song, H., Wang, B., Zhou, Q., Xiao, J., Jia, X.: Preparation and tribological properties of MoS 2 /graphene oxide composites. *Appl Surf Sci.* 419, 24–34 (2017). https://doi.org/10.1016/j.apsusc.2017.05.022
44. Colas, G., Saulot, A., Philippon, D., Berthier, Y., Léonard, D.: Tribochemical competition within a MoS2/Ti dry lubricated macroscale contact in ultrahigh vacuum: a time-of-flight secondary ion mass spectrometry investigation. *ACS Appl Mater Interfaces.* 10, 20106–20119 (2018). https://doi.org/10.1021/acsami.8b02999
45. Furlan, K.P., de Mello, J.D.B., Klein, A.N.: Self-lubricating composites containing MoS_2: A review. *Tribol Int.* 120, 280–298 (2018). https://doi.org/10.1016/j.triboint.2017.12.033

46. Gong, Z., Jia, X., Ma, W., Zhang, B., Zhang, J.: Hierarchical structure graphitic-like/MoS_2 film as superlubricity material. *Appl Surf Sci.* 413, 381–386 (2017). https://doi.org/10.1016/j.apsusc.2017.04.057
47. Mutyala, K.C., Wu, Y.A., Erdemir, A., Sumant, A. V.: Graphene - MoS_2 ensembles to reduce friction and wear in DLC-Steel contacts. *Carbon N Y.* 146, 524–527 (2019). https://doi.org/10.1016/j.carbon.2019.02.047
48. Lin, H., Jian, Q., Bai, X., Li, D., Huang, Z., Huang, W., Feng, S., Cheng, Z.: Recent advances in thermal conductivity and thermal applications of graphene and its derivatives nanofluids. *Appl Therm Eng.* 218, 119176 (2023). https://doi.org/10.1016/J.APPLTHERMALENG.2022.119176
49. Maurya, U., Vasu, V.: Boehmite nanoparticles for potential enhancement of tribological performance of lubricants. *Wear.* 498–499, 204311 (2022). https://doi.org/10.1016/J.WEAR.2022.204311
50. Maurya, U., Vasu, V., Saini, V.: Nano-innovations in lubrication: Unveiling the synergistic boundary lubrication potential of zinc titanate with phosphonium ionic liquid. *Tribol Int.* 199, 109987 (2024). https://doi.org/10.1016/J.TRIBOINT.2024.109987
51. Saini, V., Bijwe, J., Seth, S., Ramakumar, S.S.V.: Unexplored solid lubricity of Titanium nanoparticles in oil to modify the metallic interfaces. *Appl Surf Sci.* 580, 152127 (2022). https://doi.org/10.1016/J.APSUSC.2021.152127
52. Dai, W., Kheireddin, B., Gao, H., Liang, H.: Roles of nanoparticles in oil lubrication. *Tribol Int.* 102, 88–98 (2016). https://doi.org/10.1016/j.triboint.2016.05.020
53. Pham, S.H., Choi, Y., Choi, J.: Stimuli-responsive nanomaterials for application in antitumor therapy and drug delivery. *Pharmaceutics.* 12, 630 (2020). https://doi.org/10.3390/pharmaceutics12070630
54. Yin, N., Yang, P., Liu, S., Pan, S., Zhang, Z.: AI for tribology: Present and future. *Friction.* 12, 1060–1097 (2024). https://doi.org/10.1007/s40544-024-0879-2
55. Jyothish, P., Maurya, U., Vasu, V.: Advancements in metalworking fluid for sustainable manufacturing. In: Navneet Khanna, Kishor Kumar Gajrani, Khaled Giasin, J. Paulo Davim (Eds.), *Sustainable Materials and Manufacturing Technologies.* pp. 58–81. CRC Press, London (2023).
56. Upendra, M., Vasu, V.: Ionic liquids and its potential in lubricants: A review. *SSRN Electron J.* (2018). https://doi.org/10.2139/ssrn.3320367
57. Upendra, M., Vasu, V.: Synergistic effect between phosphonium-based ionic liquid and three oxide nanoparticles as hybrid lubricant additives. *J Tribol.* (2020). https://doi.org/10.1115/1.4045769
58. Cai, M., Yu, Q., Liu, W., Zhou, F.: Ionic liquid lubricants: When chemistry meets tribology. *Chem Soc Rev.* 49, 7753–7818 (2020). https://doi.org/10.1039/d0cs00126k
59. Zhou, Y., Qu, J.: Ionic liquids as lubricant additives: A review. *ACS Appl Mater Interfaces.* 9, 3209–3222 (2017). https://doi.org/10.1021/acsami.6b12489
60. Maurya, U., Vasu, V., Kashinath, D.: Three-way compatibility study among nanoparticles, ionic liquid, and dispersant for potential in lubricant formulation. *Mater Today Proc.* (2022). https://doi.org/10.1016/J.MATPR.2022.03.329

61. Maurya, U., Vasu, V., Kashinath, D.: Ionic liquid-nanoparticle-based hybrid-nanolubricant additives for potential enhancement of tribological properties of lubricants and their comparative study with ZDDP. *Tribol Lett.* 70, 11 (2022). https://doi.org/10.1007/s11249-021-01551-6
62. Swami, S., Suthar, S., Singh, R., Thakur, A.K., Gupta, L.R., Sikarwar, V.S.: Potential of ionic liquids as emerging green solvent for the pretreatment of lignocellulosic biomass. *Environ Sci Pollut Res.* 31, 12871–12891 (2024). https://doi.org/10.1007/s11356-024-32100-y
63. Salimon, J., Salih, N., Yousif, E.: Biolubricants: Raw materials, chemical modifications and environmental benefits. *Eur J Lipid Sci Technol.* 112, 519–530 (2010). https://doi.org/10.1002/ejlt.200900205

Index

Note: **Bold** page numbers refer to tables and *italic* page numbers refer to figures.